T0277003

TOXIC
SAFETY
—

ALISSA CORDNER

TOXIC
SAFETY

—

Flame Retardants,
Chemical Controversies, and
Environmental Health

COLUMBIA UNIVERSITY PRESS
NEW YORK

Columbia University Press
Publishers Since 1893
New York   Chichester, West Sussex
Copyright © 2016 Columbia University Press
All rights reserved

Library of Congress Cataloging-in-Publication Data
Cordner, Alissa.
    Toxic safety : flame retardants, chemical controversies, and environmental health /
Dr. Alissa Cordner.
        pages cm
    Includes bibliographical references and index.
    ISBN 978-0-231-17146-5 (cloth : acid-free paper); 978-0-231-54138-1 (e-book)
    1. Fireproofing agents—Environmental aspects.   2. Fireproofing agents—Toxicology.
I. Title.
    TD428.F57C67 2016
    363.17'91—dc23

                                                                    2015024698

Columbia University Press books are printed on permanent
and durable acid-free paper.
This book is printed on paper with recycled content.
Printed in the United States of America

Cover design: Diane Lugar

References to Internet websites (URLs) were accurate at the time of writing.
Neither the author nor Columbia University Press is responsible for URLs
that may have expired or changed since the manuscript was prepared.

CONTENTS

ACKNOWLEDGMENTS

I am deeply indebted to many people who have helped me conduct my research, develop these arguments, and complete this book.

First, my deepest thanks to all those whom I interviewed and spent time with over the course of my research, especially those at FlameCorp, the EPA's Design for the Environment Program, the EPA's Environmental Public Health Division, an unnamed environmental health organization, and an unnamed academic chemistry laboratory. People at each of these institutions welcomed me and answered innumerable questions. Though I do not identify individuals by name, I hope they all find their words and actions to be accurately portrayed.

I have benefited immensely from wonderful academic mentors. Many thanks to Phil Brown, who has endlessly encouraged and guided my research. I could not ask for a more supportive mentor or a better role model of how to conduct environmental health research that makes a difference. Timmons Roberts provided invaluable suggestions throughout my research and writing process, directing me to important empirical insights. Nitsan Chorev encouraged me to consider new and more significant theoretical implications. Gianpaolo Baiocchi was a constant support, through this research and through the other Project that dominated my intellectual life for the past several years.

Others have supported my work in various ways. The Contested Illnesses Research Group at Brown University provided a nurturing space for developing ideas, rewriting proposals, and revising chapters, as did the Social Science Environmental Health Research Institute at Northeastern University. Special thanks go to David Ciplet for reading drafts and providing feedback, and to Mercedes Lyson and Margaret Mulcahy for the interviews and research they conducted on the larger flame retardant project. Margot Jackson and Dennis Hogan provided insightful comments and clear suggestions for next steps. Susan E. Bell and Nancy Riley influenced my work as an undergraduate student, and I feel fortunate to count them as colleagues years later. My deep appreciation also goes to Rebecca Gasior Altman, Catherine Bliss, Rachel Morello-Frosch, Ruthann Rudel, and Sara Shostak for offering feedback, ideas, and support throughout my research and writing process.

I am extremely grateful for the collaborative environment at Whitman College. I have been lucky to count Bill Bogard, Keith Farrington, David Hutson, Michelle Janning, Helen Kim, and Gilbert Mireles as departmental colleagues. Thank you for your support of my work at Whitman and for our sociology writing group. Laura Ferguson, Adam Gordon, Emily Jones, and Lisa Uddin shared excellent feedback and good food during our summer writing group. Helen Kim and Lydia McDermott organized wonderful writing retreats where many of these pages were written or revised.

Innumerable thanks go to Laura Ferguson for Bennington runs, Sarah Koenigsberg for climbing and Burwood, Jen Cohen for trivia night, and Bridger for tennis ball time. Thank you for your friendship and encouragement. I am so lucky to count Elizabeth A. Bennett, Peter T. Klein, Erica Jade Mullen, Mim Plavin-Masterman, Stephanie Savell, Trina Vithayathil, and Mujun Zhou as dear friends and colleagues. Extra thanks to Trina, Chris, Markose, and Mathia; Pete and Steph; Dave, Jenn, Eliza, Cora, and Marlon; and Meghan Kallman for great visits to Providence while I lived all over the country. Elizabeth A. Bennett, how amazing that we both ended up in the Northwest. Thanks to you, Ben Falk, Dalva, and Rhubarb for being the best. Mandy Marvin, thank you for reminding me to seek out spontaneous fun every day. Adrian Velasco, thank you for the wild adventures we had together.

My research was supported by the EPA STAR Fellowship program (FP 917119). Additional work on the social implications of flame retardants, including some of the interviews I conducted for this project, was also funded by NSF (SES-0924241). This work has not been reviewed by the EPA or NSF, and does not reflect EPA or NSF policy in any way. I also received funding from Whitman College's Professional Development Account, the American Sociological Association, Brown University's Graduate School, and the Brown University International Affairs Travel Fund for research expenses and conference travel. I would also like to thank Chemical-Watch, the Environmental Law Institute, and the Fire Safety Bulletin for providing free or reduced-cost access to useful resources. Parts of chapter 3 were previously published in the *Journal of Environmental Studies and Sciences*, and parts of chapter 4 were previously published in *Science, Technology, & Human Values*.[1]

I was fortunate to work with Kathryn Schell at Columbia University Press, who was endlessly supportive of this project. Thanks to Patrick Fitzgerald and Phil Brown for making that connection, and to Lyndee Stalter and Vickie West for their help during copy editing and production. The comments of two anonymous reviewers have deeply strengthened this final book. Thank you to Jane Henderson for a beautiful index, and for being such a close family friend all these years.

Finally, my greatest thanks and love to my family. Susanne Cordner, I treasure what our sisterhood has become these last few years. I am so proud of you and excited for your next steps. Extra thanks for your help with 1980s legal decisions! Barbara Tyler, thank you for being a role model of strength and kindness, and for being the first sociologist in the family. And endless thanks to Ian Cordner and Virginia Tyler for a lifetime of love and support, and for teaching me the value of making a difference. Without my family, I would never have started on this path, let alone have ended up at this next juncture.

ABBREVIATIONS

ACC	American Chemistry Council
BFR	brominated flame retardant
BPA	bisphenol-A
BSEF	Bromine Science and Environmental Forum
CBI	confidential business information
CfFS	Citizens for Fire Safety
CPSC	Consumer Product Safety Commission
DBDPEthane	decabromodiphenyl ethane
DecaBDE	decabromodiphenyl ether
DES	diethylstilbestrol
DfE	Design for the Environment (EPA)
EDSP	Endocrine Disruptor Screening Program
EPA	Environmental Protection Agency
EWG	Environmental Working Group
FDA	Food and Drug Administration
FOIA	Freedom of Information Act
GLP	Good Laboratory Practices
HBCD	hexabromocyclododecane
HCRA	Harvard Center for Risk Analysis
HESI	Health and Environmental Sciences Institute
HP	Hewlett-Packard
IAFF	International Association of Fire Fighters
ICL-IP	Israeli Chemicals Limited Industrial Products
IRIS	Integrated Risk Information System

NAFRA	North American Flame Retardant Alliance
NASFM	National Association of State Fire Marshals
NGO	nongovernmental organization
NHANES	National Health and Examination Survey
NIEHS	National Institute of Environmental Health Sciences
NOAEL	no observed adverse effect level
NPSS	New Political Sociology of Science
NTP	National Toxicology Program
OctaBDE	octabromodiphenyl ether
OIRA	Office of Information and Regulatory Affairs
OPPT	Office of Pollution Prevention and Toxics (EPA)
ORD	Office of Research and Development (EPA)
PBB	polybrominated biphenyl
PBDE	polybrominated diphenyl ether
PBT	persistent, bioaccumulative, and toxic
PCB	polychlorinated biphenyl
PentaBDE	pentabromodiphenyl ether
PIRG	Public Interest Research Group
PMN	Pre-Manufacture Notice
POPs	persistent organic pollutants
R&D	research and development
REACH	Registration, Evaluation, Authorisation, and Restriction of Chemicals (Europe)
SNUR	Significant New Use Rule (EPA)
SST	strategic science translation
TB117	Technical Bulletin 117 (California)
TBBPA	tetrabromobisphenol-A
TCEP	tris(2-chloroethyl) phosphate
TDCPP or TDCP	tris(1,3-dichloro-2-propyl) phosphate
TPP	triphenyl phosphate
TSCA	Toxic Substances Control Act (U.S.)
VCCEP	Voluntary Children's Chemical Evaluation Program (EPA)
VECAP	Voluntary Emissions Control Action Program

TOXIC
SAFETY

—

1

UNCERTAIN SCIENCE AND THE FIGHT FOR ENVIRONMENTAL HEALTH

S everal years ago, I sat in the sunny Berkeley, California, office of an environmental health advocate, talking about environmental health problems. "It's an uncontrolled chemistry experiment," he told me. "Chemicals are widespread in the environment and in our bodies, and we don't know how they interact with each other or with other social factors." As an organizer, he had worked to restrict many different types of chemicals, and he talked specifically about his organization's work to restrict flame retardants. I asked why his organization had campaigned to restrict these chemicals. "The evidence is strong and compelling," he said. "We can see these chemicals are bad and that they have health effects." Although these chemicals are widely used in consumer products to slow fire ignition, a growing body of research has shown that exposure to certain flame retardants can cause reproductive and neurological problems, especially in children.

As I was writing this book, a group of unexpected allies took to the streets to protest the widespread use of flame retardant chemicals. Firefighters, who are most directly and personally on the front lines of fire safety, gathered for a day of action to call attention to the role these chemicals play in their profession's high rates of cancers and other occupational illnesses.[1] In San Francisco, they placed two-hundred pairs of empty boots on the steps of City Hall to memorialize firefighters who had recently lost their lives to cancer. In Spokane, Washington, they draped an environmental organization's banner over a classic fire truck at a press conference. And in Augusta, Maine, they partnered with local legislators and activists to show a documentary on the flame retardant industry's distortion of science. These unlikely supporters joined their voices with a growing chorus of activists, scientists, and regulators calling for a different approach to fire safety, a strengthened system of chemicals regulation, and a more precautionary way of protecting public and environmental health.

A decade ago, little was known about the health effects of flame retardants for humans, but concern was growing. Modern flame retardants have been widely used in consumer products in the United States since the 1960s, and their history involves a fascinating progression of industrial accidents, public awareness, and regulatory battles. For example, many Americans know that in the late 1970s a toxic flame retardant was added to polyester children's pajamas. Arlene Blum, then a chemistry Ph.D. student, coauthored several articles showing that the chemical, brominated tris, was mutagenic in laboratory studies and that children absorbed it after a single night's sleep.[2] Alarmed by this research, pressure from consumer groups, and emerging cancer research, the U.S. Consumer Products Safety Commission banned the compound in children's sleepwear.[3] Brominated tris was removed from commerce in the United States, though the ban was later overturned in federal court. Clothing manufacturers quickly remanufactured the children's pajamas with a similar flame retardant, chlorinated tris, but this chemical too was shown to be mutagenic and was removed from sleepwear.[4]

For several subsequent decades, Tris flame retardants received little scientific, regulatory, or activist attention. In 2007, after spending several decades as a professional mountaineer, Dr. Blum met a furniture manufacturer at a California symposium on chemicals policy. She learned that furniture manufacturers complied with California's flammability standard by adding flame retardant chemicals—including the chlorinated tris she had studied in the 1970s—to polyurethane foam used in furniture.[5] In response, she established the Green Science Policy Institute, a nonprofit environmental health advocacy organization in California, and began collaborating with other environmental groups, academic researchers, and regulatory officials on research and policy related to reducing the use of flame retardants in consumer goods. Today, chlorinated tris is regulated in California as a human carcinogen, and some companies are phasing it out of production, but exposure remains widespread.[6] As a result of this growing attention to and concern about flame retardant chemicals, several states have passed regulations that change flammability standards, restrict certain chemicals, and require greater disclosures by companies using the chemicals in their products.

This is but a sliver of the full flame retardant story that provides a rich backdrop for this book. Flame retardant chemicals have emerged as a highly public example of the problems with how chemicals are used, researched, and regulated in the United States. Exposures from products ranging from computers to couches are widespread, but the consequences of these exposures are poorly understood. A handful of these chemicals have been phased out or banned, but they have been replaced by other chemicals whose risks are largely unknown. Interactions between an energized environmental health community on the one hand, and a multi-billion dollar chemical industry on the other, are highly contested, with each side accusing the other of self-interest and scientific impropriety. Scientists interested in flame retardants have expanded the scope of their research to more chemicals, different health end points, and additional comparisons, yet their work is often inconclusive. It seems that more research is always needed. Furthermore, translating this science into policy is a difficult proposition: as a scientist at the Environmental Protection Agency (EPA) told me, "Even when you know everything, deciding what to do is not a science question."

Flame retardants have been the subject of a remarkable amount of activist outcry, scientific research, regulatory attention, and industry defense, making them one of the most prominent environmental health controversies of the early twenty-first century. These chemicals are widely used in products as diverse as smartphones, couches, and airplanes, and are produced and marketed by a powerful and profitable international industry. Flame retardants have been the focus of congressional inquiries, scientific conferences, and journalistic exposés. The cause of flame retardant regulation has been taken up by a striking coalition of activists, including firefighters, children's health clinicians, environmental health scientists, furniture manufacturers, and environmental organizations. Long-pursued efforts toward reforming federal regulation of industrial chemicals have been rekindled in no small part due to recent activity around flame retardants. Yet no academic scholarship has examined the full field of stakeholders engaged with these controversial chemicals, or explored how flame retardants have become a hot topic of inquiry across so many institutions.

This book examines how environmental health risks are defined and contested in the face of unavoidable scientific uncertainty and competing, powerful stakeholders. It tracks the *social discovery* of flame retardants, a term used by social scientists to explain the growing awareness of a previously unrecognized or poorly understood social problem, disease, condition, environmental hazard, or social phenomenon.[7] In particular, *Toxic Safety* answers two far-reaching questions: how do stakeholders develop different definitions of risk and different interpretations of science, and why do these differences matter?

The short answer to these questions is that all individuals and institutions interested in contested environmental health issues use science to make competing and strategic claims in pursuit of institutional and regulatory goals. They do so on an uneven playing field, and their divergent goals and tactics have identifiable consequences for public health and environmental protection. Environmental health policy is supposedly science based, yet research on environmental risks is inevitably uncertain, resulting in policy decisions that are driven by nonscientific forces. These controversies—labeled as scientific but also social, political, and economic in nature—play out as conflicts over how decisions should be made in the face of uncertain knowledge and unknown consequences. Each group of stakeholders develops and fiercely defends different definitions of risk because those definitions have serious and identifiable consequences for chemicals regulation and public health. Each group of stakeholders also strategically translates the scientific evidence in ways that support their goals, though this translation does not happen on an even playing field.

To date, social science explanations of these strategic uses of science tend to fall into two camps. On the one hand, the political sociology of science highlights the institutional features of scientific and regulatory practices that produce uncertainty, showing that areas of ignorance and uncertainty reflect both inevitable conditions of scientific practice and uneven distributions of political, social, and economic resources. On the other hand, researchers focusing on corporate manipulation of uncertainty show that ignorance is often intentionally cultivated by those who stand to benefit economically or politically from the lack of public consensus, scientific paralysis, and regulatory delays that result. My research bridges these two fields to offer a more nuanced and encompassing description of how environmental health risks emerge and are fought over in the public sphere. The multifaceted uncertainties connected to environmental risks are both inevitable and manipulated, unintentional and strategic.

To introduce this case study and set the stage for a detailed foray into the world of scientific uncertainty and policy contestation, I begin with an overview of contemporary environmental health controversies as global, ubiquitous, and scientifically uncertain problems.

ENVIRONMENTAL HEALTH CONTROVERSIES

Environmental health hazards are all around us. We are constantly exposed to a chemical cocktail of industrial pollutants in the air we breathe, the food we eat, the water we drink, and the products we handle every day. Preventing exposure is all but impossible, and the consequences of this exposure are poorly understood. As a result, current and future generations face potential reproductive, neurological, and other health problems. For every company representative claiming that their products are safe, there is an activist stating the opposite, a scientist conducting new experiments, and a regulator advocating for restrictions. Central to these controversies is the question of whether chemicals pose a risk to people and the environment.

Environmental health is a broad term that generally refers to the effects of the natural and built environment on human health outcomes. Today's world faces serious environmental health risks that are global, highly contested, and frequently uncertain. Some scholars call this the *risk society*: modern society produces technological risks that are scientific, incalculable, and potentially catastrophic.[8] This unknowable but pervasive nature of modern technological risks leads to more widespread awareness of risk as more people experience the consequences. Yet even as environmental risks are increasingly global in their reach, the impact of environmental hazards remains unequally distributed, exacerbating rather than transcending existing inequalities.[9]

The ubiquity of industrial chemicals in our environment is both a type of environmental degradation and a potentially serious threat to human health. Most chemicals used today are understudied: the vast majority of the more than 84,000 industrial chemicals registered in the United States with the EPA lack any data on how people are exposed to them, at what levels, and with what consequences.[10] Although there is a common expectation in the United States that the EPA fully assesses the safety of chemicals that are in use, in practice this is far from the case.

Exposures to chemicals are complex, ubiquitous, and poorly understood.[11] This is particularly true for low-level, long-term, everyday exposures, as opposed to acute, high-level exposures resulting from dangerous workplace practices or chemical disasters. An individual at home may be simultaneously exposed to flame retardants from a computer; petrochemicals from a plastic water bottle; formaldehyde from particleboard furniture; lead from old paint on the walls; polychlorinated biphenyls from a floor finish applied decades ago; phthalates from an air freshener; fine particulate matter from the traffic outside; and mercury from the power plant across town, in addition to thousands of other chemicals. These exposures interact in the human body in a complex *exposure pathway* from a product source to exposure and absorption into the human body, through interactions with a unique body chemistry and genetic makeup, leading to predisease outcomes such as gene activation and eventually, potentially, to illness.[12] Particularly concerning are chemicals that may disrupt hormonal systems—*endocrine disrupting chemicals*—because they can induce significant effects at very low levels of exposure.[13] At least eight-hundred chemicals are believed to interfere with hormonal systems, and evidence is mounting that endocrine-related diseases are widespread, though significant knowledge gaps exist.[14]

Many of today's most salient environmental health problems are related to chronic, low-dose exposures to commonly used chemicals whose health outcomes are often uncertain. A generation ago, the most prominent environmental health threats in the United States were tied to the discovery of contaminated communities or sudden toxic disasters.[15] But accumulating research highlights the potential dangers from long-term, low-level exposure to the toxic and hormone-disrupting chemicals that are broadly used in household products. Beyond the flame retardants that are the subject of this book, other examples abound. Exposure to low levels of bisphenol-A (BPA), a chemical widely used in hard plastics and food containers, is associated with health problems ranging from reproductive cancers to diabetes.[16] Fine particulate matter from transportation and industrial pollution sources is associated with higher levels of cardiopulmonary disease.[17] And the levels of industrial chemicals found in sea mammals in the Arctic have been high enough to trigger consumption advisories.[18]

Environmental health research is increasingly interdisciplinary and multidisciplinary, and has gained significant traction among health professionals. The 2010 President's Cancer Panel highlighted environmental causes of cancer, demonstrating widespread scientific and regulatory acceptance for the idea that toxic chemicals can harm human health.[19] Environmental health is also an important topic of exploration for spatial studies of inequality and is recognized as important for children's health. A growing body of research shows that health effects are linked not only to specific doses of chemicals, but to the timing of exposures and the vulnerabilities of the individuals who are exposed.[20] All this means that knowing the chemical levels that people carry in their bodies at any given time is not enough information to understand the connections between their chemical exposures and any current or future health outcomes.

As environmental contamination has increased globally, the poor, communities of color, and those living in the Global South face greater health burdens from chemical exposures.[21] A television bought in the United States contains several pounds of petroleum-based chemicals in the plastic casing, which will invisibly degrade over the course of its useful life. That television also contains *embedded hazards* from pollution that occurs throughout its life cycle: land and water pollution from metal and mineral mining, greenhouse gases and other pollutants from all moments of transportation, the global movement toward the polar regions of mercury and other heavy metals released as coal is burned to produce electricity to turn on the television, and direct exposures for e-waste recyclers as they dismantle the used television by hand.[22] The environmental life cycle of consumer products is long and far-reaching, and falls disproportionately on individuals and communities with less power and fewer resources.

In short, environmental health hazards in modern society are ubiquitous and uncertain, and the chemical cocktail of exposures we each encounter every day has unknown reproductive, neurological, and other health problems for current and future generations. As a result, a growing chorus of activists and some scientists advocate for a system that regulates environmental hazards using the *precautionary principal*, which shifts the balance in decision making toward preventing harm, even in conditions of uncertainty.[23]

ENVIRONMENTAL RISK AND THE NEW POLITICAL SOCIOLOGY OF SCIENCE

The flame retardant story has been characterized as a problem of enduring environmental risks and regrettable substitutions from one chemical to the next.[24] One activist explained it this way: "It's a really good example of what we call the 'whack-a-mole' problem. You know, you get rid of one bad chemical only to have another one crop up. Then you try to get rid of that one and then another one, and it's like they're equally as bad, or we don't know whether or not it's safe." Chemical regulation moves too slowly to protect public health from emerging chemicals of concern, and as one chemical is banned, another takes its place, usually without adequate evidence that it is in fact a safer choice.

Sociology provides several explanations for this whack-a-mole model of risk and policy. The *ecological modernization* perspective argues that environmental factors are increasingly taken into account in modern industrial processes.[25] Proponents of this perspective would suggest that the flame retardant industry will move to safer chemicals as ecological considerations become more tightly coupled with economic factors, leading to safer chemistry and more efficient chemical production and use. This perspective certainly echoes the flame retardant industry's own rhetoric. For example, the main website of a major flame retardant producer features the slogan "Grow the future through chemistry: safer, greener formulations and applications," the white-colored text standing out on top of a bright blue sky and a healthy wheat field.[26]

There is no doubt that many industries have made significant environmental progress in the last hundred years, but these changes rarely result in environmentally neutral production processes, and usually are motivated more by a strengthened regulatory apparatus than by industry's own reflexive behavior or self-policing. Regrettably, few incentives exist to encourage companies to put environmental and human health protection ahead of profits, and the U.S. chemical industry continues to release billions of pounds of toxic chemicals into the environment each year.[27] Though major companies participate in responsible conduct programs through their trade associations, these programs emphasize standardized plans, voluntary compliance, and internal information-sharing in order to build a more positive collective corporate image, rather than progressive policies on health and safety.[28] They also tend to rely on self-enforcement instead of independent evaluation, contributing to their credibility problems.[29] Many newly developed products are predicted to be safer than the banned chemicals that they are replacing, but these expectations rely on modeled data with a history of inaccurate or incomplete risk estimates. The industry also funds expensive lobbying campaigns to challenge proposed health and safety requirements, and regularly defends existing products—including identified mutagens or chemicals with high levels of human exposure—rather than voluntarily removing them from commerce at the earliest evidence of risk.

The *treadmill of production* perspective offers a very different take on the greening of capitalism. Scholars in this tradition argue that there is an enduring and unavoidable conflict between the capitalist mode of production and environmental protection, because regardless of the actions of individual companies or steps taken by the state or social movements, capitalism's need for growth and profits will always lead to greater environmental destruction.[30] Many elements of the flame retardant story confirm the treadmill of production expectations. The multi-billion dollar industry spends millions to lobby against regulations, continues to manufacture products that are potentially dangerous, and has consolidated over the past few decades. Activists celebrate small policy victories and market changes, such as single-chemical bans, but the chemical manufacturing system as a whole remains intact. Indeed, some observers call this pattern of regrettable substitution the "toxic treadmill." However, the treadmill of production theory assumes that the public and consumers exert little influence over decisions about the allocation of technologies, the use of labor, and volumes of production, and that all environmental contests are fundamentally about resource scarcity.[31] This perspective also takes a simplistic view of science, seeing it only as a tool to service industry, not as a tool used by other stakeholders to contest industry practices.

Although these macrolevel theories of political economy and environmental risk explain some aspects of the flame retardant story, they are unable to critically examine and explain decision making across multiple relevant institutions, and they place significant emphasis on a single overwhelming logic to explain all outcomes. They often take a simplistic view of science as a source of endless technological advancement or a tool to service industry. Instead, greater attention is needed to the role of science and of scientific uncertainty in particular.

SEEING SCIENCE AS POLITICAL

In the United States, science is often a primary guide and justification for environmental policy decisions. For example, the EPA's Risk Assessment Portal says that "Risk assessment is, to the highest extent possible, a scientific judgment."[32] However, risk-based decision making also involves social, economic, and political forces. As an industry representative told me, science is never "perfect." While chatting in a New York City café, he described the industry's preferred way of assessing chemical safety, involving a full evaluation of exposure information and hazard or toxicity information. "If I knew everything about every chemical, I would agree" with this method of assessment, he explained, "but we don't." Data sets are always incomplete, studies can be manipulated, and findings are rarely clear-cut. This is the blurry, contested boundary between science and policy, or between risk assessment as a science and risk assessment as a social process or value judgment.

Recent work on the new political sociology of science (NPSS) allows for a better explanation of the role of scientific uncertainty in conflicts over environmental issues.[33] The NPSS framework assumes that science is inherently political and explores the networks, institutions, and power structures that are both visible and hidden in contemporary scientific practices and organizations. This theoretical perspective draws from several areas of science studies, with an understanding that risk includes both a reality that is based in the physical world and an experience that is socially constructed. NPSS scholars recognize the importance of power and interests in controlling the production and consumption of science, but they move beyond the interpretations available within the treadmill of production theory to show that, although economic interests matter, power does not emerge or exist autonomously; instead it is contained and expressed through networks and institutions.[34] This general framework recognizes that nonexperts can generate and interpret state-produced knowledge, that corporate and state power are used to legitimate and delegitimate knowledge, and that social movements contest the authority of not just the state but also corporations and other institutions.

Scientific logic is central to contemporary policy debates about chemicals, and science is increasingly required in the regulatory, legal, and social movement arenas.[35] This process of *scientization* is critiqued for excluding lay voices and structuring debates in a way that privileges corporate interests over public ones.[36] But while the public authority of science as an institution remains strong, the authority of individual scientists declined significantly in the second half of the twentieth century as scientists took public and controversial positions linked to the military, social movements, and industry.[37]

Today, scientific systems and practices are regularly questioned by nonscientific actors, through what sociologist and leading NPSS scholar David Hess calls *epistemic modernization*.[38] Examples of epistemic modernization include the inclusion of AIDS patients in the design of drug trials; the growth in community-based participatory research and citizen-science alliances; *democratizing social movements* that contest, reframe, and produce scientific knowledge around contested issues; the production of *public* or *advocacy biomonitoring* research where participants share their results publicly; the development of an alternative expertise network on autism that includes parents and nonpsychologists; and collaborative stakeholder processes such as the EPA's Design for the Environment alternatives assessment program.[39] Additionally, some social movements are responding to neoliberalism and scientization by developing consumer-based campaigns, improving their scientific arguments, and conducting their own research projects.[40] Finally, there is recognition that not all questions can be resolved by expertise or the development or application of technological and scientific solutions. Scientific consensus proceeds at a different—usually slower—speed than do policy demands.[41]

The NPSS approach is particularly useful for understanding the scientific uncertainty at the heart of the flame retardant story. On the one hand, uncertainty "is an inherent property of scientific data."[42] Scientific uncertainty is particularly salient in emerging or policy-relevant areas of exposure research, toxicology, and epidemiology because of the complexity and length of exposure pathways leading to disease outcomes and the methodological and analytical uncertainties and disagreements involved in identifying those pathways. But on the other hand, scientific uncertainty can be exacerbated by the social, political, and economic factors that motivate the actions of interested stakeholders and institutions. Environmental contaminants themselves are seen as "political issues, involving economic and societal choices."[43] The production of science and the constitution of legitimate and accepted knowledge take place in a field defined by numerous, unequal power relationships.[44] Scientific uncertainty includes such entanglements as identified and discussed data gaps, contradictory or inconsistent findings, suggestive but not conclusive findings, and identified areas of *undone science*.[45] Yet existing theories of uncertainty do not allow for the simultaneous examination of these multiple strategies, or for thinking coherently about how scientific uncertainty crosses disciplinary and stakeholder boundaries.

By developing concepts that integrate multiple, supposedly incompatible views on scientific uncertainty, my analysis offers distinct theoretical and empirical contributions to this existing body of literature on uncertainty and environmental risk assessment. Some scholars argue that scientific uncertainty or ignorance is the inevitable consequence of the features of institutions such as regulatory science. In contrast, others emphasize the strategic manipulation of evidence in the pursuit of stakeholder goals. Yet typically these analyses focus on a single set of stakeholders rather than investigating how the full range of interested institutions engage in debates over a single environmental risk.[46] Instead of assuming that uncertainty takes on only one role in any given example or interaction, I show that scientific uncertainty in risk assessment practices is both a constant feature of scientific institutions and a feature of scientific practice that can be (and often is) manipulated in the pursuit of many, divergent goals. That is, ignorance can be strategic or conspiratorial, and simultaneously ignorance can be seen as a general feature of institutions that produce and use scientific evidence.[47]

This book fills this important theoretical gap by developing the concept of *strategic science translation* to examine the ways in which stakeholders intentionally package scientific evidence to support nonscientific goals. It also fills an empirical gap, through a multisited analysis of how science is produced, interpreted, and strategically translated by a full range of stakeholders across four institutions: the chemical industry, the regulatory sector, academic research, and the social movement sector. This allows me to identify an important site of comparison across these very different institutions: although they sometimes define risk or the components of risk similarly, they regularly construct *conceptual risk formulas* to define the boundaries of risk assessment, and they mobilize strategic science translations of the empirical evidence that allow them to pursue divergent goals. Engaging in the simultaneous analysis of competing science- and risk-based arguments allows me both to pull out the commonalities in their strategies and to identify important differences in the types of scientific arguments they use, as well as the consequences of those arguments for environmental and public health. In a society where science is a requirement for environmental policy making, it is not surprising that most stakeholders frame their claims about chemical risk as debates over the state of the scientific evidence, even as they acknowledge that risk assessment is an art as well as a science, or that politics plays a role in corporate decision making. The contestations over flame retardant chemicals that are playing out today are not just scientific. They are also social, political, and economic in nature.

I explored these topics ethnographically for more than four years using a mix of qualitative research methodologies. Between 2009 and 2014, I completed 116 in-depth interviews with the full range of stakeholders involved in flame retardant debates: representatives of the flame retardant industry, product manufacturers, and trade associations; scientists from many disciplines working in academia, industry, and consulting or nonprofit research firms; government scientists working in regulatory agencies, research institutes, and state offices; state and federal regulators and legislators; activists from environmental, health, and occupational organizations; fire scientists and fire safety researchers; and professional and retired firefighters. I do not identify any of my participants, and only describe people by name when referring to public activities or published documents. I conducted a year of participant observation across five sites: a flame retardant Research and Development (R&D) lab, two offices at the EPA, an academic environmental chemistry laboratory, and an environmental health nonprofit. My analysis also relies on analysis of publicly available documents and websites, the scientific literature, and internal documents. The Appendix of this book contains a detailed description of my research methods, as well as a reflection on conducting this type of multi-sited research.

In the next chapter, I provide an overview of the history of flame retardant use, regulation, research, and activism in the United States. Broadly introduced to commerce in the 1960s following the spread of highly flammable consumer products and the rapid development of the chemical industry, flame retardant chemicals have become a multi-billion dollar international industry. Today dozens of flame retardants are widely used in products ranging from TVs to building insulation to furniture foam. Chapter 2 details the fascinating history of these chemicals, from early controversies over toxics used in children's pajamas, to the discovery that some chemicals were found at alarmingly high levels in women's breast milk, to dynamic activist campaigns to ban dangerous chemicals, to the lobbying efforts by well-funded chemical industry front groups.

Chapter 3 turns to the issue of risk definition. Environmental health problems are frequently studied and regulated in terms of risk. In the most general sense, *risk* refers to the danger of something, a quality that incorporates both its inherent *hazards* (how dangerous something is) and the likelihood that the hazard will occur (whether something is danger-ous to a given population, often equated with *exposure*). Quantitative risk assessments are regularly used to evaluate chemical safety and develop regulations or restrictions of toxic substances, but a long history of social science research has shown that risk is a social experience of an uncertain reality, not an absolute scientific value.

Instead of focusing on how experts or social science observers define and calculate risk in the case of flame retardants, I look to the competing groups of stakeholders who use risk-based arguments to pursue certain scientific and policy goals. I move away from scholarly debates about technical, rationalist, and constructivist risk definitions that dominate academic discussions of risk. Instead, I present six *conceptual risk formulas* used by stakeholders to delineate the components that go into evaluating risk and the relationships between those components: the classic risk formula, the emerging toxicology risk formula, the exposure-centric risk formula, the hazard-centric risk formula, the exposure-proxy risk formula, and the either-or risk formula. Each is a strategic definition of risk used by stakeholders and institutions to support their policy goals. Using different risk formulas, the chemical industry can con-tinue marketing their products as "safe," the federal government can impose restrictions on chemical use, and environmental health activists can argue that any chemicals detected in breast milk should be banned. By closely analyzing moments when stakeholders interact around chem-ical assessments, I also show that institutions engage in *anticipatory* risk definition when they define the parameters of their assessments, and *strategic* risk definition when they present their results.

Risk definitions are needed in cases of scientific uncertainty, an un-avoidable condition of most areas of environmental health research. In chapter 4, I discuss the importance of scientific uncertainty in science policy debates and show that these uncertainties allow different stake-holders to interpret the same scientific findings differently. Scientific uncertainty in risk assessment is both an inevitable feature of scientific institutions and a feature of scientific practice that can be (and often is) manipulated in the pursuit of many, divergent goals. All stakeholders engage in what I term *strategic science translation* (SST), the process of interpreting and communicating scientific evidence to an intended audience for the purpose of advancing certain goals and interests. *Selective* SST involves the selective use of evidence, summarizing the full scientific corpus down to a small collection of relevant facts, findings, and conclusions. *Interpretive* SST involves emphasizing one argument over another in the case of inconsistent or inconclusive findings. The third category, *inaccurate* SST, involves incorrect communication, false attributions and summaries, or improper characterization of research strengths and weaknesses. Seeing these diverse forms of scientific manipulation as SST allows for comparative analysis across institutions. Because all groups of stakeholders engage in SST in contested environmental policy fields, their actions in those fields and the resulting policy outcomes often reduce not to questions of scientific truth but to the resources deployed to put forward their preferred translation of the science.

Chapter 5 describes science's complicated role in environmental policy debates and shows how the development of science-based decisions, ranging from the production of new chemical products by chemical manufacturers to chemical assessments at EPA, involves social acts of negotiation between multiple interpretations of scientific evidence, participation and input, and political and economic concerns and constraints. Assessments at the EPA, including those done by the New Chemicals Program and the Design for the Environment Alternatives Assessment branch, are presented as scientific documents, but in fact the data included, the evaluations made, and the conclusions presented reflect stakeholder input, both informal and codified evaluator schemas, and an often deliberate balancing of interests and perspectives. Further complicating the supposed neutrality of government assessments is the revolving door that moves industry representatives and EPA officials back and forth between the two institutions. Thus, the decisions presented as being largely scientific in nature are in fact heavily influenced by social, political, and economic considerations. Indeed, although scientific assessments must wrestle with how to interpret and use uncertain and contradictory science, scientific uncertainty is as much a *result* of these risk assessment processes as it is an *input* into these processes because of the ways that stakeholders strategically interpret any given conclusion.

Controversies over risk definition are not confined to regulatory and scientific arenas. Some of the most heated contemporary debates over chemical risk and safety occur in the public spheres. In chapter 6, I describe how controversies over flame retardants have played out for the public and through the media—from the California statehouse, to the halls of the Senate Building, to the front page of national newspapers. A broad coalition of environmental activists, public interest nonprofit organizations, and firefighters has campaigned at the state and federal level to restrict the use of certain chemicals and to revise flammability standards. Members of this coalition use science on multiple levels by translating it for their audiences, collaborating with scientists, and conducting their own research. In contrast, a small number of well-funded industry front groups have fought these restrictions, acting as the public face of the chemical industry and working to institutionalize a very different understanding of environmental risk. This chapter also leverages the concepts developed in earlier chapters to show what conceptual risk definition and SST look like in the media, on websites, and in activist conversations.

Improving chemicals regulation and protecting public health are possible only if all parties accept that scientific uncertainty is inevitable and develop tools and processes that allow for the best possible actions under conditions of imperfect knowledge. In the case of flame retardants, early evidence of hazard and exposure concerns has led to only partial restrictions and changes in the market, and high-stakes debates over chemical safety continue at a furious pace. Once we acknowledge that science is undeniably necessary and yet inevitably uncertain, we can make value judgments and policy preferences explicit in regulatory decisions. Do we want to minimize disruption to industry and maximize the availability of products? If so, an exposure-centric assessment that demands a strong weight of scientific evidence is ideal. If, however, our primary goal is to protect public health and the environment, a different pathway is needed, one that accepts the limitations inherent in environmental health research, demands reasonable evidence of safety rather than absolute proof of harm, and allows some action to be taken based on incomplete evidence.

In the conclusion I talk about these scientific and policy difficulties. I offer suggestions for regulators, environmental scientists, activists, and the chemical industry, and describe several examples of how some decision makers are moving forward to improve chemicals management in the face of scientific uncertainty. In particular, science policy processes should be designed and implemented to take uncertainty under consideration without being paralyzed in the absence of indisputable findings. Tools such as comparative chemical alternatives assessment and high-throughput chemical screening, combined with precautionary evaluations that evaluate a range of evidence and err on the side of protective assessments, could be incorporated into existing regulatory apparatuses. In the United States, our chemical regulatory system is in serious need of reform, as it is currently unable to provide adequate protections for human and environmental health. The chemical industry has much to contribute to these processes and could develop additional hazard and exposure data to ensure that their products are as safe as possible. The new ways of thinking about environmental risk that I offer in this book directly inform these broad suggestions for improving environmental policy, and I detail specific recommendations for each group of stakeholders in the conclusion.

The varying positions within the flame retardant chemical debate reflect competing institutional goals and interests, and result in divergent visions of what the future of environmental health should be. As Daniel Sarewitz writes, "Any political decision (indeed, any decision) is guided by expectations of the future."[48] The case of flame retardants shows that because of the uncertain and constantly evolving nature of environmental health science, risk management activities need to be nimble, able to incorporate new science and emerging concerns. This works against industry's primary goal of securing market stability, and to some extent it works against academic science's goal of minimizing false-positive results, but it supports erring on the side of caution when it comes to protecting public health and the environment. In this way, risk definition has profound consequences for the ways we regulate potential hazards and thus for public health.

2

HOT TOPICS

— ■■■■■ —

Flame Retardants in the Public Sphere

The public controversy over flame retardants first attracted my attention in 2008. Researchers had recently found that Californians carried high levels of polybrominated diphenyl ether (PBDE) flame retardants in their bodies, likely as the result of a well-intentioned flammability standard. I read review articles concluding that exposure to certain flame retardants was ubiquitous, had increased dramatically since the 1970s, and was associated with reproductive and neurological effects in animals and possibly humans as well.[1] A number of states had recently passed or proposed bans on certain flame retardant chemicals, supported by a broad network of environmental and health organizations working with new allies from the fire service and business sectors.

I quickly learned that many stakeholders saw flame retardants as representative of the failings of the United States' regulation of chemicals. As an EPA representative explained, flame retardants are "a poster child of how things can go wrong, or about how complicated the issue is." Or in the words of an environmental health advocate, "they're representative [of] a broad class of chemicals. Their routine release into the environment, routine exposure, environmental and health exposure, all of that becomes representative . . . The fact that a lot of these chemicals are in our homes and everyday products and materials [is] alarming." For others, flame retardants became "the new lead," with ubiquitous exposures and links to reduced IQ in children.[2]

This chapter provides an overview of flame retardant chemicals as a case study of environmental health controversies. First introduced to commerce in the 1960s alongside the spread of highly flammable petroleum-derived consumer products and the rapid growth of the chemical industry, flame retardant chemicals have become a multi-billion dollar international industry. Today scores of flame retardants are used in furniture, electronics, cars, insulating foams, carpet cushioning, baby products, and industrial and construction materials such as building insulation and electrical wiring. As this chapter makes clear, flame retardants are widely used, but they are incompletely studied and inadequately regulated in the United States.

At the center of this story is a tension between precautionary and reactionary regulation of chemicals. Federal chemical regulation in the United States is generally *reactionary*, meaning it takes effect only after risks are well documented and more or less conclusively proven. However, this approach, often described as "risk-based" because it is anchored in formal risk assessments, is frequently criticized. As a recent National Academies panel concluded, the risk assessment process is "bogged down" by challenges to its timeliness and credibility, a lack of adequate resources, a lack of engagement by key stakeholders, and "disconnects between the available scientific data and the information needs of decision makers."[3] Risk-based standards require levels of scientific certainty that are difficult to achieve, and as biologist and scientific advocate Sandra Steingraber notes, "uncertainty is too often parlayed into an excuse to do nothing."[4]

As a contrasting model, the *precautionary* principle suggests that when an activity or substance poses a significant potential risk, action should be taken even when uncertainty exists.[5] The more severe and irreversible the potential consequences, the lower the weight of evidence required before taking action to restrict or regulate the risky activity. As explained in a 1998 consensus statement developed by scientists and advocates, "Where an activity raises threats of harm to the environment or human health, precautionary measures should be taken even if some cause and effect relationships are not fully established scientifically. In this context the proponent of the activity, rather than the public, should bear the burden of proof."[6] Thus the precautionary principle suggests that those undertaking or profiting from a risky activity should demonstrate that the risks are not unreasonable, even if those risks are not fully known.

For some, flame retardants represent a rare triumph of precautionary action in the United States. As a public health scientist told me, several flame retardants "were essentially banned from manufacturing very early in the game in terms of the scientific understanding . . . [Their levels] are going up in people, they're clearly toxic to animals . . . they seem to be endocrine disruptors, they're causing developmental neurotoxicity . . . We don't need to spend another 20 years studying this to know this is probably not good, not a good a thing, and we should stop making them." For others, however, flame retardants are a poster child for the shortcomings of reactionary chemicals regulation. This chapter tells the story of how certain flame retardants have been used, fought over, restricted, and replaced with other chemicals, a story that matters greatly for public and environmental health, and that provides important clues to how we as a country regulate and evaluate chemical risk and safety.

MATERIALS AND MARKETS

Modern life is highly flammable. As a fire scientist explained, "polyurethane foam is the most flammable item you have in your house . . . In most cases, if you don't put out that polyurethane foam within the first five minutes after it ignites, you will lose the house." He told me that arson investigators used to confuse the rapid combustion from furniture foam with fires started using accelerants like gasoline. Additionally, the high-density foam insulations that are central to the green building movement require high levels of flame retardant chemicals to meet building standards.[7] And today's light automobiles contain an average of 384 pounds of plastic, making them highly flammable.[8]

Experts estimate that at least 175 flame retardants are commercially available today.[9] The growth of chemical flame retardant use has coincided with state and national policies, educational campaigns, and behavioral changes among U.S. residents, especially a more than 50 percent decrease in smoking rates. Thanks to modern fire safety measures, home fires in the United States are less common and less deadly than they were in the past, and fire is often seen as an unlikely and intangible future risk by Americans. As another fire scientist said in an interview, "humanity is the only species on the planet that deliberately sets fires. We have for the most part lost our fear of fire." But as I explain toward the end of the chapter, debate exists about whether declines in fire danger and mortality are attributable to flame retardants.

Before diving into the history of this class of chemicals, a few descriptions and definitions are in order. Flame retardants are designed to slow ignition, not prevent it entirely, and their mechanisms of action vary based on their chemical composition.[10] As Dr. Sergei Levchik, a chemist who specializes in flame retardant chemistries, explained at a scientific conference, the benefit of the chemicals "is in preventing small accidental flames from growing into large uncontrollable fires . . . [The] use of flame retardants doesn't make material . . . fire-proofed."[11] Flame retardants are highly application specific, meaning they are developed to be used in particular types of products and to meet specific flammability standards. For example, a specific chemical might work in polyurethane foam to meet California's furniture flammability standard, but might not meet a different flammability standard for the same product type (such as the British furniture standard), and likely could not be used in a different product (such as plastic used in electronic housings).

Flame retardants can be divided into two general groups based on how they are used in a product. They can be *additive*—meaning they are mixed into the product but are not chemically bound to the chemistry—or *reactive*—meaning a chemical reaction joins the flame retardant to the product material itself. Reactive flame retardants are less likely to leave the product, decreasing the potential for exposure during normal product use, but this type of chemistry is not available for all types of consumer products. For example, no reactive flame retardants are on the market for use in furniture foam, one of the most controversial uses of the chemicals.

Flame retardants can also be divided into categories based on their chemical composition: halogenated, inorganic, phosphorous based, and nitrogen based.[12] The most commonly discussed division is between halogenated chemistry, based on their position on the periodic table of elements, and everything else. *Halogenated flame retardants* (also called organo-halogens) are made with carbon and chlorine or bromine (or more rarely fluorine or iodine), which line up vertically one column from the far right of the periodic table. Halogenated flame retardants are widely used in electronics, plastics, and foams because of their low cost and high efficiency.[13] *Nonhalogenated chemicals* include chemicals made with phosphorous or with minerals such as nitrogen or aluminum. For example, the leading global flame retardant by volume in 2011 was aluminum hydroxide, though it is not practical for all products because it must be used at very high levels (up to 65 percent of the product).[14] Additionally, some flame retardants have to be used in combination with a synergist chemical to be effective. For example, many brominated flame retardants are used with the synergist antimony trioxide to increase the efficiency of the flame retardant.[15]

Most research, activism, and regulations have focused on certain halogenated flame retardants that have been found to be *persistent* (meaning they continue to exist in the environment without breaking down), *bioaccumulative* (meaning they accumulate in body tissues), and *toxic* (meaning they can cause harm to body tissues). This combination of traits is referred to as *PBT*: persistent, bioaccumulative, and toxic. Some PBT chemicals have been included in the Stockholm Convention on Persistent Organic Pollutants (POPs) and receive special attention by regulatory bodies, including the EPA and the European Union.[16] The importance of these three characteristics comes to the fore when stakeholders work to define risk differently, as I discuss in the next chapter. In general, *toxicants* are chemical substances that induce a toxic, or harmful, effect; *toxins* are biologically occurring toxicants; and *toxics* are chemical or synthetic toxicants that do not occur in nature.[17] Thus snake venom would most correctly be called a toxin, and a flame retardant chemical would be called a toxic chemical or a toxicant, though these terms are frequently used imprecisely.

Today the halogenated flame retardant industry in the United States is dominated by three companies: Albemarle, Chemtura (and its flame retardant division, Great Lakes Solutions), and ICL Industrial Products (ICL-IP, the U.S. division of Israeli Chemicals Limited).[18] In my research, I spent time in a research and development laboratory at one of these companies, which I call "FlameCorp."[19] All three companies manufacture halogenated, nonhalogenated, and blended flame retardants, and they also manufacture other chemical products in addition to flame retardants. These companies are represented in the United States by the American Chemistry Council (ACC), the trade association for the American chemical manufacturing industry, and its flame retardant advocacy panel, the North American Flame Retardant Alliance (NAFRA). In Europe, the industry is represented by the Bromine Science and Environmental Forum (BSEF) and the European Chemical Industry Council. Despite the restrictions on certain flame retardant formulations that have been enacted in recent years, the sector remains a profitable and growing international industry.[20] The U.S. demand for flame retardants is predicted to reach nearly a billion pounds in 2016, and global revenues are predicted to top $10.3 billion by 2019.[21] The international nature of the industry is also important. The Asia/Pacific region accounts for half of all global demand, and the spread of flammability standards throughout the developing world is a significant driver of growing demand for the chemicals.[22]

EARLY FLAME RETARDANT HISTORY

Flame retardants are "some of the earliest examples of modern chemistry," in the words of a fire scientist. Both the ancient Egyptians and the Romans used alum and vinegar to make wood less flammable.[23] The first patent for a flame retardant was granted in England in 1735.[24] The manufacture of modern flame retardants began in the 1960s as part of the post–World War II expansion of the chemical industry.[25] The infamous legacy chemicals asbestos and polychlorinated biphenyls (PCBs) were used as flame retardants, among their many other uses.[26]

The initial use of flame retardants in consumer products such as furniture, electronics, and even children's pajamas was in response to well-intentioned fire protection measures. The fire incidence rates in the United States were especially high in the middle of the twentieth century. Over 40 percent of U.S. adults were smokers, and 45 percent of households still heated with wood or coal stoves.[27] Furthermore, the growth in plastics and polyurethane products meant that materials in the home were more flammable than ever. So-called "torch sweaters" made of brushed rayon or cotton and children's costume cowboy chaps made of fuzzy rayon fibers woven to resemble fur caused an unknown number of deaths and serious injuries, and led to over 100 lawsuits in the 1940s and 1950s.[28] At least 6,400 Americans died each year between 1950 and 1970 from fire, flames, and smoke, based on an analysis of death certificates.[29] Flexible polyurethane foam—the soft, petroleum-derived foam used in upholstered furniture—was introduced into the furniture markets in 1957.[30] This foam is so flammable that it is commonly referred to as "solid gasoline." This confluence of fuel and fire sources contributed to high rates of home fires: there were almost 750,000 home fires and 6,000 civilian deaths in 1977, more than twice today's levels.[31]

In response, state and federal governments developed flammability standards and building codes to improve fire safety. After years of industry opposition following the torch sweater and cowboy chap incidents, the Flammable Fabrics Act was passed in 1953, which included testing provisions specifically targeting the most flammable cotton and rayon fabrics. A set of revisions in 1967 and 1972 expanded coverage to children's sleepwear.[32] Typically flammability standards like these do not directly require the use of flame retardants; however, for some materials, including polyurethane foam, flame retardants are the only cost-effective way to reduce flammability. As a result, these standards are very consequential. As a flame retardant industry representative argued, "Our flame retardants industry lives and dies by regulation." These standards are typically product specific, such as the Consumer Product Safety Commission's regulations for children's sleepwear, the State of California's standard for upholstered furniture, or the U.S. Department of Transportation standards for automobile interiors.[33] Often these performance-based standards and fire codes are developed and revised through fire safety organizations and international regulatory bodies using lengthy, consensus-based processes involving numerous stakeholders.

California's Technical Bulletin 117 (TB117) was especially important in driving the use of flame retardants in furniture. First enacted in 1975, it required furniture foam to withstand a 12-second open flame. Adding flame retardants to the foam was the most common and cost-effective way for manufacturers to meet the standard; exposure scientists have found that the vast majority of all upholstered furniture—purchased in

California or elsewhere in the United States—contains flame retardants.[34] After a lengthy regulatory process and significant debate between environmental health advocates and the chemical industry, TB117 was updated to a smolder standard that can be met using external fabrics instead of added flame retardants. The manufacturing company Chemtura sued to stop the regulation, but the lawsuit was thrown out of California Superior Court in August 2014. The new standard has been adopted by other municipalities, and major companies are now producing lines of furniture without added flame retardants.

THE 1970S: FLAME RETARDANT EXPOSURES AND DISASTERS

Though modern flame retardant use began in the 1960s, the chemicals received little public attention until two high-profile events occurred in the 1970s. The first involved polybrominated biphenyls (PBBs), halogenated flame retardants so hazardous that two major chemical companies, Dow and DuPont, decided not to proceed with manufacturing after early toxicity testing.[35] In 1973, several thousand pounds of PBBs sold by Michigan Chemical Corporation under the trade name "FireMaster" were mistakenly mixed with animal feed in place of a nutritional supplement named "NutriMaster." The contaminated feed was distributed around the state and fed to millions of livestock.[36] Cattle were acutely poisoned, and many became very sick and were slaughtered. Low-level contamination followed of all types of livestock on over five-hundred farms, along with the poisoning of people who had consumed the animal products. Farmers dumped the contaminated milk and buried the dead animals in their fields, not knowing that they were adding a persistent and bioaccumulative toxicant to their pastures. Instead of taking responsibility for the mistake, Michigan Chemical Corporation and the Farm Bureau accused farmers of poor animal husbandry and management.

It took nearly a year to identify the source of contamination, develop testing methods, and establish and refine safety levels. During this time, the animal products made their way into the food supply, and thousands of people consumed contaminated milk and meat. PBBs were identified as the source of contamination only through a remarkable series of events: one of the first farmers whose cows were sickened happened to be a former chemist, and he and others reached out to a few dedicated individuals who pursued multiple avenues of investigation, including scientists who doggedly investigated the mysterious results from their chemical analyses.[37] Joyce Egginton, a journalist who wrote a book on this incident, calls this contamination episode "probably the most widespread, and the least reported, chemical disaster ever to happen in the western world."[38]

PBBs are no longer manufactured in the United States and are restricted through the European Union's Restriction of Hazardous Substances law.[39] Epidemiological studies of people who ate contaminated meat and dairy are ongoing.[40] Researchers have found associations between PBB exposure and reproductive disorders, altered hormone levels, and higher rates of lymphoma and digestive-tract cancers, with some health effects occurring in the next generation.[41] For example, the daughters of women with high levels of PBB exposure have elevated rates of miscarriage.[42]

A second key event in the 1970s also brought widespread public attention to flame retardants. Following the implementation of new flammability standards for children's sleepwear in 1973, clothing manufacturers chose to add the cheap, readily available chemical tris(2,3-dibromopropyl) phosphate (also called brominated tris) to polyester pajamas.[43] Several years later, researchers Arlene Blum and Bruce Ames published a paper showing that brominated tris was potentially mutagenic—meaning it could cause changes to genetic material such as DNA.[44] The discovery that the chemical was mutagenic was alarming, as metabolites of the chemical could be measured in children's urine after only one night of sleeping in treated pajamas.[45] The Consumer Products Safety Commission (CPSC) was already aware of preliminary findings from a National Toxicology Program (NTP) study showing that brominated tris was a potential carcinogen, and the new publication added to research findings from the National Cancer Institute and mounting pressure from environmental organizations. Within months, the CPSC banned the use of the compound in children's sleepwear.[46] Though this ban was overturned that same year in federal court on procedural grounds, the CPSC maintained that they retained enforcement authority against products containing brominated tris through the federal court system, and the commission pursued eight of these lawsuits in 1978.[47] Brominated tris is no longer used in the United States, and it is restricted by the EPA under a Significant New Use Rule (SNUR).[48] Companies with an unwanted surplus of Tris-containing pajamas sold them to wholesalers for pennies on the dollar, who shipped them overseas to foreign markets.[49] To top off this implausible series of events, President Ronald Reagan signed legislation in 1982 allowing companies to seek reimbursement from the federal government for losses suffered as a result of the brominated tris ban.[50] One manufacturer did file a multi-million dollar suit to claim damages, but it was dismissed on the grounds that the CPSC and its members could not be held liable when they developed regulations within their congressional authority; the dismissal was affirmed on appeal in 1983.[51]

After the removal of brominated tris from the market, clothing manufacturers briefly switched to tris(1,3-dichloro-2-propyl)-phosphate (TDCPP, TDCP, or chlorinated tris), but it was soon shown to be a mutagen like its chlorinated counterpart.[52] The industry stopped using TDCPP in children's pajamas voluntarily in the early 1980s.[53] Today children's sleepwear must either be made with fabric that passes flammability tests or be "tight fitting."[54] According to the CPSC, most of the market for children's sleepwear is tight fitting for younger children. For older children, sleepwear is made out of fabrics such as polyester that are inherently flame resistant without added chemicals, though some companies do treat cotton sleepwear with flame retardants. But although TDCPP was removed from children's sleepwear, it was not removed from the market altogether. By the mid-2000s, it was a leading chemical in furniture foam.[55]

THE 1980S: HOUSE FIRES, FLAME RETARDANTS, AND THE TOBACCO INDUSTRY

Following the two high-profile incidents in the 1970s, flame retardants received little public attention for several decades. However, behind the scenes, important changes involving flame retardant advocacy and research set the stage for the chemicals' high profile later on.

The first of these changes involved advocacy work around fire standards, largely directed by the tobacco industry. Concerned that high numbers of cigarette-caused house fires would prompt restrictions on cigarettes or force changes in cigarette formulations, industry leaders at the Tobacco Institute developed a strategy to shift the blame for house fires away from cigarettes as the cause of ignition and onto furniture as the object that caught on fire. As early as the 1930s, researchers had developed a "fire-safe" cigarette that would self-extinguish, meaning the cigarette would not smolder all the way down without active inhalation by the smoker, but the tobacco industry was concerned that this type of reformulation would decrease sales and repel customers.[56] To deflect attention away from cigarettes' role in house fires, the tobacco industry pushed for increased flammability standards and invested heavily in relationships with the fire service, giving hundreds of thousands of dollars in grants to fire departments around the country.[57]

The key players in this work were Peter Sparber, a former tobacco industry lobbyist and executive at the Tobacco Institute, and the National Association of State Fire Marshals (NASFM), a nationwide association of senior fire officials working to this day on fire codes and standards, fire prevention, and fire incident investigation. In 1989, Sparber helped to found NASFM, and he served on their executive board for over a decade. For years, NASFM received funding from the tobacco industry and flame retardant manufacturers.[58] Sparber was even paid by the tobacco industry for his supposed volunteer duties at NASFM.[59] Though NASFM denies that Sparber unduly influenced their agenda, his connections with the tobacco industry and his assistance with NASFM's legislative, organizational, and advocacy work is undeniable.

NASFM played a key role in preventing the enactment of a nationwide requirement for fire-safe cigarettes, echoing the tobacco industry's critiques of the test method. As *Chicago Tribune* reporters uncovered as part of a 2012 investigation, one of the first regulatory acts taken by NASFM was to support federal legislation requiring further research on fire-safe cigarettes, instead of a competing piece of legislation that would have required the cigarettes to reformulate more quickly.[60] Into the 1990s, NASFM remained supportive of flammability standards and critical of fire-safe cigarettes and the "Gann method" used to evaluate their safety, reflecting the tobacco industry's positions. As outlined in R.J. Reynolds' 1996 strategic plan, "fire officials must keep pressure on the [Consumer Products Safety] Commission to focus on the *fuels* rather than *ignition sources*."[61] Involvement by NASFM was listed as the first "Project" in the tobacco company's strategic plan, which noted that "NASFM is the one fire group to actively and knowledgeably object to the Gann test method."[62] But pressure to require fire-safe cigarettes continued to mount across the country, and between 2000 and 2010 all fifty states passed fire-safe cigarette legislation. All cigarettes in the United States are now self-extinguishing.[63]

The Tobacco Institute closed in 1999, and by that point Sparber was lobbying on behalf of the advocacy organizations representing the flame retardant chemical industry, often on the same issues.[64] From 1998 to 1999, Sparber and Associates was paid $80,000 for lobbying fees by BSEF, the European chemical trade group, and another $80,000 by Great Lakes Chemical (now part of Chemtura).[65] Working alongside Peter Sparber at Sparber and Associates was Peter O'Rourke, who, according to his current employment profile, represented NASFM for fifteen years.[66] O'Rourke was listed as a NASFM "staff member" in a 2008 newsletter.[67] Sparber's connection with the flame retardant industry endured. From 2005 to 2008, Chemtura was Sparber and Associates' only named client, though no lobbying fees were disclosed.[68]

Throughout this period, NASFM supported strengthened furniture flammability standards, again addressing fuels instead of cigarette ignition. They also testified against flame retardant bans in Maine, Washington state, and California, arguing that environmental concerns should not trump fire safety. NASFM continues to support a federal furniture flammability standard for small open flames, like the earlier version of California's TB117, though they now also support restrictions of "chemicals of concern" in furniture foam.[69] Sparber no longer publicly works with NASFM, though he was awarded the NASFM "Hall of Fame" award in 2008.[70]

Thus, the work of NASFM and proindustry advocates laid a foundation for the regulatory conflicts to come. On the scientific front, the work of environmental scientists studying other POPs laid the foundation for the scientific research on PBDEs, a group of flame retardants classified based on the number of bromine atoms around two carbon rings. Three main commercial PBDE formulations were sold in the United States: pentabromodiphenyl ether (PentaBDE), which was mostly used in polyurethane foam; octabromodiphenyl ether (OctaBDE), which was mostly used in plastics for electronics; and decabromodiphenyl ether (DecaBDE), which also was mostly used in plastics for electronics.[71] PBDEs are persistent and bioaccumulative, and their chemical similarities to other known toxicants soon attracted researchers' attention.

PBDEs were first measured in the environment in a handful of studies in the late 1970s and early 1980s, often alongside of other organo-halogen chemicals of concern.[72] A 1979 study found PBDEs, PBBs, and other brominated chemicals near a manufacturing plant.[73] Researchers tested for the PBBs that had poisoned Michigan farms and farmers, and for PCBs, widely used industrial chemicals that were banned when the Toxic Substances Control Act (TSCA) was enacted in 1976.[74] The similarities between PBDEs and these other organo-halogens extend beyond their alphabet-soup names to their chemical structures, and researchers who had been studying PCBs often looked to PBDEs as a next chemical of interest. Researchers in the early 1990s noticed that, unlike other persistent chemicals that showed steady or declining levels in environmental samples, PBDE levels were increasing dramatically.[75] This move from PCBs to PBDEs was an easy and desirable one for several reasons: interest in PCB research was declining after global bans, funding sources were drying up for the same reasons, and PBDEs are structurally quite similar to PCBs.

In addition to this exposure research, there was some attention to PBDEs in the late 1980s at the EPA because of concerns that they became contaminated with dioxins during manufacturing. The EPA put in place a Test Rule to study dioxin contamination in PBDEs and other chemicals.[76] After this Test Rule, "the issue faded from attention," according to an EPA scientist. Flame retardants remained a niche topic. A longtime researcher said that when he started studying brominated flame retardants in the late 1990s, "you could fit all the research into two three-inch binders."

THE 1990S: "EYE-OPENING" BIOMONITORING RESEARCH

The next major shift in the timeline of flame retardant research occurred when a team of Swedish researchers studied PBDEs using *biomonitoring*, research that tests for the presence and accumulation of chemicals or chemical metabolites in the human blood. According to researchers active in the field, Åke Bergman, Koidu Norén, and Daiva Meironyté had studied other POPs, and in the mid- to late-1990s noticed mysterious peaks on readings from their chromatograph (a laboratory machine that separates and identifies chemical components in a mixture). They adapted their analytical methods from their work on organo-chlorines in order to measure the PBDEs, and they analyzed stored samples of breast milk collected from Swedish women dating back into the 1970s.[77] At a scientific conference in 1998, they presented explosive findings: PBDE levels in Swedish women's breast milk had increased exponentially between 1972 and 1997. These results were published in 1999.[78] A follow-up study the next year compared increases in PBDE levels with declines in other organo-halogens, concluding that "the exponential increase of PBDEs in breast milk is alarming" and calling for measures to stop exposure to PBDEs.[79] They were not the first to measure PBDEs—their article cites over a dozen previous studies in environmental media, wildlife, and humans. However, the documentation that these chemicals were increasing so dramatically in breast milk sent waves throughout the international scientific community.

Their work directly inspired researchers in the United States and around the world who were stunned by the exponential increase. In interviews, numerous scientists traced their work on flame retardants or the work of their colleagues back to this single Swedish study. Some had themselves attended the 1998 conference, while others heard the news from colleagues or through scientific networks. One exposure scientist who had been in attendance said the Swedish paper "electrified many of us." A toxicologist agreed: "We went to Sweden to a Dioxin meeting in '98, in Stockholm, and—I'm never going to pronounce his name right—Meiron had that famous [paper], so we were just in the audience, and it was bang! We're going back and getting this stuff." Scientists said that the graph showing an exponential increase in PBDE levels prompted endless scientific questions. One researcher talked about a colleague who had a similar "aha" moment:

> When they presented that at the conference [my colleague] was at, he was like: "wow!" You know, he just had this 'wow' moment, where he was like, "this is so interesting that this is happening" . . . There's evidence that there's a big problem . . . something is going on, and we don't really know much about it at all, which creates huge opportunities for interesting research. So I think as soon as you see that, it's like, oh man, this is, there's a lot of interesting stuff that we could do here. You know, you could just put out that one figure up in a room full of scientists and it generates lists and lists of questions that you would be interested in answering. And all those questions become research opportunities.

Another toxicologist echoed this perspective: "It's a 1998 study which initially created a lot of buzz in Sweden, but then it just rippled out from there. And that's what prompted everybody else to say, well, what about environmental media? What about wildlife? What about health effects? What about people in other countries?" This one well-conducted, timely study with remarkable findings inspired a significant number of scientific follow-up questions and projects.

The ripple effects of this study reached the EPA as well. One prominent EPA researcher described attending a scientific meeting on flame retardants in 2001, and emerging "shell-shocked" because of data on PBDE increases in breast milk and emerging toxicological research showing neurodevelopmental effects from PBDE exposure.[80] After returning from the conference, this researcher "began to kind of raise a ruckus [because] this was clearly an area that we needed to know more about." A series of talks and conference calls organized by the Office of Research and Development (ORD) and Region 9 (the EPA region in California) created a "groundswell" of interest and activity at EPA.

These early efforts to mobilize activity around PBDEs, however, were resisted by some in the EPA's Office of Pollution Prevention and Toxics (OPPT), perhaps because of the insinuation that the OPPT was failing in its mission to appropriately regulate chemicals. I was told that Region 9 "went off the reservation a little," meaning that they pursued activities that challenged the official EPA agendas and sanctioned risk management efforts. One researcher involved in raising awareness early on said she became *"persona non grata*, because they didn't want to hear this." But the body of scholarship on PBDEs conducted by different offices at the EPA, other federal agencies, and academic researchers continued to grow through the early 2000s.

The volume of scientific research on flame retardants has increased dramatically since those early publications. Figure 2.1 shows the number of articles in the PubMed database annually from 1995 to 2012 that resulted from searches for flame retardants, brominated flame retardants, or PBDEs.[81] In addition to the hundreds of articles on flame retardants now published every year in the peer-reviewed literature, this research area is significant enough to have its own Annual Workshop on Brominated and Other Flame Retardants (commonly called the BFR Workshop), to figure prominently in other large academic conferences, and to receive millions of dollars in federal grants annually from the EPA and the National Institutes of Health.

In interviews with leading flame retardant researchers from the past decade, only one scientist mentioned work on flame retardants prior to 1998. This researcher had studied dioxin contamination in PBDEs, not the toxicity or exposure of PBDEs on their own, and the Swedish study inspired her to return to the issue. She attended a BFR Workshop in the early 2000s, and described how the findings brought the issue back into focus:

> My focus had been the brominated dioxins and furans, which in fact are contaminants of the brominated flame retardants. And when I had done work on those, I felt like unless there was some new information or unless people started finding them out in the environment or in people, it just wasn't an issue . . . But at this meeting what I hadn't realized was the tremendous growth in the BFR industry that started kind of in the late '80s and had just peaked. And this hit home really hard when they showed blood data, or, I can't remember if it was serum or milk data from Sweden, showing linearly increasing levels in the population. And that was very concerning to me.

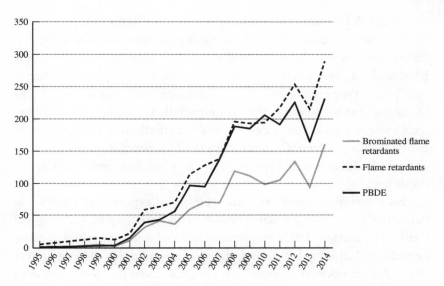

FIGURE 2.1. Number of PubMed Articles, 1995–2014

Note: PubMed searches for terms and date completion conducted March 23, 2015.

Thus, even for researchers who had worked on flame retardants prior to 1998, the Swedish study represented a momentous shift and inspired great scientific interest and concern.

THE 2000S: RESEARCH AND REGULATION OF PBDES

Very little was known about human health effects in the early years of PBDE research. What was known was that PBDE exposure was widespread. Researchers at the Silent Spring Institute in Massachusetts conducted the first household exposure studies of PBDE levels in the United States and found high levels of PBDEs and other flame retardants in house dust, kicking off a substantial amount of exposure research in the United States.[82] Small studies of U.S. adult populations found widespread exposure in blood and breast milk, and a nationally representative study by the Centers for Disease Control and Prevention confirmed that over

97 percent of U.S. adults had detectable levels of PBDEs in their blood.[83] Evidence was mounting that exposure was ubiquitous and varied greatly: PBDE exposure was distributed log normally, meaning that most people have relatively low levels in their bodies, but the people at the upper "tail" carry very high levels.[84] As a toxicologist who researched PBDEs in the early 2000s told me, this was especially troubling: "Even at low-level increased risk, you're dealing with population-based numbers in public health . . . Even if you just take the U.S. population, so you have 300-plus million people exposed, even at a low risk, well that's really important."

This growing information about widespread exposure to PBDEs was particularly concerning when combined with the emerging toxicological research in animals: PBDE exposure in rats and mice was found to cause neurological deficits, behavioral changes, reproductive disorders, liver damage, disruption of immune function, hormonal problems, and even cancer.[85] Observers also noted that the industrial markets for PBDEs and other flame retardants were large, profitable, and growing. The first regulations followed shortly. Sweden took action to regulate PentaBDE in 1999 after publication of the Swedish breast milk study.[86] The European Union completed a formal risk assessment in 2000, and banned the marketing and use of PentaBDE and OctaBDE two years later.[87] All of this prompted the first attempts to regulate PBDEs in the United States. Starting in the early 2000s, environmental activists pushed for PBDE bans, focusing on the two main commercial formulations: PentaBDE and OctaBDE. The third formulation, DecaBDE, was used in different types of products (especially electronics and wiring, not furniture foam), and at the time was (falsely, we now know) believed to be a large enough molecule that it could not be readily absorbed by people.

Flame retardant chemicals received significant attention from the EPA during this time, with numerous scientific, regulatory, and voluntary projects including a PBDE "Project Plan" in the mid-2000s and "Action Plans" for several flame retardants announced at the end of 2009.[88] As one senior person in the New Chemicals Program told me, flame retardants were "a good thing" for the agency to focus on because they attracted so much attention that they forced the EPA to "poke our head out of the fox hole" to see what the whole chemical landscape looked like. Despite the early scientific and regulatory interest in flame retardants at the EPA, everyone knew that federal regulation by the EPA was unlikely. The EPA has the authority to regulate industrial chemicals through the TSCA, enacted in 1976.[89] However, TSCA is widely criticized as being insufficiently protective of public health and the environment even by EPA administrators; it has so little authority to restrict the "existing chemicals" that were in use when the statute was enacted that it banned only five chemicals in commerce before 1976.[90] (I will discuss TSCA and its limitations in greater detail in chapter 5.)

Because of the shortcomings of TSCA, the EPA pursued alternative but limited means of controlling flame retardant chemicals. As a senior EPA representative told me about the agency's work on PentaBDE, the chemical was "a problematic flame retardant: health concerns, bioaccumulation concerns, persistence concerns, and doubling in breast milk every five years in the U.S." The Voluntary Children's Chemical Evaluation Program (VCCEP) was one of the key features of EPA's early work on PBDEs.[91] VCCEP was launched in 2000 under President George W. Bush to identify and fill data gaps related to health and exposure concerns for children through voluntary rather than regulatory actions. The EPA identified twenty-three chemicals, including the three PBDEs that had been detected in biomonitoring research, that were actively used in the United States and were thought to be of particular concern for exposure in children. Twenty of those twenty-three VCCEP chemicals were then sponsored by companies or industry consortia who agreed to participate in this voluntary program. PentaBDE and OctaBDE were sponsored by Great Lakes Chemical Corporation (now part of Chemtura), the only U.S. company to manufacture the chemicals, and DecaBDE was sponsored by the ACC's Brominated Flame Retardant Panel, consisting of Great Lakes Chemical Corporation along with Albemarle Corporation and Ameribrom (now part of ICL-IP). These companies completed a "peer consultation process" to assess PentaBDE and OctaBDE in June 2003, which included a review of all available data as well as an assessment of ongoing data needs.[92]

In the meantime, inspired by the PBDE bans in Europe and the mounting scientific evidence that these chemicals were of pressing concern, several states began to take action. This work was led by a broad coalition of environmental and health activists, including traditional environmental groups, public health associations, disability advocates, environmental health and justice groups, consumer rights organizations, and the fire safety community. (I discuss this coalition, their activities, and the industry-funded opposition in greater detail in chapter 6.) In 2003, California pushed for the first state-level ban on PentaBDE and OctaBDE. This regulatory action was directly linked to advances in the science on PBDE exposure: inspired by the 1998 presentation by Swedish researchers about PBDE levels in breast milk, a team of researchers in California modified the methods of their ongoing study of organo-halogens in breast cancer tissue to include PBDEs. In the words of one of these researchers,

> We were lucky at the time we were doing a breast cancer study, so we had the adipose tissue from breast cancer survivors and [our local university]. And the grant was to study dioxins and PCBs, and they were linked to breast cancer. Well, we modified the methods and included PBDEs, and our levels were 30 times higher than what the Swedes had been seeing! And we thought there was something wrong: check your calculations, are you sure? But it was correct.

These findings and subsequent studies showed high levels of PBDEs in sea mammals and humans in California, higher than the national average.[93] These emerging scientific advances were covered in local media and led to an increased awareness among regulators and environmental activists in California, who organized against strong industry opposition to pass the PentaBDE and OctaBDE ban in 2003, to go into effect in 2006.[94] Governor Gray Davis signed the legislation just weeks before he lost his reelection bid to Governor Arnold Schwarzenegger.

California's legislative precedent inspired similar actions around the country. By 2006, Hawaii, Illinois, Maine, Maryland, Michigan, Minnesota, New York, Oregon, Rhode Island, and Washington state had all passed similar bans on the production, use, sale, and or/distribution of PentaBDE and OctaBDE, and a number had turned their attention to the last PBDE standing: DecaBDE.[95]

Also by the early 2000s, Great Lakes Chemical Corporation had developed Firemaster 550, a functional alternative for PentaBDE's primary use in furniture foam. This chemical was reviewed by the EPA's New Chemicals Program and approved for use in 1997, with a Consent Order requiring additional testing. Chemtura conducted fifteen studies on the new brominated component of Firemaster 550, and the EPA found that the chemical was "less persistent and less likely to bioaccumulate" than PentaBDE.[96] In combination with this regulatory review, the agency conducted its first alternatives assessment through the Design for the Environment (DfE) program to evaluate alternatives to PentaBDE's use in furniture foam. As a DfE representative explained it, once Great Lakes developed Firemaster 550 and decided to withdraw PentaBDE from the market, the "EPA was concerned that we might be jumping from the frying pan to the fire and wanted to understand the alternatives . . . We felt that the responsible course was to understand what would happen if we blocked further production of PentaBDE after this phase-out." They identified functional alternatives for PentaBDE, and after a multiyear effort, the report concluded that, in the words of an EPA scientist involved in the assessment, "the alternatives were not associated with the same level of concern as PentaBDE, but that all of the alternatives were associated with some level of concern, and further work was needed." The report was published in 2005, and was updated in 2015 to reflect the latest research and regulatory activities.[97]

This combination of growing evidence of hazard and exposure concerns, domestic and international regulatory pressures, and the opportunity to market a newly developed alternative led Great Lakes to negotiate a voluntary withdrawal of PentaBDE and OctaBDE in November 2003, effective at the end of 2004.[98] As one long-time EPA employee said,

> They [Great Lakes] came up with [a chemical] that would work and they felt that they could cannibalize their market. And so they wanted to quick switch out and minimize liability associated with having produced this pentabromodiphenyl ether . . . They were concerned that having produced it even when it was pretty broadly understood that it was a problematic chemical, that they could be vulnerable to class-action.

The EPA added onto this by adopting a SNUR in 2006 for PentaBDE and OctaBDE to prevent another manufacturer producing or importing the chemicals without additional review by the EPA.[99] Informally, EPA representatives call this type of regulatory action a "dead chemical SNUR," because it was put in place for a chemical that was leaving the market.

Building on the successes of regulating PentaBDE and OctaBDE, activists pushed for state-level restrictions of DecaBDE starting in 2004. Though the science on the chemical's toxicity and exposure concerns was not as clear-cut or voluminous as with PentaBDE, concerns grew regarding the toxicity of DecaBDE as well as its ability to break down into lower congener PBDEs.[100] The industry's voluntary VCCEP assessment of DecaBDE began in 2003, but in 2005 the EPA responded that the industry's assessments "are not sufficient to adequately characterize its risks to children . . . The fate of decabromodiphenyl ether in the environment is largely unknown."[101] The EPA and the industry consortium were unable to agree on the test methods needed to fill data gaps within the voluntary program, and the agency pursued a regulatory Test Rule requiring a developmental neurotoxicity study through the Existing Chemicals Program. In a 2008 letter to the ACC, EPA Assistant Administrator James B. Gulliford expressed his "regrets" that the voluntary approach had failed, and said the agency would "move forward to use the mandatory means within its authority to obtain data to adequately characterize the risks of this chemical to children, adults, and the environment," thus formally signaling the agency's intent to require additional testing through TSCA.[102] In response, the ACC consortium offered to comply with an Enforceable Consent Agreement to develop the data, but EPA objected to the specific test methods the ACC proposed to use.[103]

Before the EPA could move ahead with the Test Rule, the flame retardant companies announced a voluntary phase out of DecaBDE in the United States, to take effect by 2013.[104] According to EPA insiders, the companies decided on their own to phase out DecaBDE because they did not want to do the more than $1 million of testing that would be requested of them under VCCEP.[105] The entire VCCEP program was discontinued in 2010, and the Office of the Inspector General subsequently found that the program had failed to provide adequate or timely data on chemical risks.[106]

The EPA pursued additional risk management activities through the 2009 DecaBDE Action Plan, one of ten Action Plans issued by EPA as part of the Obama administration's early first-term efforts to use TSCA's existing authority.[107] The Action Plan outlined a series of risk management actions, such as including PBDEs on the agency's proposed "chemicals of concern" list, implementing a dead chemical SNUR and Test Rule for DecaBDE, and supporting the phase out of PBDEs with DfE alternatives assessments.[108] The Test Rule and SNUR for DecaBDE were proposed in 2012, and the rules designated any manufacture, import, or processing of DecaBDE for any nonongoing use as a significant new use, placing the burden on the supply chain industries to request exemptions.[109] The proposed DecaBDE SNUR covered all uses of PBDEs, even though the EPA was aware of ongoing uses in the transportation and aviation sectors. As of April 2015, these proposed regulations have not been finalized, leaving open the possibility that they have not taken effect as intended.[110] Concerns about DecaBDE's use by the aerospace industry and its presence in recycled plastics may be holding up finalization of the rules.[111]

Concurrently with EPA's ongoing negotiations and regulations over DecaBDE, several states passed their own bans of DecaBDE in certain product uses, sometimes as a preemptive or symbolic ban in sectors where they are not currently used, or dependent upon the identification of safer alternatives. Europe has also strengthened regulation of DecaBDE. The European Union's Registration, Evaluation, Authorisation, and Restriction of Chemicals (REACH) law requires industry to demonstrate the safety of new and existing chemicals, and in 2009 identified other flame retardants for future risk assessment.[112] Recently, REACH added DecaBDE to its list of "Substances of Very High Concern" because some of its breakdown products are considered to be very persistent, very bioaccumulative, and toxic.[113]

Some stakeholders have argued that much of the PBDE regulation was precautionary in nature. When PentaBDE and OctaBDE were removed from commerce in 2004, there was little epidemiological evidence that they had caused harm to humans. As a policy expert explained to me in an interview in 2009, PBDEs are a "good example" of how precautionary regulation can be successful: "We don't have a lot of human evidence, but . . . we know that they build up in the environment and in bodies, we know that vulnerable populations are exposed, we know that there are alternatives, and they may not even be necessary. So that should lower our threshold of how much we actually need to know before we can do something about it." The ubiquity of people's exposure to the chemicals, combined with evidence that they were bioaccumulative and toxic to laboratory animals, provided a strong enough foundation for restriction. Scientists who had studied flame retardants for a decade or more told me that PBDEs, especially PentaBDE, became more problematic the more closely researchers looked at them. One toxicologist described his early research: as he reviewed the toxicological and exposure research on PBDEs, "it became more and more of a story . . . The deeper you got, the more interesting the science was. The more it piqued your interest that really this may be an environmental challenge."

Scientific advances since the early 2000s have shown that the early regulation of PentaBDE was prudent. Over the past decade, researchers have found that PBDEs are ubiquitous and rapidly accumulate in the environment, wildlife, and people, who are exposed through household dust, physical contact, ingestion, smoke, and contaminated air.[114] Even though the chemicals are no longer manufactured, exposure is nearly impossible to avoid. The 2008 to 2009 National Health and Examination Survey (NHANES) study detected a dominant component of PentaBDE in "nearly all of the NHANES participants."[115] Studies have found levels of PBDEs in highly exposed women, young children, and infants that are at or above the levels that cause neurological and reproductive harm to laboratory animals.[116] Certain occupations face high exposure, including some manufacturing workers, workers at recycling plants and airports, flight attendants and pilots, and firefighters.[117] Additionally, minorities and the poor have higher body burdens of PBDEs.[118] However, PBDE exposures may be decreasing following the chemicals' removal from commerce. A recent biomonitoring study on PBDE exposure in pregnant women found lower levels than in previous studies, leading the authors to conclude that "lower levels may be in response to legislation restricting the production, sale and use of these compounds."[119]

Scientists have learned more about the toxicity of PBDEs as well through laboratory and animal studies. Brominated flame retardants as a class of chemicals disrupt the endocrine system, interfering with hormones and potentially harming reproduction and development.[120] Because

PBDEs can cross the placenta and blood-brain barriers, researchers are particularly concerned about potential neurological and developmental effects and thyroid hormone disruption; pregnancy increases the demand on the mother's thyroid hormone system, and thyroid hormones are closely connected to fetal and child development.[121] In animal studies, PBDEs have been shown to be neurological and reproductive toxicants, and DecaBDE is possibly carcinogenic.[122] Not all studies have found significant effects; for example, a multigenerational study of DecaBDE in rats funded by the flame retardant industry found no developmental neurotoxicity effects.[123]

Additionally, epidemiological studies in human populations have confirmed earlier findings from animal studies.[124] In epidemiology, chemical toxicity is often discovered following occupational exposures or industrial disasters that cause significant human exposure, and animal studies then confirm the findings and identify toxicity mechanisms. According to a toxicologist, PBDEs are "actually one of the few chemicals where the animal studies actually showed effects and the human studies are kind of confirmatory. You get things like lead, you get mercury, you get some of these other neurotoxic chemicals. Unfortunately they're found first in humans, and then the animal studies are done to kind of confirm that and to look into mechanisms and that sort of thing. But PBDEs kind of went the other way around." With PBDEs, the animal toxicological studies preceded epidemiological studies by years.

Some of the greatest concern is around neurological development in children and reproductive health in adults. Responding to research in laboratory animals suggesting that PBDE exposure during pregnancy might lead to developmental or behavioral problems in childhood and later in life, a number of epidemiological research teams have initiated cohort studies tracking PBDE exposure in pregnant women and the health and development of their babies.[125] One longitudinal study found that a tenfold increase in prenatal PBDE exposure was associated with decreased IQ and increased hyperactivity by the time the child turned five, confirming results from earlier studies.[126] PBDEs are also thyroid disruptors, interfering with a complex hormonal system that impacts body functions ranging from metabolism to reproduction. Researchers have found that higher PBDE exposure is associated with hormone level changes and lower semen quality, longer time to pregnancy in women, and lower thyroid hormone levels during pregnancy.[127]

TODAY: BEYOND PBDES

A growing body of recent research examines other flame retardants, some that have been used for decades and others that have been recently developed as replacements for PBDEs following regulation. These flame retardants are often called "new" or "emerging," though these terms can be misleading because many of them have been in use for decades. Starting in 2011, the annual Workshop on Brominated Flame Retardants became the Workshop on Brominated *and Other* Flame Retardants (emphasis added) to include the growing body of research on chlorinated and phosphorous flame retardants. The flame retardant industry has quickly filled the gaps created by PBDE bans with other chemicals, some longtime chemical workhorses (such as TDCPP, in use since the 1970s) and other newly developed chemicals. Scientists working on flame retardants called this a "moving target," a "scenario of scientific catch-up," and a "toxic shell game."

Many researchers who studied PBDEs have since conducted research on other flame retardants. Sometimes they essentially replicate their early research projects with the next "hot" flame retardant, using similar scientific methods or analyzing saved samples of dust or body fluids from earlier projects. Part of this movement represents the trajectory of environmental exposure and toxicity research. An exposure scientist described the work of a chemist who develops methods that allow researchers to test for chemicals:

> She figures out how to measure Deca[BDE], and that's novel and brand new . . . But for her then to just use her lab as this sample processing facility at that point gets old or redundant, right. So she says, "here's the method. Now commercial labs can do that. I'm doing the methods for TBBPA now." . . . You just keep going down the line . . . She's brilliant. It wouldn't be exciting if she didn't move on.

Sometimes it is no small feat to identify the next generation of flame retardants of concern. One researcher explained how he collaborated on a project where analytical chemists worked for nearly a year to identify "some unknown peaks in [the] chromatograph" by ordering chemicals from flame retardant manufacturers to "match up and figure out what it was." Because manufacturers can claim that basic information about their products, including chemical structure and identity, is confidential for business purposes, it is often difficult or impossible to find out which flame retardant chemicals are used in which consumer products. As a public health scientist said, "The identity of the brominated flame retardant is trade secret, so they don't even have to tell you the identity of the compound. I think that's incredible." By the time researchers identify the chemicals, he explained, "we're five years behind the game already."

To combat this delayed timeline, some scientists and environmental activists cultivate relationships with chemical or supply manufacturers, who may confidentially provide information about the chemicals in use. For example, on an environmental advocacy conference call, a participant mentioned that he knew that a major furniture distributor was using TDCPP, but wondered what a well-known baby product manufacturer was using. Others take a more direct tack and test products to see what flame retardants can be measured in products. One researcher called this "doing it the hard way." Other scientists then use this information to choose their next research projects because they want to be working on current-use flame retardants. Many lamented the inevitable time lag involved: as an epidemiologist noted, "part of it is a little bit frustrating [because] we're ten years behind whatever they produce now." This time lag means that it is unknown whether newer flame retardants are safer than those they have replaced.

The PBDEs are widely known—the celebrities of the flame retardant world, as it were—but others are rarely mentioned, even if they are used in great volume. For example, aluminum hydroxide, the most widely used flame retardant by volume in 2011, was not mentioned by name in any of my over 110 interviews (though it did come up repeatedly at DfE). But a handful of other chemicals have also received significant regulatory, scientific, and activist attention. Tetrabromobisphenol-A (TBBPA) is used as a reactive flame retardant in electronics components. The EPA conducted an alternatives assessment of TBBPA's use in printed circuit boards, primarily motivated by end-of-life concerns: when printed circuit boards treated with flame retardants are burned to recapture valuable metals, they release higher levels of dioxins and furans, known carcinogens.[128] Hexabromocyclododecane (HBCD), used in building insulation, is a large, brominated molecule that is subject to an EPA Action Plan and was added to the Stockholm Convention on Persistent Organic Pollutants, with a limited exemption for use in building insulation.[129] Several new flame retardants have been developed as replacements for HBCD, including Emerald Innovation™ 3000 developed by Dow Chemical Company and licensed to Albemarle, Chemtura, and ICL-IP.[130] Though all currently available alternatives to HBCD's use in insulation are brominated, the large size of these chemicals suggests they may not be bioavailable to humans and thus would be of less concern for exposure and toxicity. However, some environmental advocates note that these chemicals may break down into smaller components that could pose their own risk.[131]

Finally, organo-phosphate chemicals used in furniture foam, including several chlorinated Tris formulations, have received significant attention. Part of this is likely due to the historical connections between TDCPP and children's pajamas: the same chemical that was used in children's sleepwear in the 1970s became one of the major flame retardants in furniture foam. As I described in chapter 1, Dr. Arlene Blum reengaged with flame retardants after a multidecade absence upon learning that the TDCPP she had studied in children's pajamas was widely used in furniture foam. The flame retardant Firemaster 550 is also used as a replacement for PentaBDE in furniture foam, and is a formulation of four other brominated and nonhalogenated chemicals; a recent study found that levels of Firemaster 550 in house dust had increased between 2006 and 2011, as the market shifted away from PentaBDE.[132]

Toxicological research suggests that some of these emerging flame retardants may be of significant concern for exposure or toxicity. Several chlorinated Tris chemicals, including TDCPP, tris(2-chloroethyl) phosphate (TCEP), and triphenyl phosphate (TPP), have been found in furniture, house dust, and baby products.[133] A toxicological study found that

TDCPP may be more toxic than some pesticides that are known neuro-toxicants.[134] Exposure to TDCPP and TPP is also associated with changes in hormone levels and decreased semen quality.[135] Recent studies have found that Firemaster 550 exposure is associated with changed hormone levels, reproductive and developmental alterations, and weight gain in rats.[136] Firemaster 550 also damages genetic material in minnows.[137]

As a chemical class, flame retardants remain in the public spotlight. Activists have pursued bans of several types of Tris chemicals at the state level. In particular, TDCPP was listed in 2012 on California's Proposition 65 based on evidence of carcinogenicity, the chemical has been added to Washington's state's list of chemicals of concern in children's products, and a flame retardant manufacturer recently announced it would phase-out production of TDCPP by 2015.[138] In 2013, the EPA announced that twenty flame retardants, including HBCD, TCEP, and two components of Firemaster 550, would be the focus of in-depth assessment activities.[139] This accumulating evidence supports assertions by some scientists and advocates that, although known toxic flame retardants should be phased out, the identified replacements may not be safer, and thus additional research and possibly regulation is needed.

LOOKING AHEAD

The science on flame retardants continues to grow and evolve. New studies regularly appear in the peer-reviewed literature, confirming the health risks from PBDE exposure and complicating the story about PBDE replacements. Flame retardants encapsulate the complexities of how industries, researchers, regulatory bodies, and activists act and interact around chemical risk and safety. The story of flame retardants comes down to a not-so-simple question: in the face of scientific unknowns and uncertainties, what should be done? The paths from knowledge to action, or from public health goals to public health policies, are frequently detoured by scientific complications, values, politics, and economic interests.

In spite of the regulatory and activist pressures and growing scientific concerns I have described, flame retardants remain a growing and profitable industry today, though the industry is more on the defensive than perhaps ever before. The three main flame retardant manufacturers compete against each other in the marketplace and sometimes work independently on advocacy efforts, but they often combine forces for product defense work. Significant tensions also exist between the halogenated and nonhalogenated flame retardant sectors. This is an important divide in the flame retardant world, as some argue that nonhalogenated chemicals are environmentally preferable.[140] As a *Plastics Engineering* article on flame retardants stated, "Halogenated or non-halogenated? That is the debate."[141] Additionally, companies along the supply chain have moved away from some types of flame retardants, a trend that is likely to spread in the furniture industry in particular after California's passage of a labeling requirement for furniture treated with flame retardants.[142]

In addition to concerns about the safety of the chemicals, an important contemporary debate involves whether the chemicals are beneficial in real-life fire scenarios. The effectiveness of flame retardants in laboratory tests is hard to dispute, but competing research studies demonstrate the conflict over flame retardant efficacy. As part of an ongoing debate over a proposed federal furniture flammability standard, researchers at the CPSC conducted flammability tests with flame retardant foam and with barrier fabrics, and concluded that "the fire-retardant foams did not offer a practically significantly greater level of open-flame safety than did the untreated foams."[143] In addition to these concerns about chemical efficacy, some fire scientists argue that flame retardants actually increase fire danger. When materials containing flame retardants burn, the smoke can be more toxic: because the chemicals inhibit full combustion, the emitted smoke contains higher levels of carbon monoxide, hydrogen chloride or bromide, and partially burned hydrocarbons.[144] This was known as early as 1973, when a fire safety report published by the National Commission on Fire Prevention and Control noted that "ironically, efforts to make materials fire-retardant . . . may have increased the life hazard, since the incomplete combustion of many materials treated to increase fire retardancy results in heavy smoke and toxic gases."[145] Because most fire deaths result from inhaling carbon monoxide and other toxic gases, not directly from the fire itself, flame retardants may increase risks from fires in other ways, even if they slow ignition.

■ ■ ■

This chapter has summarized the history of flame retardant controversies, explained the animal and human health effects, and described the experiences of scientists and regulators who argued that flame retardants were too hazardous to use. At the same time, it has shown that chemical companies continually reformulate their flame retardant products, allowing them to claim greater safety. What are the logics that allow these opposing parties to make generalizations about risk and hazard? How do "experts" like EPA regulators and chemical developers, and "outsiders" like citizens, firefighters, and legislators make reasoned arguments about risk and hazard? These questions will be addressed in the next chapter.

3

DEFENDING RISK AND DEFINING SAFETY

I n the fall of 2011, I worked as a research fellow in the EPA's Design for the Environment (DfE) program, part of the office at the EPA that evaluates industrial chemicals. The DfE program conducts alternatives assessments for chemicals that have been identified by the EPA as a priority for action.[1] These assessments are developed through voluntary collaborations with groups of stakeholders, and their principle goal is to inform chemical decision making. That is, if a company wants or needs to stop using a certain chemical because it has been restricted or because of health or environmental concerns but needs a replacement chemical that fulfills the same function, these *comparative chemical hazard assessments* provide guidance on which less hazardous chemicals might be good replacements. When I visited the program in 2011, three of the five active assessments were of flame retardants: deca-bromodiphenyl ether (DecaBDE) in electronics, hexabromocyclodo-decane (HBCD) in building insulation, and tetrabromobisphenol-A (TBBPA) in printed circuit boards. Unlike traditional risk assessment, these reports evaluate the chemical's toxicity and other characteristics such as chemical persistence, but do not characterize actual levels of exposure.

About a month into my fellowship at the DfE program, I joined James and Sarah, two DfE employees, to listen on speaker phone as they led a Partnership call for one of the ongoing assessments. About forty participants had joined the call, representing chemical companies, supply chain manufacturers, state regulators, other federal agencies, and nongovernmental organizations (NGOs). Sarah whispered to James, "that's pretty good attendance," and James nodded in agreement.

Sarah started by reading the project goal that had been developed at a previous meeting and asked for comments. A participant from the chemical industry responded immediately: "This is focused on hazard. Why not risk?" Sarah responded quickly: "DfE focuses on risk, recognizing that hazard is one component of risk." After she finished, James whispered to her, "Throw 'functional use' in there if they ask again," referring to the assumption that exposure scenarios are likely to be the same for a given product regardless of which chemical is used.

Another industry representative added to this first comment, saying, "It's very concerning that hazard would be taken in isolation." He suggested the report should include "cautionary notes" about the problems with hazard assessments. Sarah opened her mouth, but before she could respond, another person on the call joined the discussion. This representative of an environmental NGO said, "It's very clear what DfE's goal always is. It's clear from the kick-off meeting that this is a *hazard* assessment." A consultant spoke up next, saying that DfE's principles and goals are "designed to inform substitution" within functional use categories. As the consultant spoke, James looked at me and whispered, "Revolving door," suggesting this person had recently moved between EPA and the private sector. James leaned into the speaker phone and said diplomatically, "We do hear you on this . . . This is not a one-dimensional problem."

Sarah continued down the agenda, turning to the project scope. She summarized: the scope is a hazard assessment, and DfE will use the same criteria as in previous assessments. She mentioned the criteria document that DfE had developed a few years ago which clearly explained how different hazard determinations would be made.[2] Another industry stakeholder spoke up: "Hazard must be balanced with other considerations." He noted that some of the chemicals included in the assessment had significant data gaps, and he wondered whether there would be a minimum set of data that qualifies a chemical for inclusion: "absence of data shouldn't certify a chemical as being safe." Sarah responded promptly: "First, I want to make it clear that we don't certify that chemicals are safe. We don't rank them." She explained that the program evaluated the hazards of chemicals using existing data, expert judgment, and modeling. Each hazard evaluation, she said, is "a snapshot in time" to inform decision making using the best-available information. "In a perfect world we would have full testing data, but unfortunately that's not the reality. So you make the best decision you can with the information you have." James nodded, adding, "We can't exclude chemicals because they're not fully characterized." The NGO representative who had spoken before chimed in again: "That's why this process is so important. It's not regulatory, but it encourages more information and more data." On this point, she agreed with the industry representative who had spoken earlier: "We can't allow [a lack of] data to be equated with safety." The call continued, and participants talked through questions about specific chemicals and discussed the timeline for the process.

The stakeholders involved with this DfE alternative assessment supported radically different policy positions, and diverged in their interpretations of whether the existing scientific data were sufficient to demonstrate that the chemicals under review were dangerous to humans or the environment. The NGO representative supported DfE's focus on hazard, whereas industry stakeholders wanted DfE to evaluate exposure as well. These differences are at the heart of the questions motivating this chapter: How do we know if something is dangerous, and how do we know whether it is so dangerous that we need to take action? In short, how do we define risk?

In this chapter, I show how competing institutions define chemical risk to variously emphasize chemical exposure, hazard identification, and levels of scientific proof. These divergent *conceptual risk formulas* set the stage for and outline the parameters of future risk management practices. Such formulas matter when it comes to operationalizing risk assessment practices because before actors can disagree over whether flame retardants or any chemicals pose a risk to human health or the environment, they have to identify how that risk should be calculated.

SOCIAL AND SOCIOLOGICAL DEFINITIONS OF RISK

Environmental health problems are frequently studied and regulated in terms of risk. In common parlance, *risk* refers to the danger of something, a quality that incorporates both its inherent *hazards* (how dangerous something is or how severe the consequences would be) and how likely that hazard is to occur (whether something is dangerous to a given population, often equated with *exposure*).

Risk assessment is the practice of perceiving, interpreting, and calculating adverse effects. Thus, in risk assessment a chemical is typically evaluated based on a combination of exposure and hazard or toxicity information.[3] These assessments function as rules that dictate how chemical safety should be assessed. *Risk management* refers to the evaluation and implementation of actions taken to address the risk as a result of the assessment process. Delimiting the risk formulas used in any risk management strategy sets the boundaries for related risk assessments, with significant implications for chemicals policy, chemical production, and environmental health.[4]

Yet significant disagreement surrounds the assessment of environmental health risks. Imagine this scenario: A scientist publishes research findings from a study measuring a flame retardant in people's bodies, and in the concluding section expresses concern about possible human health consequences. An activist responds to this research with a call to action: "the trespass is the harm" and the chemical should be withdrawn from commerce. An industry group responds that the study only measures exposure, and does not address hazard or risk. What explains these very divergent interpretations of risk? This scenario demonstrates that risk is a social experience of an uncertain reality, not an absolute scientific value.

The sociology of risk demonstrates that perceiving, interpreting, and acting upon risk involve social factors beyond the calculated risk.[5] Risk and risk assessment have received significant attention in sociology, other social science disciplines, and the interdisciplinary field of risk analysis, which has its own professional society, conferences, and journals.[6] Crucially for my analysis, social and technical definitions of risk can vary across contexts, disciplines, and institutions.[7] For example, sociologist Kelly Joyce investigated how risk assessments of mercury in fish differed between the EPA and Food and Drug Administration (FDA). Although these two agencies issued a joint advisory on fish consumption, their "different forms of expertise and institutional cultures" led to different understandings and framings of the risks associated with fish consumption.[8]

The social definition of technological risk matters because future risks are at least somewhat unknowable in the present, and thus acting on risk requires distinguishing between multiple possible courses of action.[9] Definitions are central to how contemporary society experiences, debates, and responds to risk. As risk society theorist Ulrich Beck argues, risks are "products of struggles and conflicts over definitions with the context of specific relations of definitional power."[10] Though these contests are particularly visible when stakeholders debate the conclusions of risk assessments or risk-based regulations, conceptual risk definition must first establish the legitimate bounds of those assessments and regulations. When the stakes are high—when profits or public health are threatened, for example—risk definition becomes a site of intense competition between stakeholders.

Though risk assessment is a well-developed field of scientific study and practice, the conceptualization of risk itself has received less attention.[11] Scholars have taken different approaches toward investigating or acknowledging these differences. Those from the interdisciplinary field of risk analysis see risk as calculated future outcomes based on the magnitude and likelihood of predicted events, and they typically assume that uncertainty about risk can be decreased with more and better information.[12] Risk scholar Catherine Althaus defines risk as the "ordered application of knowledge to the unknown," and then examines how risk definition varies across disciplines ranging from medicine to law.[13] She notes, for example, that science and medicine pursue an objective, measurable risk under the assumption that uncertainty relates to not-yet-known values that are capturable through better, advancing science.

In contrast, social scientists typically emphasize social and contested elements of risk. For example, sociologist Sheldon Krimsky and global risk scholar Dominic Golding distinguish between realist, subjectivist, psychometric, technocratic, and sociocultural approaches to risk.[14] Building on early work by William Freudenburg, others have highlighted how risk calculations and risk management decisions are socially influenced and often reflect stakeholders' values and economic and political interests.[15]

My perspective on risk definition differs in two ways. First, I am less interested in how commentators or observing academics talk about risk than I am in how practitioners and stakeholders themselves talk about and operationalize certain risk formulas in the midst of unfolding events, in hopes of influencing those events. Second, I argue that risk formulas are themselves worthy of empirical study. Before stakeholders can argue over the outcomes of certain risk assessment or management decisions, risk formulas impact how they will develop arguments and assemble evidence to make their arguments, because different formulas allow or require different types and amounts of scientific data and describe different relationships between bodies of evidence.

By studying how multiple institutions compete to define the risks of the same chemicals, I am able to see risk as a concept whose definition travels, changes, and operates within and across institutions.[16] It is in these institutions that science is made, interpreted, and strategically used to influence policy.[17] The institutionalization and codification of risk formulas act as rules that govern the assessment of chemical safety, setting the boundaries for subsequent risk assessments and risk management activities. Thus, the adjudication of these rules has significant implications for chemicals policy, chemical production, and environmental health.

RISK DEFINITION IN THE EPA'S RISK GLOSSARY

As I have noted, disagreement over how to define risk and other key terms in the chemicals arena is common. The EPA, for example, uses many risk management tools and approaches the project of assessing chemical risk and safety from a variety of ways across programs. As one DfE person said of another EPA office, "They have a different base of knowledge, and they speak a different language."

These divergences are exemplified by the "Risk Glossary" project at the EPA. I first learned that this glossary existed during a conversation with someone on the Risk Assessment Forum, an EPA committee that promotes consensus on risk assessment issues. Contacts in the Office of Pollution Prevention and Toxics shared a glossary with six-hundred different definitions for risk-related terms, ranging from acceptable risk to weight of evidence. In an interview, Steven, an EPA scientist working on this glossary, explained that the project grew out of a 2010 conference on human health risk assessment that brought together 120 risk assessors and experts from across the agency to talk about the newly released *Science and Decisions: Advancing Risk Assessment*, a comprehensive look at the future of risk assessment by the National Research Council informally called the "Silver Book."[18] At this meeting, he said, participants lamented the lack of consensus definitions in risk assessment, the multiple definitions for the same term, and the tautological or cyclical nature of some EPA definitions.[19]

Steven has conducted risk assessments at EPA for over twenty years, but he told me that "risk assessment documents are all over the place" and "it's basically chaos out there." When the Risk Assessment Forum started thinking about how to advance risk assessment after the release of the Silver Book, coherent definitions were something that quickly came to mind because a single set of consensus definitions could reduce the variability in risk assessors' work. They compiled statute and programmatic definitions for risk terms across the agency in a wiki site, and the differences were revealing.

Of the eight distinct definitions of *risk*, some are short and sweet: "The probability of deleterious health or environmental effects."[20] Others are more complex: "The probability that damage to life, health and/or the environment will occur as a result of a given hazard (such as exposure to a toxic chemical). Some risks can be measured or estimated in numerical terms (e.g., one chance in a million)."[21]

Likewise, the ten definitions of *risk assessment* vary greatly. Some are procedural:

> Risk assessment can be divided into four parts: identification of hazards, dose response (how much exposure causes particular problems (i.e., cancer, convulsions, death), exposure assessment (determining how much exposure will be received by people during particular activities), and risk characterization (determining a probability that a risk will occur).[22]

Others define risk according to hazard and exposure: "A qualitative or quantitative evaluation of the environmental and/or health risk resulting from exposure to a chemical or physical agent (pollutant); combines exposure assessment results with toxicity assessment results to estimate risk."[23] Still others are based on policy activities:

> A tool created to compare and rank environmental problems based on the potential for environmental and public health impacts. Traditionally, risk assessments draw together a number of experts in fields such as toxicology, economics and natural resources. These experts are expected to use "pure science" to assess the risk to public health from contaminants and identify appropriate resource investment or mitigation measures. This approach does not generally allow for public participation or input into the process.[24]

These differences in risk definition reflect and reinforce subtle differences between how EPA programs operationalize risk assessment.

Although risk assessment is frequently discussed as though it was in significant part a scientific process, the act of assessing risk involves numerous social and scientific decisions that impact how it is conducted. Risk assessments must also deal with the data gaps and scientific uncertainties that are inevitable in the environmental health sciences, particularly around emerging environmental health threats or when assessing newly developed technologies. As noted science and technology studies scholars Holger Hoffman-Riem and Brian Wynne write, "In risk assessment, one has to admit ignorance."[25] Risk assessment incorporates this ignorance in dramatically different ways, assuming safety and charging ahead, or fearing danger and pulling back.

CONCEPTUAL RISK DEFINITIONS

When it comes to environmental health controversies, competing institutions develop and defend *conceptual risk formulas* that place different emphases on chemical exposure and hazard identification. Before the risks of flame retardants can be assessed as higher or lower, conceptual definitions of risk set the ground rules and boundaries for the assessment. Different risk definitions can thus favor the goals of different actors or institutions.

For the remainder of this chapter, I describe six conceptual risk definitions that play out in the environmental health sphere. They range from
the classic risk formula—that risk is a function of hazard and exposure—
to what I call the either-or risk revision—that a risk can be identified
through hazard *or* exposure, not necessarily both. I also describe the ways
in which each formula acknowledges and incorporates uncertainty, and
how and why each risk formula is favored by certain stakeholder groups.
Although all stakeholders engaged in flame retardant research and policy
are focused on risk and use words like risk, hazard, and exposure quite
frequently, in fact they have very different meanings behind the terms.

The Classic Risk Formula

An EPA chemist described his view of risk using an unusual metaphor:

> When we go to the zoo, you know, there are hungry bears and . . . hor-
> rible carnivores there. Why are we not concerned when we go to the zoo?
> Because, well, first of all, these bears and other stuff like that, there is a
> potential hazard there, a very grave, huge hazard. But there's a very low
> exposure. So the ultimate risk . . . I almost think of them as two num-
> bers being multiplied together. There's the hazard component, and
> there's the exposure component, and if the exposure component is zero,
> which it is ideally—the fence is tall enough to hold in the bear—then the
> exposure is zero. Even though the hazard is a really big number, a big
> number, obviously, times zero is a zero for the product of those two
> numbers, which is the risk.

The chemist explained that even with a hazard as "grave" and "huge" as
a hungry bear, if exposure is predicted to be nonexistent, then the risk
must also be nonexistent. This represents what I call the *classic risk formula*, which states that risk is a linear function of hazard and exposure.
Hazard is understood to be the danger or severity of something; *exposure*
refers to its likelihood or probability; and the *function* incorporates uncertainty about the two factors and the relationship between them.[26] This
formula is widely used to describe so-called high-consequence, low-
probability events such as a failure of a nuclear reactor: the hazard or

severity of the incidence would be quite high, but because the expected exposure or probability of the event is quite low—say, one in a million—the risk can be calculated as being low as well.[27]

The classic risk formula identifies a multiplicative or linear relationship between exposure and hazard. That is, it assumes a toxicological *dose–response:* the response (or hazard) should increase as the dose (or exposure) increases, so the risk increases as the dose goes up. Chemical risk assessments use this model to arrive at determinations of risk: if a predicted or measured exposure level is below the level believed necessary to cause a negative effect, then the risk is assumed to be low or nonexistent. This reflects a foundational tenet of toxicology, that "the dose makes the poison," meaning that the effect should increase with dose, and, consequently, that at a sufficiently high dose any substance will have an effect.[28]

The classic risk formula, used to describe dose–response relationships and identify reference doses or safety factors in formal risk assessment practices, is commonly discussed by regulators, industry representatives, and some scientists. In many contexts, the EPA defines risk as a function of hazard and exposure. Official EPA risk assessments use the definition explicitly, and EPA scientists state that both the hazard and exposure components are important. The website providing an overview of the New Chemicals Program states in italic letters, *"Hazard × Exposure = Risk."*[29] As another example, the EPA's Integrated Risk Information System defines risk as "the probability of adverse effects resulting from exposure to an environmental agent or mixture of agents," and it uses this model to calculate a reference dose for a given health end point.[30]

EPA scientists in the Office of Pollution Prevention and Toxics, the office that evaluates and regulates industrial chemicals, echoed this definition in interviews. Like the scientist talking about the zoo, another EPA chemist shared this emphasis on the importance of exposure: "without the exposure, it's not worth our [EPA's] time . . . Let's focus on the stuff where we suspect there's risk." This chemist also emphasized dose–response: "The simplification is, the dose makes the poison . . . Too much of anything—water, oxygen, nitrogen, right—is harmful." These examples show that EPA representatives often describe chemical risk as a function of hazard and exposure.

Representatives of the flame retardant industry—like the chemical industry more broadly—also regularly invoke the classic risk formula. This industry perspective came up in many contexts, ranging from confidential interviews with FlameCorp employees to corporate press releases. As one industry scientist told me in an interview, "the common definition of risk . . . [is], what is the probability that there will be a problem, that something will happen, and what is the seriousness of the effect." He continued, explaining that the risk was a function of those two elements. A high-hazard chemical could pose a risk even under low-exposure scenarios because of its toxicity: "If you were working with a chemical which is very toxic, but it is extremely unlikely that anybody will ever be in contact with it, there is the fact that may still be significant, because you have to consider both the seriousness and the probability." But a chemical that was low hazard and low exposure would pose no risk: "If you have a chemical which has relatively low toxicity and then the probability that anybody will be exposed to it is very low, there is no risk." These examples demonstrate his faith in the multiplicative risk formula—and his rejection of other risk formulas that might evaluate a chemical primarily based on hazard or exposure alone, rather than a combination of the two.

This perspective was also shared by some scientists. For example, an exposure scientist argued that exposure data needed to be combined with information about dose–response and toxicity in order to understand risk: "Just because I can find five molecules of compound X in you . . . you're not going to probably get cancer or some other risk." Similarly, several chemists invoked the classic risk formula to downplay a chemical's risk. One chemist told me that because chemists learn how to limit their personal exposure to chemicals in laboratory settings, they tend to be less concerned about chemical risk generally: "chemists tend to be very non-afraid of things" because "you work with chemicals, and you learn how to deal with them." Another said that he was not concerned about his exposure to the controversial chemical BPA because he had assessed its human health toxicity to be very low. "I would eat BPA," he told me with a straight face. These two examples demonstrate chemists' faith in the multiplicative risk formula: because any number multiplied by zero is zero, a nonhazardous chemical or a chemical with no exposure must by definition pose no risk.

Similarly, many toxicologists embrace the classic risk formula and their disciplinary tenet that the dose makes the poison. As one toxicologist told me, she would place her work at the intersection of hazard and exposure "because almost everything we do, we're looking at dose-response." A regulatory toxicologist offered a precise definition of the classic risk formula: "When you talk about risk, it's the intrinsic toxicity plus exposure. So the hazard can be very high, but if you're not exposed to it, then obviously there's no risk. Conversely, you can have a very low hazard, but if you're eating a pound of it every day, the risk can be very, very high." Another toxicologist, one who had worked in both government and industry settings, explained dose–response using the example of a common medication: "The dose makes the poison. And you know, you can take two Tylenol every day . . . and have very little chance of having a toxicity. But you take eight, it's going to blow out your liver." However, as I will describe, toxicology's reliance on a linear dose-response model has been significantly challenged by research in recent years showing that the dose–response relationship is influenced by other factors, such as timing of exposure.

The classic risk formula typically deals with uncertainty through the use of standard *uncertainty factors*. Regulatory risk assessments adjust measured toxicological values by various uncertainty factors, including those that account for the difference between animal species and humans (typically a value of 10), the assumption that some people will be more susceptible than others (also typically a value of 10), and the assumption that children are more susceptible than adults (typically a value of 3). Thus, the classic risk formula accounts for scientific uncertainty and data gaps by making the assumed reference dose—or "risky" level of exposure—much lower than the level actually measured and found to be problematic in any existing studies. This also means that the classic risk formula relies on concrete scientific data and default uncertainty values that can be plugged into a multiplicative formula.

The classic risk formula that is widely used in technical risk assessment processes typically favors industry stakeholders.[31] The formula requires a great deal of data on exposure and hazard, and without this data a risk assessor can only use speculation or information about similar chemicals to calculate risk. Thus, the absence of data regarding either exposure or hazard can be interpreted as affirming the absence of risk.[32] Outside of the standard uncertainty factors, there is little room for scientific uncertainty in the classic risk formula, or for incorporation of the complicating factors related to emerging areas of toxicological research. This benefits the chemical industry because estimations of low risk allow chemicals of concern to stay in commerce.

The Emerging Toxicology Risk Formula

The classic risk formula assumes a linear dose–response: the relationship between the level of exposure and the level of effect is assumed to be constant as doses increase or decrease. However, a growing body of scientific evidence suggests that for some toxicants the assumption that "the dose makes the poison" is fundamentally flawed and incomplete. Instead, the relationship between hazard and exposure can vary based on the amount of exposure, the timing of exposure, and the characteristics and genetic makeup of the exposed individual. In the words of Linda Birnbaum, director of the National Institute of Environmental Health Sciences (NIEHS), "We now know it's not only the 'dose that makes the poison,' it's timing, it's inherent susceptibility."[33]

These revisions are making their way into discourses and practices in various institutions through what I call the *emerging toxicology risk formula*. This risk formula diverges from the classic risk formula in several key ways.

First, some dose–response relationships may be nonlinear, in that the relationship between exposure and effect does not monotonically increase—or may even decrease or change shape—from low to high doses.[34] Vitamins are a classic example: many vitamins are beneficial at low doses but can be harmful and even intensely toxic at high doses. Scientists call this *hormesis*, a biphasic dose–response relationship in which low-dose responses differ in significant ways from high-dose responses.[35]

In particular, a growing body of research suggests that hormone-disrupting chemicals induce significant effects at very low levels of exposure.[36] The potential of chemicals to cause health problems by interfering with the hormone system has been controversial since the theory was first proposed by scientists decades ago, but an emphasis on nonlinear dose–

response relationships has gained regulatory traction in the past few decades.[37] The EPA's Endocrine Disruptor Screening Program (EDSP) evaluates chemicals for their potential to cause health problems by interfering with the hormone system.[38]

Second, the timing of exposure also impacts health effects. In particular, fetal and early childhood exposures to some chemicals may induce strong later-life effects.[39] Health effects can also span generations. As a well-known example, diethylstilbestrol (DES) was a drug used by millions of women in the mid-1900s in the mistaken belief that it would improve pregnancy outcomes. Exposure to DES led to greatly increased risks of cancer and reproductive effects in children who were exposed to the drug in utero.[40] The field of transgenerational effects has documented examples of harmful reproductive effects persisting for as many as four generations after initial exposure.[41]

Many researchers and regulators are particularly concerned about exposures during childhood. As researchers regularly note and the EPA's Office of Children's Health Protection has codified, children are not just small adults, due to their behaviors, physiological differences, metabolic systems, and windows of susceptibility.[42] Children are worthy of special attention, not just because of their social value but also because, as one environmental health advocate explained in an interview, "children are at the top of the food chain" when they are in the womb or when they are breastfeeding.

This perspective has gained significant traction among environmental health scientists and activists. As a slide at a 2013 scientific meeting on evidence-based toxicology stated, "Response Is Dependent on Dose and Time."[43] Scientists often discuss the timing of exposure alongside an emphasis on nonlinear effects of hormone-disrupting chemicals. For example, one public health scientist told me that "with the flame retardants . . . we know that these are potent thyroid disruptors in humans. We know that in animals . . . exposures especially *in utero* can have big impacts on the developing brain." Thyroid disruption is believed to be one of the mechanisms behind observed associations between polybrominated diphenyl ether (PBDE) exposure and negative neurodevelopmental outcomes in children.[44] Thus, the timing of exposure and low-dose exposures are both important under the emerging toxicology risk formula.

Finally, developments in genetic toxicology, epigenetics, and the environmental health sciences point to variation in individual response and susceptibility to exposure.[45] The growing field of epigenetics examines alterations in gene expression resulting from an environmental exposure, including genetic changes that can be passed between generations.[46] On the surface, this bears some similarity to the intraspecies uncertainty factor commonly used under the classic risk formula. However, the contemporary toxicology revision of that formula takes a more nuanced and exploratory perspective, seeking precise answers to how, why, and to what degree individuals vary in their susceptibility to environmental exposures.

The Exposure-Centric Risk Formula

The next risk definition I call the *exposure-centric risk formula*, because although it does recognize the importance of evaluating hazard, it prioritizes a multifaceted understanding of exposure that expands the classic risk formula in two ways. First, the exposure component of the risk formula becomes complex, requiring high levels and multiple forms of data. Second, the formula is strictly multiplicative, meaning that if any component (or subcomponent) is estimated to be zero, the risk equation as a whole will equal zero, regardless of the magnitude of the other components in the model. This formula is most widely used and favored by industry stakeholders. Importantly, the exposure-centric formula tends to minimize risk, assuming that exposure data are sufficiently certain and comprehensive for accurate risk characterization, and that exposures at levels of concern are unlikely.

The exposure-centric risk formula separates the exposure component of the risk formula into four pieces: the potential for exposure based on the *physical-chemical properties* of the chemical, the potential for exposure based on *use scenarios*, an established *exposure pathway*, and *levels* of documented exposure in people or the environment. Each of these components can be assigned a value and "zeroed out" in the risk calculation process.

Under the exposure-centric formula, exposure potential depends on the physical-chemical properties of the flame retardant. A product development manager at FlameCorp told me that "reactive" flame retardants, which are chemically bound inside consumer products, have no exposure potential outside of occupational exposures: "If you develop flame retardants that are reactive . . . they will never, never expose anybody." However, this perspective ignores exposures that occur during chemical and product manufacturing and end-of-life disposal. The flame retardant industry is also developing polymeric products, large-molecular-weight compounds that are believed to be too large for the molecules to be absorbed by organisms. In a recent industry-funded assessment of alternatives to DecaBDE, polymeric flame retardants were found to be of low concern for exposure, generally because they were assumed to be of low bioavailability.[47]

Second, exposure potential is based on the chemical's use scenarios. A FlameCorp advocacy representative explained that flame retardants used in home insulation can be assumed to have a low exposure potential, flame retardants used in furniture or electronics have a moderate exposure potential, and flame retardants used in cell phones have a high exposure potential. Another FlameCorp scientist provided a nice summary of the industry's emphasis on how the product will be used: "For us [industry], it's a balance of exposure as well as toxicity . . . If you're applying a chemical to a fabric that is going to be used as apparel, you have a much bigger concern in the future that you have to deal with . . . Now, if it's in a banner that's being used as an awning or something, it's not going to come into human contact often, so it's not as much of a concern." Thus, a product's intended use delineates its exposure potential.

Exposure assessment in the exposure-centric risk formula also involves identifying and documenting an exposure pathway.[48] For exposure to occur, the chemical must make its way out of a source, through a medium of exposure (like air or dust), and through an absorption pathway into the human body. This component of exposure was often mentioned by exposure researchers, whose work involves identifying levels of the chemicals in different environmental media, measuring the presence and concentration of chemicals in people, and clarifying the exposure pathway. As a public health professor who has researched the flame retardant exposure pathway told me, "There could be nasty stuff in the dust in this room, but that doesn't mean that I have exposure to it." This individual continued, explaining that documenting how exposure happens is very important: "Just because we find a relationship, what's in dust and what's in people, doesn't say whether [it] is dermal exposure or whether [it] is from incidental ingestion or . . . inhalation." As a toxicologist explained, a misspecified exposure pathway can lead to incorrect risk calculations: "Someone might have predicted that dust isn't an important route of exposure . . . but if you actually go and measure it, then it's not true"—because people are significantly exposed through house dust.[49] Inaccurately characterizing the exposure pathway can lead to inaccurate risk calculations.

The final component of exposure definition is that the chemical must be actually measured in people or the environment. Without these data, even if there is exposure potential based on the physical-chemical properties, use patterns, or an exposure pathway, the exposure-centric risk formula can be used to argue that exposure is not a concern. For example, after environmental chemist Heather Stapleton and her colleagues identified certain flame retardants in baby products using analytical chemistry methods, the American Chemistry Council (ACC) issued a press release, stating, "This study attempts to examine the existence of certain flame retardants in a small sampling of children's products; it does not address exposure or risk."[50] Although it would seem that the presence of nonreactive flame retardants in baby products would have high exposure potential, the trade association argued that the research did not address exposure because it did not test for levels of the chemicals in people or the environment, and did not address risk because it did not connect these levels to a dose required to lead to a toxic effect.

Even if the data needs of these multiple exposure criteria are met, the strict multiplicative nature of the exposure-centric risk formula means that known toxic chemicals can be evaluated as low risk given limited exposure. Many scientists at FlameCorp said that flame retardant exposure levels are so low for the average consumer that they are not a concern. As one research and development (R&D) scientist explained, "If it's at parts per trillion, what is the actual dose level which creates an issue, a human health issue? It could be a part per million or something even more than that. So then the question is, so what?" Industry scientists told me that analytical chemistry methods had developed so much that chemists could detect "the most infinite particle" at toxicologically irrelevant levels of chemicals in the human body: "Just because they [chemists] can detect it doesn't necessarily mean it's a bad thing." Another R&D scientist noted that toxicological studies finding effects in animals may be "a thousand times higher, ten thousand times higher" than the levels found in people, and said that given low exposure, "probably there is no effect at all." These comments demonstrate the widespread nature of the argument that low exposures, even of toxic substances, are not risky.

As an extension of this emphasis on the dose–response relationship, industry respondents repeatedly argued that everything is toxic at some dose. In the words of a FlameCorp manager, "Everything is toxic. If you eat one hundred grams of salt, table salt, you will be sick or probably, maybe dead." Another FlameCorp scientist told me that flame retardant exposure "is like anything else . . . You can abuse anything. You can drink too much water and die." In addition to these last two quotes mentioning table salt and water, industry representatives also compared the risks of exposure to flame retardants to sugar, gasoline, peanut butter, motor oil, and Thanksgiving dinner. These arguments represent a type of *commensuration*, in which different components are put into a common metric for purposes of calculation and evaluation.[51] However, the toxicological principle that everything is toxic at some high dose does not necessarily imply the correlate that everything is safe at some low dose.

Another key focus of industry's attention to exposure is the expectation that exposure—and therefore risk—can be controlled through proper handling and manufacturing techniques. For example, a senior R&D scientist described a hypothetical chemical: "The human health [evaluation] looks really good, but it could be . . . toxic to fish or something . . . So you put a label that [says] don't release to waters." In this case, controlling aquatic exposure means the chemical's documented aquatic toxicity no longer presents a risk.

To control manufacturing emissions, the brominated flame retardant industry coordinates the Voluntary Emissions Control Action Program (VECAP).[52] This program, which started in Europe and has since spread to the United States, works to identify and reduce point-source emissions of flame retardants from manufacturing and processing facilities through education and audits. These strategies are premised on the idea that most or all exposures to flame retardants come from manufacturing plants, an assertion that is not supported in the peer-reviewed literature. As an industry representative explained, companies want to minimize point-source emissions because flame retardants are "stable by design" and therefore are persistent. VECAP assumes that exposures result from mishandling or production errors (the improper disposal of packaging waste, for example). Additionally, the calculations provided by VECAP materials presume that incineration or land landfilling leads to zero emissions, but this perspective has been critiqued by others. An industry consultant said that this argument is "nonsense."

Identifying multiple components of exposure and assigning each of them a multiplicative role means that the risk formula can more easily be zeroed out in final calculations, arriving at a determination that the chemical or product in question presents no risk. At FlameCorp, for example, the toxicity testing conducted during new product development is frequently "balanced" with or analyzed alongside of information or expectations about exposure.

The exposure-centric risk formula also tends to dismiss or downplay scientific uncertainty. The formula assumes that without each required exposure component, any concerns or uncertainties about hazard become less important. Though numerous EPA representatives and scientists told me that exposure was extremely difficult to accurately or completely measure, industry respondents rarely mentioned the uncertainty associated with exposure research. In contrast, as I show in the next section, uncertainty provides significant motivation for those supporting a hazard-centric risk formula.

The Hazard-Centric Risk Formula

An EPA representative explained her approach to risk reduction in an interview. She began by distinguishing between risk and hazard: "Hazard is related to the intrinsic properties of a chemical that make that chemical capable of harm, whereas risk is a probability of harm given a scenario. So risk, of course, is a function of both hazard and exposure." But she continued by arguing that hazard assessment was necessary because of the difficulties in evaluating exposure: "Exposure is notoriously difficult to estimate. Also, exposure controls can, will, and do fail. It's one of the reasons why we exist as an Agency. So ultimately pollution prevention . . . is best when done upfront . . . We need to reduce the number of hazardous chemicals, highly hazardous chemicals . . . that we're using."

This *hazard-centric risk formula* modifies the classic risk formula with the assumptions that hazard is complex and multifaceted, exposure is hard to measure, and some elements of exposure can be understood as hazard characteristics. In some ways, the hazard-centric risk formula parallels the exposure-centric formula I have described: it defines hazard as multifaceted, and maintains the multiplicative relationship between hazard and exposure, with the expectation that decreasing one component—hazard, in this case—will decrease risk overall. Beyond this parallel, however, the formulas are dramatically different.

Unlike the exposure-centric formula, the hazard-centric formula assumes that uncertainty around chemicals is often so significant that it precludes effective risk assessment. From this perspective, exposure is insufficiently researched and hard to fully characterize, and thus the best way to reduce risk is to reduce the hazard of chemicals that will inevitably, uncontrollably, and often invisibly make their way into the environment. For those embracing the hazard-centric formula, hazard should be put front and center when it comes to chemical decision making as a "performance attribute." Perhaps most significantly, proponents of the hazard-centric risk formula redefine several chemical properties typically connected to exposure as hazard characteristics.

The hazard-centric risk formula sees hazard as multifaceted. In some ways, the EPA's DfE program embodies the hazard-centric risk approach. Described briefly at the start of this chapter, the DfE program conducts comparative chemical hazard assessments of chemicals of concern targeted by the EPA for other risk management activities. These assessments highlight the multifaceted and complex nature of hazard assessment: they evaluate fifteen end points for each chemical, ranging from carcinogenicity to skin sensitization, and assign a hazard value of very low to very high for each end point.[53]

The hazard-centric risk formula assumes that assessments should be protective, relying on the most conservative end point available. This is because the hazard-centric formula acknowledges and emphasizes the uncertainty that surrounds hazard data, even as it works to fully characterize a chemical's expected hazard. This is evident at DfE as well: for example, if multiple neurotoxicity studies of sufficient research quality exist for a chemical, DfE will follow the study that found effects at the lowest dose.[54] In the program's recent DecaBDE assessment, several chemicals were predicted to be of "moderate" concern for bioaccumulation, and an EPA researcher explained to me that moderate is a conservative measure based on their experience with chemicals like DecaBDE, where scientists do not fully understand the mechanisms behind bioaccumulation.

In spite of the hazard-centric formula's focus on hazard, it still maintains the multiplicative assumptions of the classic risk formula. That is, the classic risk formula is maintained in principle, but for supporters of the hazard-centric model the best way in practice to reduce risk is to reduce hazard. This is based on the assumption that exposure considerations are unlikely to vary greatly across the same use of a chemical: your exposure to the flame retardant in your couch is likely to be the same no matter which flame retardant was used. A DfE representative called this the "functional use" approach: "If you have a flame retardant that's used for plastic X in this part of an airplane, and you're going from flame retardant A to flame retardant B, then the exposure concern for those two things should be the same. And if risk is a function of hazard and exposure, you would expect that being able to render that exposure variable null allows you a much simpler way to manage risk." Thus, "hazard assessment lets you take exposure assessment out of the equation." As another DfE representative told me, "If you can move from a hazardous chemical to a less hazardous chemical that does the same job, it seems to me that you are managing and reducing risk." This approach still defines risk as a function of hazard and exposure, but favors hazard reduction over exposure reduction.

The hazard-centric risk formula places a great deal of emphasis on scientific uncertainty in assessing chemical hazard. Even if toxicity data exist, proponents of the hazard-centric risk formula recognize limitations related to characterizing toxicity pathways and mechanisms of action, low-dose and non-monotonic effects, and the reliability of animal studies, thus connecting to the emerging toxicology risk formula. In the first place, toxicology may identify effects but not provide information about why a chemical is toxic: as one researcher said, "We don't understand *why* [many chemicals] are toxic. We know that they are, but not the mechanism [of action]." A second uncertainty involves the difficulties of understanding low-dose effects based on animal research typically involving relatively high doses of a chemical. As another toxicologist explained, "I can tell you if there's a 5 percent increase in the tumors . . . But EPA regulates at one in ten-thousand. I'm at five in a hundred. One in ten thousand is not necessarily an unreasonable number, but I'm not getting there with an animal." Animal models almost inevitably involve much higher doses of a toxicant than is typical for human exposures, necessitating high-to-low-dose extrapolation. Additionally, animal models are imperfect predictors of illness in humans, just as results using different animal species do not always align perfectly.[55]

In response to these uncertainties, some toxicologists conduct research at "environmentally relevant" exposure levels: rather than testing whether high doses of the chemicals lead to serious outcomes like death or cancer, researchers use smaller doses of the chemical—similar to those measured in the environment or in people—and look for more subtle effects such as neurological outcomes or chromosome damage. Other times, scientists will call attention to relatively small differences between the levels measured in the environment or in people and the levels found to cause harm in animal studies. For example, at a scientific conference, an environmental scientist pointed out that the levels of a chemical she had measured in household dust were within a 100-fold safety factor of the effects that her collaborators found in a toxicological study, suggesting that was close enough to be concerning.

Thus, the hazard-centric risk formula advocates reducing hazard as the best way to reduce risk, while maintaining the multiplicative relationship between exposure and hazard. In these ways, the hazard-centric model aligns closely with the principles of green chemistry, especially the need to think about hazard from the beginning of chemical design and production.[56]

The Exposure-Proxy Risk Formula

During my time at the EPA in Washington, D.C., I gave a presentation on my research to a room of scientists and regulators from the New Chemicals Program. At the end of my talk, conversation turned to how chemical evaluation had changed under the Obama administration. One exposure scientist who had worked at the EPA for years spoke about an initiative by then-Assistant Administrator Paul Anastas (considered one of the founders of green chemistry) to include chemical persistence in hazard evaluation. At first, this exposure scientist thought this change was very unscientific, but eventually he decided that it made sense. Another scientist chimed in: at the EPA, they think of the combination of persistence and bioaccumulation as a "proxy for exposure, because we assume that if a chemical is persistent and bioaccumulative, we will be exposed." A DfE representative agreed, and added that this way of thinking about persistence and bioaccumulation moves the agency from risk assessment to risk management.

As I noted above, the classic risk formula accounts for uncertainty by dividing known dose–response levels by default uncertainty factors. In several programs, the EPA goes farther by adopting what I call the *exposure-proxy risk formula*, which includes persistence and bioaccumulation alongside of hazard and exposure. Environmental persistence refers to how long the chemical will stay in the environment, and is typically measured by the chemical's half-life before it is transformed or broken down by chemical or natural processes. Bioaccumulation is the absorption of a chemical into a human or other organism by all possible routes of exposure in the environment, including inhalation, ingestion, and dermal exposure, as well as how the chemical is transformed within the organism. Not all chemicals persist or bioaccumulate. Some break down quickly or are passed through the body without being absorbed.

The exposure-proxy formula assumes that exposure is likely to occur if a chemical is persistent and bioaccumulative, even if the chemicals themselves have not yet been measured scientifically. In essence, this formula adds a third component to the classic risk formula, so the formula now reads: risk is a function of hazard *and* exposure *and* potential to persist and bioaccumulate. An EPA chemist used the exposure-proxy risk formula to reflect on past lessons learned: "If you look at the chemicals that EPA has had a problem with in the past . . . it was because they were very persistent and bioaccumulative."

Persistence and bioaccumulation are particularly relevant for many flame retardants, which are often designed to maintain their chemical efficacy in products such as furniture or cars with life spans of several decades. For example, 30 out of the 32 flame retardants recently evaluated by the DfE assessment of alternatives to DecaBDE were assessed as highly persistent, and 11 were highly bioaccumulative.[57] Several respondents specifically mentioned DecaBDE as a cautionary tale. DecaBDE was long believed to be too large of a molecule to be absorbed by living organisms, but research now shows that the chemical is highly persistent and bioaccumulative.[58] According to EPA scientists, although scientists now know that DecaBDE is absorbed, they still do not fully understand how it is absorbed or all of its breakdown pathways. If the exposure-proxy risk formula had been applied to it before extensive measurements of exposure were developed, perhaps different risk management practices would have been implemented earlier, reducing widespread exposure. As one researcher told me, "We cannot apply exposure controls to deca[BDE], for example . . . The damage is done."

Because persistence is so common among flame retardants, the chemical industry has taken numerous active measures to redefine hazard so that persistence is seen from a more "neutral" perspective. For example, a hazard assessment methodology proposed by the industry does not include persistence as a distinguishing characteristic. As a FlameCorp representative explained to me, this is because "persistence is part of flame retardant design." As another FlameCorp scientist explained, "We make molecules which are very, very stable. We call them stable. NGOs call them persistent." But even if persistence is not a distinguishing characteristic, because so many flame retardants share this trait, removing persistence from evaluation altogether is misleading and may give the impression that chemicals have fewer exposure-related concerns than they in fact do.

The exposure-proxy revision allows regulatory bodies to make protective determinations of risk in the absence of measured exposure. The EPA's New Chemicals Program uses a version of the exposure-proxy risk formula to determine whether newly developed chemicals pose an "unreasonable risk" to human health and the environment.[59] When a new chemical is developed by a manufacturer and submitted to the EPA for review, it is reviewed by EPA experts for hazard, exposure, and persistence and bioaccumulation potential, among other characteristics such as expected usage. This departure from the classic risk formula is one way that the EPA accounts for uncertainty. If chemicals will persist in the environment and accumulate in the body, the potential consequences of underestimating hazard and exposure are more substantial. This mirrors work in Europe under Registration, Evaluation, Authorisation, and Restriction of Chemicals (REACH) guidelines to prioritize "very persistent, very bioaccumulative" chemicals.

The exposure-proxy risk formula was also used by the EPA in 2012 to prioritize the Toxic Substances Control Act (TSCA) "Workplan Chemicals" for additional review.[60] The agency identified a list of 345 candidate chemicals based on existing concerns or use patterns (for example, chemicals used in children's products). Based on a review of the scientific literature, consultation with experts, and structure–activity relationship evaluation, the EPA assigned each chemical three scores: a Hazard Score, an Exposure Score, and a Persistence/Bioaccumulation Score. These three numbers were added together: Chemical Score Calculation = Hazard Score + Exposure Score + Persistence/Bioaccumulation Score.[61] The combined scores were normalized and grouped as low, moderate, or high priority. This shows how the exposure-proxy formula is operationalized by the EPA.

The Either-Or Risk Formula

Finally, some stakeholders reject the multiplicative risk formula altogether. As an environmental health activist explained, her work on flame retardants was motivated both by hazard and by exposure, but not by a risk-based approach to either. "One of the reasons that we chose to work on flame retardants was because of the widespread use in consumer products. Flame retardants were in everybody's homes, they were in virtually everybody's body . . . And the science was really developing, as to the harmful nature of these chemicals. How harmful is this to people? How prevalent, how widespread it is?" Though she mentioned both harm and prevalence, she did not feel the need to integrate the two into some "safe" exposure level, and she explicitly rejected a "risk-based approach . . . It's flawed in many ways."

This final redefinition of the risk formula I call the *either-or risk formula*. According to this risk formula, a chemical can be found to pose an unreasonable risk based on hazard or exposure data, not necessarily on the combination of the two. That is, a risk can be actionable based on documented hazard but no proven exposure, or documented exposure (especially to children or in utero) but no proven hazard.

This perspective was most common among the environmental activists, and very few activist respondents spoke of anything like the classic risk formula. One exception was an environmental health organizer who said that biomonitoring allows researchers to take information about "our actual exposure" and compare it with "the exposure level where harm occurs . . . , and see, you know, do we think harm is occurring to the general population now or not?" This person saw exposure data as being useful when combined with hazard data to arrive at a determination of risk.

Much more common among activists, however, was explicit critique of the classic risk formula and traditional risk assessments. In the words of an environmental health organizer, risk assessment "basically says . . . this is how high the stack of bodies can be." They would agree with former EPA Administrator William Ruckelshaus, who famously wrote, "Risk assessment data can be like the captured spy: if you torture it long enough, it will tell you anything you want to know."[62] Others noted that risk assessment does not adequately protect vulnerable populations. As an environmental health activist commented, quantitative risk assessments are based on:

a healthy, white, 30-year-old, 150-pound male. They don't take into account adequately the more vulnerable among the population, you know, the developing baby or child, people with other health problems, the elderly . . . They don't take into account cumulative and synergistic effects that chemicals have based on multiple exposures to multiple sources of chemicals, which is much more based in reality.

The "acceptable levels" that are built into risk assessment modeling, she argued, are problematic "because that risk happens to be higher in certain groups: certain minority populations, certain socioeconomic populations.

So it's not an evenly distributed burden of risk." This is a common argument in environmental health social movements.[63]

Activists were not alone in critiquing the classic risk formula or its multiplicative variations. When scientists and regulators critiqued risk assessment, they tended to emphasis scientific shortcomings: inadequate data, incomplete models, or inappropriate parameters. In contrast, while activists were aware of faulty scientific assumptions or data gaps, they mostly emphasized social problems or forms of injustice perpetuated by risk assessment. Environmental activists also emphasized a need to avoid "dueling science"—that is, comparing stacks of scientific evidence regarding flame retardant safety—because they felt that this played into the chemical industry's tactics.[64]

The either-or risk formula differs from the classic risk formula in important ways. According to this perspective, decision making can proceed based on hazard data alone, and hazard is sufficient to prioritize and regulate chemicals. As one environmental organizer explained it, "We prefer a hazards-based approach, which is, 'let's look at the chemicals. If it's hazardous, don't use it.'" Thus, the argument is that hazardous chemicals should not be used, regardless of exposure. This is particularly important because it is difficult to predict or control the full range of exposures likely to result from consumer products. To give just one reason for this, many consumer products are eventually sold secondhand, so labeling requirements may not reach buyers of used goods. As another activist explained, "parents go and buy things at garage sales and maybe didn't get the flyer that said this has lead paint, this is going to cause brain damage in your child. So, you just ban it." Following the either-or risk formula, regulating hazardous products becomes the safest course of action because exposures cannot be predicted or controlled. Additionally, activists have noted that consumers, especially parents, frequently make purchasing decisions based on hazard. As a market campaigns activist said, "No mom wants a toxic chemical in their children's product, in their pacifier, in their home. If you have a choice between a toxic product or [a nontoxic product], you know, which are you going to choose?" This anecdote aligns with survey research showing that those across the political spectrum think chemicals used in consumer products should be more tightly regulated.[65]

The biggest distinguishing characteristic of the either-or risk formula is its assertion that exposure on its own is actionable for chemical risk. That is, in some cases documented exposure provides sufficient justification for regulation, without the need for hazard data. Activists have emphasized that exposures are inevitable, ubiquitous, and often more significant than predicted. On a call about flame retardant policy, one advocate said, "I defy any of us to get through the day without being exposed to some form of toxic chemical marketed as a flame retardant." Activists also have called attention to faulty assumptions in the past that chemicals "were in the products and stay there." Instead, chemicals used in consumer products are likely to escape those products, and thus human exposure is likely. One environmental health organizer explained, "Even if we were to stop right now, some of these chemicals are very persistent, like the flame retardants, and they're not going away any time soon." Instead, some highly persistent chemicals will be around—and in human bodies—for generations to come.

This perspective aligns with the refrain "the trespass is the harm." In the words of an environmental activist whose words aligned with the either-or risk formula, this means that "you don't even have to show a health effect. If you're showing that these chemicals are getting into my body, that trespass is unauthorized." This perspective counters the chemical industry's argument that exposure levels are too low to present a risk. In the words of an environmental health activist, "companies can dish all they want about how these levels are not significant and they're not very high, and this and that. But it's just an indefensible position. These chemicals shouldn't be in our bodies, period." This then becomes an ethical argument about consent and right to know, of "environmental outrage" in addition to an argument about toxicity and health effects. This type of *exposure experience* is associated with increased knowledge about personal chemical exposure as a form of risk awareness.[66]

Other activists specifically have singled out chemicals that are found in cord blood or breast milk as deserving regulatory action. Early academic and advocacy studies that detected flame retardants in these two bodily fluids attracted significant public and media attention. One advocate described the impact of an advocacy biomonitoring study by the Environmental Working Group that found flame retardants in cord blood:[67]

when children are born, "they've never breathed the air, they've never eaten our food, never drank the water, yet they're born with these chemicals in them . . . It's a very powerful tool from the political side. I mean, it points out a flaw in the system." Others talked about how this knowledge could be used for policy. An advocate involved in the California biomonitoring program said that chemicals found in cord blood could be prioritized for further regulatory attention: "We're suggesting that biomonitoring data should be used by the state in prioritizing chemicals . . . If a chemical is found in cord blood, it should be slated for immediate action instead of having to go through some long, convoluted process." Thus, under the either-or risk formula, either hazard or exposure data can be sufficient for decision making. The two need not be combined and evaluated together for risk management activities to proceed.

RISK DEFINITION: ANTICIPATORY AND STRATEGIC ACTIONS

Why do these risk definitions matter? What difference does it make if a stakeholder prefers to complicate hazard or exposure, sees the relationship between the two as more or less multiplicative, or requires more or less scientific certainty? These risk formulas matter greatly because they represent stakeholders' *anticipatory* and *strategic* attempts to set the boundaries of risk definition and, by extension, chemicals regulation and use.

Risk definition identifies how formal and informal risk assessments should be conducted: what pieces of information are included, how they are weighted or organized, and how they combine into an overall assessment of risk. The data needs of the hazard-centric and exposure-centric risk formulas, for example, are extremely different; given the huge existing data gaps around industrial chemicals such as flame retardants, choosing one risk formula or another will lead to a different overall assessment of risk. Furthermore, once risk protocols are put in place, they act as rules that guide future risk assessment activities.[68]

Because risk formulas are drawn from the stakeholders' own words, not from the formal definitions of risk scholars or social science observers, they also provide needed analytical leverage on how stakeholders interact with each other. Returning to the description of a DfE Partnership call that opened this chapter, the DfE stakeholders and the environmental NGO leader were advocating for a hazard-centric risk formula, while the industry stakeholders spoke up in favor of the exposure-centric model. They were engaging in *anticipatory risk definition*, battling over the boundaries of the risk management practices on the table. Within the confines of the DfE alternatives assessments, most acts of risk definition have already occurred: DfE has a clearly stated mission and set of goals that focus action on hazard identification and comparison, not full risk assessment. Yet repeatedly in these assessment processes, industry stakeholders attempt to redefine the scale and scope of DfE assessments to include commentary in support of the exposure-centric or classic risk formulas because those formulas are more likely to arrive at a determination that a chemical is safe for use. Additionally, hazard assessments can be completed more quickly than full risk assessments, given the latter's greater data demands which can draw out any possible regulatory process and ensure that chemicals under evaluation stay on the market longer.

Risk definition is highly strategic, meaning it is intentionally developed to pursue a certain goal or outcome. Each stakeholder's preferred risk definition can be linked to their interests and policy goals. A few examples will make this strategic element clear.

Design for the Environment and the Hazard-Centric Risk Formula

The DfE program has partially redefined exposure as a hazard characteristic by including persistence and bioaccumulation in their comparative chemical hazard assessments and evaluating these characteristics alongside traditional hazard measures such as carcinogenicity or reproductive toxicity. DfE redefines these exposure characteristics as part of the hazard-centric risk formula for two main reasons. First, they place great emphasis on scientific uncertainty, especially data gaps around chemical exposure, because "exposure controls can, do, and will fail." They therefore see hazard reduction as the most effective way to reduce risk. From this perspective, risk assessment and risk management strategies are useful only when exposures are well known and controlled, a highly unlikely scenario. In the words of one DfE scientist, "risk-based analysis is really probably only valid if you know that there is no exposure, or the exposure only occurs in very controlled circumstances."

Second, redefining exposure characteristics as hazard enhances the legitimacy of hazard assessment. Alternatives assessments such as those conducted by the EPA are not always well received in regulatory and corporate environments, where they are critiqued precisely because they diverge from the classic risk formula and the industry's preferred exposure-centric risk formula. Including exposure characteristics in hazard assessment not only increases the scientific relevance of the documents but may provide important legitimating signals. Additionally, the inclusion of exposure characteristics buttresses DfE against accusations that their assessments ignore exposure. The contestation around risk assessment and management methodologies is clearly visible in the way DfE strategically seeks to define its hazard methodologies apart from risk assessment, in spite of programmatic similarities with some aspects of risk assessment.

Flame Retardant Manufacturers and the Exposure-Centric Risk Formula

Chemical industry stakeholders regularly embrace both the classic risk formula and the refinements present in the exposure-centric risk formula. They also frequently push back against the hazard-centric and exposure-proxy risk formulas because both formulas are more likely to arrive at determinations of significant risk.

This is not to say that hazard evaluation has no role in industry practices. To the contrary, flame retardant companies pay serious attention to hazard in product development, in balance with exposure potential and production and cost concerns. One R&D scientist emphasized the need to develop low hazard flame retardants because exposure to flame retardants is inevitable: "in our case, it is impossible to avoid the exposure because this thing goes in furniture, cars, and so on. So the probability that somebody will be exposed is relatively high . . . So we have to develop things that had very low toxicity or very low adverse effects." Despite their awareness of toxicity and the need to develop low-hazard chemicals, flame retardant manufacturers have put significant energy and resources into critiquing the scientific research showing health effects and lobbying against flame retardant regulations.

In spite of the importance of toxicity testing in new product development, representatives of the flame retardant industry told me in interviews that hazard was not sufficient for making decisions about chemical safety, especially for existing products. Sometimes hazard-based decision making was rejected outright. As a FlameCorp representative told me, "hazard must be balanced with other considerations." Key among these considerations are market concerns, use patterns, and exposure characterization. One industry representative maintained that decisions about which tests to conduct are influenced by a company's sales department and by market concerns: "A choice between two tests might not come down to which test is best scientifically." Additionally, hazard assessment was not seen as a valid way of assessing chemicals already in commerce. "You cannot just go with hazard," a new product developer at FlameCorp told me. More commonly, industry respondents expressed the opinion that even hazardous chemicals could be low risk if exposures were sufficiently controlled.

The flame retardant industry also explicitly rejects the exposure-proxy risk formula. According to an industry representative, "If you've got P [persistence] and B [bioaccumulation], is that enough? . . . I would probably say 'no' . . . If it's not toxic, then I just am not sure that it's a problem." Instead, the industry prefers the exposure-centric risk formula. This approach was criticized by EPA representatives, who said that companies often have little knowledge about the true exposure pathways for their products, and that focusing solely on consumer exposure ignores exposure during manufacturing and product disposal. An EPA exposure scientist said companies often provide only limited exposure data because "industry has less control over exposure. They're not good at knowing where chemicals are going." Scientists at FlameCorp acknowledged as much. For example, one told me that although they knew more about how their products are used in foam, "in textiles we typically know not as much . . . A lot of it is more proprietary than foam . . . People don't want to tell you sometimes where they're putting [chemicals]" because of confidential business information and corporate competition.

In these ways, the industry disregards more precautionary risk formulas in favor of the exposure-centric risk formula that, due to significant data requirements, is less likely to arrive at a determination of risk and more likely to protect existing markets. In the words of a DfE representative, an emphasis on exposure allows proponents to justify the continued use of "nasty chemicals" that are very toxic but have low predicted exposure.

CONSEQUENCES OF COMPETING RISK DEFINITION

The six conceptual risk formulas described in this chapter and summarized in Table 3.1 represent fundamentally different ways of thinking about and evaluating chemical risk and safety. They place different emphases on chemical exposure, hazard identification, and scientific proof. They also matter greatly when it comes to the operationalization of risk assessment and management activities because risk definitions can be codified into formal practices or used informally to evaluate the merit of different risk-based arguments.

Though these risk formulas are anchored in a scientific analysis of chemical risk, they are also frequently accompanied by assertions and claims that extend beyond science. The either-or risk formula is frequently accompanied by the ethical assertion that chemicals should not enter into infants' bodies or women's breast milk. Similarly, the exposure-centric risk formula is regularly accompanied by claims about fire safety and economic realities. Those at FlameCorp, for example, argued that economic or societal benefits of flame retardants deserve consideration in the risk assessment process and that fire safety should not be hampered by efforts to decrease chemical exposure. EPA representatives acknowledged that their work was biased in favor of public health protection. As a DfE scientist told me on the last day working in their office, "We're not impartial. We are biased. But we're biased by the Agency's mission, which is to protect public health and the environment." The links between stakeholders' biases and their preferred risk formulas are undeniable.

Risk assessment is a multifaceted practice or social process: it is at once a scientific practice, a decision-making tool, and a contested language. It

TABLE 3.1 Conceptual Risk Formulas

	FORMULA	KEY FEATURES	SITES AND STAKEHOLDERS
Classic	Risk = f (hazard · exposure)	*Function* is a linear and multiplicative dose–response relationship	Discipline of toxicology Regulatory bodies Discourse from industry and EPA
Emerging toxicology	Risk = f (hazard · exposure)	*Function* includes individual susceptibility and timing of exposure Rejection of linear dose–response in favor of more nuanced understanding of dose–response relationships Particular relevance with certain classes of chemical (e.g., hormone disruptors) or points of exposure (e.g., fetal development)	Low institutionalization but growing scientific and regulatory awareness and acceptance EPA's Office of Children's Health Protection
Exposure-proxy	Risk = f (hazard · exposure · persistence/bioaccumulation)	All components draw on the most protective measures available Assumes that exposure is expected if a chemical is persistent and bioaccumulative	Prioritization or screening of chemicals of concern EPA's Chemical Work Plan process

(continued)

TABLE 3.1 (continued)

	FORMULA	KEY FEATURES	SITES AND STAKEHOLDERS
Exposure-centric	Risk=f(hazard · exposure potential · human exposure levels)	*Function* includes a measured dose–response relationship (i.e., exposures below a certain threshold are of no risk) *Exposure potential* is multifaceted, including mechanism and pathway of human exposure	Prominent in chemical industry discourse and evaluation
Hazard-centric	Risk=f(hazard · exposure)	*Hazard* is multifaceted and includes characteristics of exposure potential *Exposure* is assumed to be undermeasured Reduces risk by reducing hazard within functional use categories	Low institutionalization but growing awareness Alternatives assessment programs in government and nongovernmental organizations
Either-or	Risk=f(hazard) *or* Risk=f(exposure)	Either hazard *or* exposure provides grounds for action Alignment with precautionary principle Toxic trespass perspective	Environmental advocacy Low institutionalization State biomonitoring programs

can be codified into a series of steps regarding the evaluation of the scientific literature, as it is in many regulatory programs. It serves as a decision-making tool, providing a way for the EPA or industry stakeholders to say whether a certain chemical is safe enough for intended use. Risk is also a widely contested term, because its highly variable definition and operationalization allows stakeholders to achieve certain goals.

At the root of risk definition is the question of safety: what does a safe chemical look like, and how will we know it when we see it?[69] As an environmental health activist explained, "In an ideal system, we would never bring a chemical into the marketplace without proof of safety." But the meaning of safety for any stakeholder is inextricably tied to how they conceptually define risk. Following the chemical industry's exposure-centric risk formula, a chemical is assumed to present no risk unless exposures are scientifically expected and measured in people. Critics of this contemporary chemical regulatory scheme call this the "innocent until proven guilty" approach because it allows chemicals to remain in commerce long after initial concerns are voiced.[70] The EPA is clear that alternatives assessments and reviews of new chemicals do not certify chemicals as safe, although that is often how DfE evaluations are interpreted.[71] In contrast, risk assessment aims to identify (in)tolerable levels of hazard and exposure, hopefully with a large enough margin that safety can be ensured. Because of the uncertainties and assumptions outlined here, risk assessment is, according to a high level EPA scientist, "an art, not a science." Risk-based regulations mean that action is taken with the hope that uncertainties will not play out in the direction of harm.[72]

A key feature of the classic and exposure-centric risk formulas is that it is possible to arrive at a determination of zero risk if exposure levels are not above a level of concern. In contrast, the hazard-centric risk formula makes it unlikely that an assessment will arrive at such a determination because everything is hazardous in some way at some level, as defenders of risk assessment and representatives of the chemical industry readily admit. While we were going over a hazard assessment document, a DfE scientist pointed this out to me. Notice, she said, that the hazard sheet says "low" hazard, not "no" hazard; "we will never say safe, nontoxic, or no toxicity." As another researcher at DfE told me, "It is a little more difficult to use hazard assessment to defend a chemical," because hazard assessments typically rely on the worst toxicity end point. In contrast, the exposure-centric model relies on a multiplicative formula with many exposure components, creating a greater chance that one element will be predicted to be low or nonexistent, thus zeroing out or reducing the entire risk equation.

This use of conservative estimates for multiple hazard end points may be a way to implement precautionary regulations without the explicit use of the precautionary principle, which advocates erring on the side of caution in the face of uncertainty about potential risks.[73] Although multiple EPA representatives talked about erring on the side of caution or using worst-case scenarios when they lacked measured data, none told me that their judgment calls allowed the agency to implement the precautionary principle. Indeed, "regulators, with the backing of industry, have largely rejected the precautionary principle,"[74] and several EPA scientists told me that the precautionary principle was a forbidden term within the EPA. But as the history of PBDEs discussed in chapter 2 makes clear, many see PBDE regulation as an example of the precautionary principle in action: "There were no bodies in the street when Penta was banned," as a public health scientist explained.

Creating science-based public health regulation is unquestionably difficult. Deciding on and evaluating the weight of evidence needed for action involves policy decisions anchored not just in science or existing regulatory or legal frameworks, but in normative positions, ethics and values, different levels of risk tolerance, and economic and political interests. To put it another way, political, social, and economic factors must outline the parameters of the decision-making process before science can be used to fill in any needed equations or story lines.

■ ■ ■

The divergent definitions of risk that have been the focus of this chapter establish the boundaries and practices that guide future risk management practices. But how do stakeholders justify their arguments for a given policy outcome or interpretation of the state of the science? In the next chapter, I show how scientific uncertainty creates openings for strategic (mis)interpretations and (mis)uses of scientific evidence in pursuit of decidedly nonscientific goals.

4

STRATEGIC SCIENCE TRANSLATION

I n June 2012, over a hundred scientists and environmental policy makers plus a handful of environmental activists, industry representatives, journalists, and this lone social scientist met in Winnipeg, Canada, for the 13th international meeting on flame retardants.[1] The first Brominated Flame Retardant (BFR) Workshop had been held in Sweden in 1989. After a ten-year lapse, the workshop resumed in 1999 with a small gathering in Burlington, Ontario, and it has been held almost every year since. Since these early meetings, the community of researchers studying flame retardants has grown substantially, as this class of chemicals has received increased scientific, regulatory, and public attention. At the thirteenth workshop, attendees heard two days of presentations on cutting-edge, usually not yet published science on exposure, toxicity, and epidemiology research on many different halogenated flame retardants.

On the second day of the conference, a researcher shared results from a recently completed study looking at the reproductive effects in kestrels (a type of falcon) from a flame retardant chemical called β-Tech. Her team had given the kestrels environmentally relevant doses of this chemical, levels commonly measured in the environment. They then tested for transfer of the chemical into the eggs and tissues, and for reproductive and other behavioral changes such as changes in egg-laying patterns. Although they had the analytical methods to be able to test for this chemical, they were unable to find measurable levels of β-Tech or any known β-Tech metabolites in any of the eggs or tissues they tested. However, they still observed reproductive and behavioral differences between the dosed and undosed birds. The birds exposed to β-Tech laid fewer eggs and their eggs were lighter; fewer eggs were hatched overall, and there was a sharp decline in the number of male chicks.

The researchers faced two dramatic research puzzles. First, though they knew the birds were exposed to β-Tech, they could not find the chemical or its metabolites anywhere in any of the birds' eggs or tissues. Second, even though they could not find measurable levels of the chemical, they were able to show that chemical exposure led to negative health effects in these birds.[2]

This study illustrates the high level of scientific uncertainty in contemporary environmental health research. The researchers had to figure out that β-Tech was an actively used flame retardant, often a challenge given the limits of industry disclosure requirements. They also had to find ways to detect the chemicals in the laboratory: it is one thing to want to measure a chemical, but quite another to develop the complex laboratory techniques that allow a researcher to actually find the chemicals in various types of samples. In the analysis of the data from the study, the researchers faced a surprising uncertainty: what happened to the β-Tech? It was consumed by the kestrels and led to health outcomes, but it was undetectable. Although many chemicals break down and are transformed in the environment and in living organisms, this uncertainty has serious consequences for other studies on emerging contaminants: if chemicals cannot be measured in gathered samples, there is no way to look for associations between chemical exposure and health outcomes.

Scientific practices of all types require innumerable judgments, and these acts of interpretation and translation were frequently on display throughout the two days of scientific talks, networking coffee breaks, poster sessions, and group dinners at the BFR Workshop. Social studies of science recognize the breadth of methods and disciplines captured under the label "science," and the powerful historical legacy of value-free research in which scientific authority rests on science being seen as "value-free and politically neutral."[3] Yet the field of environmental health research is highly uncertain. As a long-time science policy expert with the National Institute of Environmental Health Sciences (NIEHS) explained, making science-based decisions regarding public health requires "an acceptance by all parties that decisions need to be made in the face of uncertainty." Anything approaching certainty regarding the cause of a negative health outcome would by definition require observed negative public health impacts, yet this would suggest an observed failure by the relevant policy-making institutions. As a public health researcher put it, once researchers find population-level effects, "the horse is out of the barn."

What does scientific uncertainty look like to the scientists conducting research and the decision makers trying to translate that research into action? And what are the impacts of institutional orientations toward risk on scientific translation and interpretation? Environmental health science is uncertain in many ways that inevitably leave it open to interpretation—and, sometimes, to misinterpretation. In this chapter, I argue that because the stakes are high in environmental policy arenas, research findings will be strategically translated by interested stakeholders. After introducing scientific uncertainty as an object of social science inquiry, I introduce a typology of methodological, data-driven, and conceptual uncertainty. I then turn to how uncertainty impacts the use of research findings in scientific practices, through the concept of strategic science translation.

ENVIRONMENTAL HEALTH RESEARCH AND SCIENTIFIC UNCERTAINTY

Science carries a special authority in social and political life, and is often a primary guide and justification for environmental policy decisions. Decision makers use research findings to make and validate decisions, scientific credentials give weight to opinions and identities, and scientists generally share values such as disinterestedness and organized skepticism.[4] Yet science itself is not objective or removed from social context. Research is conducted by people, is influenced by factors ranging from funding agency priorities to systematic inequality, and involves the constant negotiation of boundaries that separate science from other social institutions.[5] Policy-relevance challenges scientific objectivity, autonomy, and neutrality, and accusations of uncertainty decrease the authority of the science in question, moving it from the realm of objective facts into areas of interpretation and value judgments.[6]

Scientific uncertainty and the production of knowledge have received extensive attention from social scientists in recent years. In common parlance, uncertainty often refers to unknown probabilities of outcomes or the "totally random unknown."[7] For risk analysis scholars, uncertainty involves conditions where outcomes are well specified but probabilities are unknown.[8] Social scientists have also focused on differences between types of uncertainties. Sociologist Matthias Gross, for example, distinguishes between *knowledge* (beliefs that are justified and accepted as true), *ignorance* (knowledge about the limits of what is known), and *non-knowledge* (knowledge about the unknown combined with a recognition that those absences are important).[9] Historians of science Robert Proctor and Londa Schiebinger separate ignorance into three types: ignorance as a "native state" or absence of knowledge waiting to be filled in, ignorance as a "lost realm" that is the product of inattention, and ignorance as an engineered or "strategic ploy" by interested parties.[10]

I see uncertainty as a general feature of the scientific landscape that identifies the locations of social, political, and economic power. In many ways, uncertainty "is an inherent property of scientific data."[11] This is particularly the case for emerging environmental health threats, for many reasons: documenting historical and current chemical exposure is scientifically complicated and resource intensive; numerous exposure pathways exist for any chemical; many of the suspected health effects of chemicals take decades if not generations to emerge; few toxicological experiments are permitted on humans, for ethical reasons; and the effects of animal studies do not always translate perfectly into human health effects. People are exposed to hundreds of chemicals on a daily basis, which may act in synchronistic, antagonistic, or cumulative ways, and chemical exposures interact with other individual factors such as genetics to influence potential health outcomes. Contradictory findings are commonplace, and there are always additional studies to be conducted.

Areas of ignorance can also be produced through structural features of social institutions that produce and use scientific evidence.[12] Uncertainty is exacerbated in emerging or policy-relevant areas of exposure research, toxicology, and epidemiology due to the complexity and length of exposure pathways leading to disease outcomes and the methodological and analytical uncertainties and disagreements involved in identifying those pathways. Environmental health research can be characterized as having four *moments of uncertainty:* choosing research questions or methods, interpreting scientific results, communicating results to multiple publics, and applying results for policy making.[13] In these moments of uncertainty, scientists find that traditional approaches, protocols, and paradigms neither adequately describe the current state of affairs nor provide clear guidelines for action.

In other cases, however, ignorance and uncertainty are themselves strategically produced and used, challenging conventional expectations that knowledge will always, inevitably communicate and be put to good use.[14] Knowledge gaps or areas of *undone science* exist because research topics are overlooked or deliberately ignored, and because the priorities of civil society or the public can differ from the priorities of industry or the government.[15] Other areas of "forbidden science," controversial topics such as intelligence–genetic links or the development of new pathogens, are off the table for social and ethical reasons, not scientific ones.[16] In science and technology studies, the field of *agnotology* examines doubt and ignorance as cultural products that can be "made, maintained, and manipulated by means of certain arts and sciences."[17] As these researchers have shown, industries ranging from big tobacco to big oil have concealed evidence, called for additional but irrelevant scientific exploration, funded research that addressed different research questions, and gathered cohorts of paid experts to cultivate public uncertainty about the health impacts of their products.[18] Additionally, research on the "funding effect" in scientific publications has shown that the studies funded by companies are much more likely than independent or government-funded studies to have results that favor industry interests.[19]

Thus, through both "intentional" and "unwitting" forms of control over the production, dissemination, and use of scientific information, stakeholders in government and industry deploy and produce ignorance through a process of unknowing that discredits or ignores exposure–health linkages.[20] Furthermore, uncertainty is both an input into risk assessment processes and often an outcome of those processes.

Recent work in this area follows Pierre Bourdieu's conceptualization of the scientific field, a social arena of "forces, struggles, and relationships that is defined at every moment by the relations of power among the protagonists."[21] Seeing science in this way, as "an arena in which actors struggle for power," allows for an attention to how participants in the field compete over scientific authority as a specific form of capital.[22] But scientific authority, or scientific capital in Bourdieuian terms, is not only powerful within scientific disciplines. Through the process of *scientization*, scientific authority is increasingly valued and required in regulatory, legal, and social movement fields.[23] Additionally, policy-relevant research inspires nonscientific actors, such as legislators and activists, to be interested in scientific findings and to appropriate those findings for multiple uses in other fields. The application of science for policy making becomes a site of contestation because stakeholders desire different policy outcomes, which they justify using different interpretations of the state of the science or pieces of scientific evidence.

The topography of the scientific field reflects distributions of power in its dips and valleys—its data gaps and areas of missing research—just as it does in its highest peaks—the topics that become the focus of popular and accepted inquiry. This reflects a "broader politics of knowledge" involving complex, multi-institutional struggles for resources and research agendas.[24] Funding priorities can be set by federal agencies or the military, and disciplines can compete for intellectual territory and scarce grant dollars.[25] Furthermore, the topography of the scientific field is also connected to political considerations, with research questions of interest to communities and nonelites often left unanswered while the questions of interest to the government or industry receive more attention.[26] This is particularly the case with environmental health topics of interest to the public and community groups but not to corporate or regulatory stakeholders.

Thus, uncertainty results both from the practice of science itself and from the political, economic, and social forces that influence the production, interpretation, and use of research findings. Yet while scholars widely acknowledge the existence of both forms of scientific influence, they have not developed conceptual tools that allow for the examination of both simultaneously. Before turning to this conceptual development, I investigate uncertainty directly, showing that it is multifaceted and includes methodological uncertainty about how risk should be measured and assessed, data-driven uncertainty about what the evidence demonstrates, and conceptual uncertainty about what risk is.

Methodological Uncertainty

At the 2012 BFR Workshop, two researchers debated the virtues of using "pure" chemical formulations instead of the commercial mixtures that are actually sold and used. A toxicologist had used the technical mixture available from the chemical manufacturer, but an environmental chemist noted that their findings might not reflect the risk of the pure chemical itself, because the impurities in commercial mixtures might be toxic themselves. The toxicologist replied that whatever is in the technical mixture is relevant, whether it is the intended product or a contaminant that causes the effect, because this is the substance that will be used in product manufacturing and thus the substance to which consumers will be exposed.

This is an example of *methodological uncertainty* in environmental health research, in that it concerns research choices. These researchers disagreed about the choice of the chemical substance used in toxicological research, but they could just have easily questioned each other about research choices such as the type of animal model chosen, the behavioral or toxicity tests used, or the timing of exposures.

The ubiquity of methodological uncertainty is widely accepted by practicing researchers. I spoke with a toxicologist who told me that "what questions you choose to ask has a lot of influence on what answers you get." Similarly, another researcher argued that differences in results are "usually" the result of different research methods. For example, in animal toxicology research, some species or strains of animals are more likely to get certain diseases than others. As a toxicologist noted, "I can give you a rat that will *always* get breast cancer, and I can give you a rat that will *never* get breast cancer. And which rat is representative of the human population?" This researcher identified other research choices that significantly impact results: "What was the diet? What food was it on? . . . What were the housing conditions that they were in? Were they singly or group housed, for example? And what were the questions that you asked? Well, obviously you can't ask a question for long-term effects of developmental exposure if you didn't start exposing them until they were adults." Thus, many methodological choices—some obvious, others more subtle—impact the results found in a study. This is similar in any research enterprise: all science involves decisions made by people, introducing the subjectivity of the researcher at every step in the research process.[27]

Many methodological choices are influenced by the structure of the scientific field, including disciplinary assumptions and traditions that identify valid and authentic methods and types of knowledge.[28] For example, a broad area of methodological uncertainty surrounds the question of whether studies must be conducted according to "Good Laboratory Practices" (also called GLP or "Guideline" studies) in order to be relevant for risk assessment and regulatory decision making. GLP guidelines were first implemented in the 1970s and 1980s by government agencies including the EPA and Food and Drug Administration (FDA) in response to instances of fraud and data misrepresentation by the pharmaceutical industry.[29] Today, Guideline studies follow rigorous data keeping and standardized procedures designed to ensure that the developed data are reliable and reproducible.[30] Supporters of GLP argue that the guidelines improve the accuracy and reliability of findings. However, opponents argue that the checks and balances contained in GLP protocols are designed to standardize practice and "are not always effective guards against biased or even bad science."[31] Additionally, GLP studies are large in scale, time consuming, and very expensive, so they are mostly conducted by industry or the federal government.

Methodological uncertainty can also function at the level of the discipline. For example, in interviews, toxicologists highlighted the limitations of studying one chemical at a time, and they emphasized that toxicological experiments are limited to outcomes people already know how to study: "There are a large number of real-world toxicological end points . . . and toxicology just can't do them all." Toxicology studies are designed to assess the impact of (usually) a single chemical on a small number of defined health outcomes. A study that looks for cancer is not suited for the assessment of neurodevelopmental or reproductive effects. This too can be a source of methodological uncertainty.

Data-Driven Uncertainty

At the EPA's office in North Carolina, I attended a meeting of ToxCast, a high-throughput screening program that tests over a thousand chemicals in hundreds of *in vitro* assays.[32] High-throughput screening involves putting large numbers of chemicals through *in vitro* assays, ranging from cell cultures to small organisms, to test for changes in gene expression, biomarkers, or behavioral effects. But this requires that the researchers know certain things about the chemicals being tested. At this meeting, ToxCast researchers were discussing at what point a chemical is not pure enough or is at too low a concentration to be included in the overall data set. As one researcher stated, "whatever *zero* is should be redacted." But there was no consensus on what that number was. Deciding whether the cutoff for "zero" would be 0.1 percent, 1 percent, or 10 percent had a significant impact on the analyses to be conducted by in-house researchers and the data that would be released to the public.

This represents an example of *data-driven uncertainty* in environmental health research, when pieces of evidence and the labels assigned to them are contested or imprecise. Debates about the growth of high-throughput screening are often debates about data-driven uncertainty. Many scientists express great optimism for the ability of this research to decrease this kind of data-driven uncertainty by providing useful, timely, and less expensive data than is available through traditional toxicological measures. Researchers recognize that high-throughput screening could open up the "black box" of toxicological research, connecting exposures and outcomes by identifying the mechanism that leads to a given health outcome.[33] But high-throughput screening may also be an example of the "molecularization" of health and environmental research that displaces attention from environmental hazards and polluter responsibility.[34] Many questions remain about how to translate findings from high-throughput screening into policy, and at present it is only used to influence how chemicals are prioritized for standard toxicological review.[35] High-throughput screening currently faces numerous types of data-driven uncertainty: researchers can only work with the assays that are already developed and validated, limiting the types of evaluation that can be conducted, *in vitro* data must be understood in the "broader context" of chemistry and biology, and not all types of chemicals can be screened on currently available equipment.

Another form of data-driven uncertainty surrounds chemical identification. As I shadowed academic and government researchers, I was surprised to learn that scientists are not always working with the chemical they think they have or the needed purity of the chemical. These problems were front and center at another ToxCast meeting. Through systematic analysis of chemical substances, researchers found that up to 6 percent of the chemicals they receive from the manufacturer are the wrong chemical in some way, due to mislabeling, mixtures, or impurities. A toxicologist explained: "we've got about a 3–6% error rate," where "either it's actually not the chemical or it's nowhere near the purity we want." This poses a significant problem because a study that is conducted using the wrong chemical will provide false, misleading, or unhelpful results. In some cases, impurities can impact health at extremely low levels. A toxicologist shared a personal experience from his own research, when a 0.1 percent contaminant in the compound he worked with "entirely screwed my study." Thus, if researchers fail to identify certain characteristics of their samples, their analyses may be seriously flawed. But few toxicological studies confirm the chemical identities of the products they work with, because researchers lack the time, resources, or equipment.

Other times data-driven uncertainty results when researchers set parameters for a study and make decisions about how to interpret raw data. Like the earlier example about identifying "zero" for ToxCast, a flame retardant advocacy representative noted that advances in analytical chemistry have shifted zero in a different way: "Zero, if you go back ten years in analytical chemistry, was a different number than it is today." In analytical chemistry, the level of detection has declined dramatically, allowing for the identification of chemicals in people and the environment at parts-per-billion or parts-per-trillion levels. This changing understanding of what "no exposure" really is has created a moving target for industry to argue that people are not exposed to their products, and for public health advocates to identify chemicals of concern due to human exposure.[36]

Conceptual Uncertainty

An environmental health activist with a Ph.D. in genetics told me that scientific uncertainty was the most difficult part of her job: "You'll never definitively prove that something is causing harm, because that's the limitation of the scientific method. You can have a preponderance of evidence that shows that something is having a negative health impact, but you can never get definitive proof." This represents *conceptual uncertainty* about environmental health research. Although regulators would often like science to provide unequivocal answers and the public often thinks that answering questions definitively is the goal of science, this is rarely if ever the case in environmental health research. An environmental health regulator told me that what counts as "reliable" science "is hugely debatable." A public health scientist said, "Science is never a slam dunk." And an environmental health researcher compared science to the analogy of the blind man feeling the elephant: "Everyone picks up on a different piece."

Conceptual uncertainty matters greatly in areas of science policy, where findings are applied to decision making.[37] For example, even if there is consensus as to whether a dose of a chemical causes a health effect, how does that translate into policy recommendations? As a representative of the EPA's DfE program told me, "The decision to say that a 'very high' level for aquatic toxicity is one milligram-per-liter or greater, I mean, that's science policy . . . There's no sort of empirical test you can do to say, 'this is very high.' That's a decision, a policy decision." Scientists or different review processes will arrive at different conclusions because, in the words of an environmental health advocate, there is "interpretation of data that's always going on throughout the whole process from the point that you're doing the lab work, to the point that you're making a final policy decision. And so at each one of those points there's this interpretation of evidence."

Identifying unknowns is a particularly difficult task, as a scientist explained about the search for new chemicals: "It isn't until they [scientists] start looking for other ones [chemicals] that they know they're out there." Additionally, stakeholders have different ideas of what constitutes an acceptable level of proof: "The bigger problem is that different people have different feelings on what that level of proof is." As a concrete example, there is uncertainty about which toxicological effects matter and should be tested. As an EPA scientist explained, "With endocrine disruption, what does that mean? I mean, that's certainly a test that people are arguing, what's a positive test?" Thus, even toxicity end points can be conceptually uncertain.

Across these three types of uncertainty, the constant is that scientific uncertainty is multifaceted and is neither produced randomly nor interpreted neutrally. On the one hand, uncertainty is an inevitable and productive component of scientific practice; on the other, it is often strategically developed or deployed. I met with two EPA risk experts for a joint interview. One said, laughing, "If you put 12 scientists in a room and asked their opinion, how many opinions do you think you're going to get?" The other replied, that's why we do peer review. Similarly, a toxicologist spoke passionately about the importance of interpretation in environmental health work:

> That's the one thing I learned through all of my years in the environmental world: there's six sides to an issue . . . There's always, you know, 20 different perspectives from 20 different scientists . . . They just have different worldviews, let's put it that way, and when you're talking about shades of grey, instead of black and white, those worldviews can push you in one way or the other to a conclusion and that's sometimes very different.

These types of epistemological, disciplinary, and sociocultural influences are certainly important and are visible in what this toxicologist calls scientists' "worldviews." But assuming that scientific uncertainty and contestation are merely tied to disciplinary training or personal risk preferences understates the social, economic, and political forces that influence the scientific field. Instead, these differences are often strategic in nature, reflecting differences in power, competing interests, and policy goals.

APPROACHES TO SCIENTIFIC UNCERTAINTY

One of the leading flame retardant researchers in the country is Heather Stapleton, an environmental chemist at Duke University who has studied flame retardants since graduate school. With collaborators in public health, epidemiology, and environmental science departments around the country, she and her laboratory group have conducted innovative research to identify flame retardants currently used in products, develop analytical methods needed to detect chemicals in the environment, trace the absorption of these chemicals into environmental media, and study the toxicity of several flame retardants in computer, molecular, and animal models. She has published over a hundred scientific papers on flame retardants that are indexed in the PubMed database, nearly a quarter of which are cited over fifty times.[38] At the BFR 2012 conference, she was a coauthor on six papers, and on the second morning of the conference she took to the scientific podium to share new research on Firemaster 550, a flame retardant developed by Chemtura in the 1990s as a replacement for pentabromodiphenyl ether (PentaBDE) in furniture.[39]

Stapleton began by sharing some of her history working with Firemaster 550: the components are proprietary, but her laboratory was able to identify the chemical composition and structure after the flame retardant manufacturer Chemtura gave them a sample of the product. Her laboratory then published the chemical identities in this previously confidential substance.[40] She continued with her description: the Firemaster 550 website says it has a "favorable environmental profile," and an industry study required by the EPA in 2005 concluded that there was "no migration" out of furniture foam. Following the classic risk formula, Chemtura and the EPA thus concluded that the risk was negligible because there is no exposure. But, she said, she consistently detects Firemaster 550 components in dust and handwipes, "so they certainly migrate out."[41]

In her talk, Stapleton shared preliminary results of a collaborative study looking at whether Firemaster 550 caused health and behavioral effects in rat offspring after maternal exposure to the flame retardant.[42] They chose environmentally relevant levels, similar to those that have been measured in household dust. Stapleton and her collaborators chose to use environmentally relevant exposure levels because of questions raised by a study funded by the flame retardant industry. This study found reproductive effects at the lower doses but not at the highest dose; however, instead of focusing on these findings, the industry study dismissed these low level effects as "spurious" because they did not follow a linear dose–response relationship.

In her study, Stapleton reported, she found changes in body weight, reproductive development, and behavior. Speaking to an audience of hundreds of flame retardant experts, she focused on the extra weight gained by the animals exposed to Firemaster 550, and speculated that some of the behavioral changes they also measured were not due to neurological changes or anxiety but occurred because the rats were "so damn fat" that they had become the rodent version of "couch potatoes"! But, Stapleton cautioned, the study used a small sample size, with only three groups per treatment level. She concluded her presentation with questions, not firm answers: "What is the true NOAEL [no observed adverse effect level] for FM550? Should FM550 or its components be considered an endocrine disruptor or chemical obesogen?" Wrapping up her talk, she said this research "raises a lot of red flags about Firemaster 550."

Questions about scientific uncertainty and the meaning of research findings for environmental health are frequently and openly discussed by researchers. In her research talk, Stapleton highlighted the scientific uncertainty in her study and in the industry-funded study. She saw low-dose effects as potentially meaningful for human health, while the industry-funded study came to the opposite conclusion—that the results were spurious and unimportant. Yet she was also critical of her own research, highlighting the small sample size and the need for additional research.

How are these types of uncertainties interpreted and discussed by multiple groups of stakeholders? Existing social science research on uncertainty is generally divided into two groups. On the one hand, agnotology researchers point to deliberate manipulation or cultivation of uncertainty by industries (and, less often, government agencies). On the other hand, scholars of the new political sociology of science highlight the institutional features of academic and regulatory science that lead to the uneven production and dissemination of scientific evidence. Both provide significant leverage for examining the flame retardant story. Yet practices are not strictly confined to one group of stakeholders or the other. Industries are not the only actors who strategically present scientific evidence to strengthen an argument, and the institutional limitations that exist in academic or government science can also be found in advocacy and industry research. Indeed, comparing the ways different institutions strategically use science can deepen our understanding of science policy, scientific uncertainty, and even industry practices.

Stakeholders' varying perspectives on scientific uncertainty demonstrate that parties can and do respond very differently to the state of the science. Though individual stakeholders may feel that there is one "most correct" interpretation of the science, these interpretations differ greatly across institutions. The malleable nature of individual data points, the decisions that underpin data sets and statistical results, and the multiple moments of interpretation that occur as "data" become "findings" leave science open to what I call strategic science translation.

STRATEGIC SCIENCE TRANSLATION

An EPA researcher described a recent chemical assessment she had completed. "It's the glass half empty and the glass half full," she began. In her review article, she described long-term health effects after exposure to a certain chemical, but another recent publication came to the opposite conclusion:

> A paper just came out that did the exact same thing, looked at all the same papers, and [the author] concludes that the weight of evidence shows that it is not a developmental neurotoxicant . . . We just don't agree on certain interpretations. But she was paid for, she was paid by the manufacturer. I was paid by the EPA . . . She pulls out the uncertainty in the studies, and I try to look across the positives in the studies . . . I'm not saying that she says that because she is paid for by the manufacturer, but . . . her years of experience have been in . . . looking at the things people are doing wrong or not quite doing so right.

The two publications analyzed the same set of findings, but they came to different conclusions and presented different recommendations that reflect divergent interests and policy goals. The EPA study identified similarities across multiple studies in order to identify areas of concern for public health protection, aligning with the agency's interest in identifying and regulating chemicals that are found to pose unreasonable risks to human health or the environment. In contrast, the industry study found differences and uncertainties across multiple studies, highlighting scientific limitations and lack of consensus in order to make the argument that the chemicals are not a concern for developmental toxicity and aligning with the industry's interest in keeping their product on the market and profitable for as long as possible. This involves interpreting and communicating scientific evidence to an intended audience for the purposes of advancing certain goals and interests, a practice I call *strategic science translation* (SST).

Each word in the concept's name is significant. First, this activity involves *science*, the systematic collection of evidence and observations that describe and explain the world. Stakeholders use science to support their arguments by deploying scientific capital as a weapon in scientized fields such as science policy or environmental activism.[43] Scientific rationality and data-based arguments are valued weapons in these fields, but they are not the only legitimate arguments.[44]

Second, this activity is *strategic*, meaning it is deliberate, purposeful, and goal oriented. In the case of SST on environmental health, an immediate goal is often advancing a certain understanding of the state of the science or an individual research finding, and a more overarching goal is a policy or regulatory outcome or orientation. This conception goes beyond the explicit "strategic functions" of coercion, resistance, dissimulation, and (de)legitimization recognized within the interdisciplinary field of translation studies.[45] SST is a more encompassing term, including both manipulative aspects of coercive translation and more benign acts of knowledge translation such as selective citation, though it is similar in its attention to the role of power in deliberately crafting and presenting an argument.

Finally, this activity involves *translation*.[46] Translation is the movement of one definition or understanding of something to another definition or understanding of that same thing, often seen as moving between different discourses or languages, and often happening between fields. I conceive of translation as something much broader than the communication of scientific findings to nonscientific audiences, especially in the context of communicating federally funded research findings to relevant communities.[47] This conception of scientific translation also goes beyond the linguistic translation of scientific material and texts between languages.

Translation necessitates choices that convey certain pieces of information and omit others, but high quality translations preserve meaning across fields with little observable fingerprint from the translator.[48] Furthermore, translations are always intended for an audience or receiver of information, and translators make assumptions about their audiences that guide the translation. Translators must have a level of knowledge and expertise that enables them to engage with participants regarding the subject matter being translated.[49] To engage in SST, stakeholders need this type of interactional expertise in multiple fields, including the field that produces evidence to be translated and the field that is the target of their translation.

Following Bourdieu's understanding of scientific fields, stakeholders engage in SST when they deploy scientific capital in nonscientific fields. Within purely scientific fields, arguments anchored in scientific rationality are the only legitimate "weapons" that can be used by actors to influence the structure of the scientific field and the actors' own authority.[50] In interdisciplinary fields, however, including many policy-relevant fields, scientific capital carries significant weight but "no single form of scientific capital enjoys a monopoly."[51] This can be extended to scientized fields such as federal regulatory processes, environmental health activism, and industry decision making, all of which value both scientific and nonscientific rationality. Engaging in SST therefore allows actors to deploy scientific capital, often alongside other types of arguments, in the most useful and targeted way possible, whether their audience is the public, policy makers, or scientists. Because SST involves translating existing findings, it is distinct from other topics central to undone science regarding the structural origins of data gaps and uncertainties. Rather, it refers to the deliberate practice of translating existing scientific findings in order to make an argument, and it goes beyond the assumption that only industry supporters are capable of using science strategically to support identifiable economic interests.

The practice of SST allows stakeholders in environmental health arenas to bolster their arguments, strengthen their authority, and inspire change regarding a policy-relevant issue, whether their goals involve public awareness, corporate behavior, or regulation. Due to the interpretive nature of science and the inevitable uncertainty and complexity in emerging and policy-relevant areas of research, some level of SST is necessary because findings must be condensed and evaluated in order to be useful for science policy and decision making. SST thus allows actors to apply science to all types of contradictory arguments in scientific, public, corporate, and policy arenas.

An Example of SST: The State of the Science on DecaBDE

The case of competing interpretations of the research on the flame retardant decabromodiphenyl ether (DecaBDE) demonstrates how science can be used strategically. In 2009, the industry announced it would phase out production and distribution of DecaBDE by 2013.[52] Even after announcing this phase out, the flame retardant companies maintained that DecaBDE was not a risk and described the decision as one motivated by market concerns, not science. In a public statement provided to an environmental radio program, an advocacy representative with the flame retardant company ICL-IP argued, "Deca-BDE, one of the world's most effective and widely used flame retardants, has been the focus of controversy over the past several years in spite of extensive studies that it offers no health or environmental risk."[53]

The assertion that DecaBDE was not risky involved two strategic science translations designed to protect the industry from accusations that it continued to market the chemical long after initial calls to cease production. First, industry representatives routinely claim that the studies that found significant health effects after exposure to the chemical were critically flawed because their sample sizes were too small, their analytical methods were unreliable, and they failed to follow necessary GLP protocols.[54] These critiques are designed to delegitimize the findings suggesting that DecaBDE is too risky to remain in commerce.

The second scientific translation involves a large GLP research project funded by the flame retardant manufacturers to investigate the developmental and neurological effects of prenatal DecaBDE exposure in rats, published in 2011 by Chemtura employee John Biesemeier and eleven coauthors.[55] According to a flame retardant representative, the industry conducted the Biesemeier study to fill identified data gaps around developmental neurotoxicity: "There were some data gaps . . . and so industry was chartered to go and understand if this was a problem or not and create the science." Although the Biesemeier study did find isolated statistically significant results (e.g., differences in startle response in the lowest-exposure group), the authors concluded that these results were not biologically significant because they did not follow a linear dose–response curve.[56]

The Biesemeier study was officially sponsored by the Bromine Science and Environmental Forum (BSEF), a lobbying organization based in Brussels that is funded entirely by flame retardant manufacturers.[57] This strategy of steering research funding through nonprofit entities instead of directly acknowledging corporate sponsors is a common industry practice, sometimes intended to create an illusion that companies "play no role in research that supports their interest."[58] In the case of the Biesemeier paper, however, this disconnect was impossible because six of the twelve authors on the study were employed by flame retardant manufacturers (Albemarle, Chemtura, and ICL-IP), and the other six were employed by the contract research organizations that conducted the study and analyzed the findings.[59]

Since its publication, the chemical industry has frequently referenced this paper in regulatory and scientific disputes as evidence of DecaBDE's safety. For example, after the EPA's DfE program released its draft assessment of alternatives to DecaBDE, flame retardant manufacturers, supply chain users, and consulting firms submitted fifteen sets of comments and critiques of the report's findings.[60] Albemarle, a DecaBDE manufacturer, submitted seventy-five pages of comments on its own and also paid for additional sets of comments by three environmental consulting groups.[61] Many of these comments focused on DfE's assessment of DecaBDE as a "high concern" for developmental and neurological toxicity assessed in the DfE report. In these comments, Albemarle called the Biesemeier

study "high quality" and "robust," and argued that since the Biesemeier study found "no neurodevelopmental effects . . . at the highest dose
tested," the assessment for DecaBDE should be low concern, not high.[62]
Other comments critiqued DfE's assessment for relying on studies that
found effects at very low levels of exposure but did not meet GLP standards. The strategic argument at play is twofold: the Biesemeier study
found "no effects" (a misleading translation that ignores several statistically significant effects that did not follow a dose–response curve), and
any other studies showing effects should carry little weight now that a
large GLP study has been conducted—that is, the industry study is
definitive. These arguments represent a strategic translation of the science in an effort to lower the assessment of DecaBDE's health effects.
And indeed, these efforts appear to have been partially successful, because in the final DfE report, DecaBDE was assessed as having only "low
concern" for neurological toxicity.[63]

SST is perhaps most obvious in industry research and arguments using
science because the industry has particularly visible economic goals and
regularly deploys significant resources to influence policy debates. But all
institutions, not just industry, engage in SST. This is visible in how other
institutions discuss—or ignore—the Biesemeier study. The initial draft of
the DfE assessment of DecaBDE did not cite the Biesemeier study until
requested to do so by the chemical manufacturers during early reviews
of the assessment.[64] Once the study was added, the assessment simply
said that the Biesemeier study had found no effects and identified a
NOAEL of 1,000 mg/kg per day.[65] The assessment continued to establish
its hazard determination using a group of studies conducted by another
group of researchers who identified a much lower NOAEL, because DfE
assessments default to the most conservative finding.

The Biesemeier study is also regularly overlooked or dismissed by activists and academics. In my research, I came across no fact sheets, blog
posts, websites, or technical summaries on DecaBDE written by NGOs
or environmental health activists citing this study to describe DecaBDE's
effects.[66] Of the dozen scholarly publications that have cited the article,
only two are academic research articles, neither of which presents the
study as definitive research: one lists it as one study about potential health
effects among many, and the other cites it to suggest that questions about

DecaBDE's risks remain unanswered. Of the remaining ten publications, one uses the article as an example of how Guideline-compliant studies dismiss detected non-monotonic low-dose responses, one is available only in Chinese, seven are letters to the editor written by industry coauthors, and the final one is a response to one of those letters to the editor.[67] Each of these citations—along with the innumerable times the article could be cited but is not—is an act of SST, strategically presenting a piece of scientific evidence to make a larger claim about the state of the science, scientific credibility, or environmental health policy.

I now describe three types of SST using the words and actions of stakeholders working at the intersections of science and policy. *Selective* SST involves the selective use of evidence. Just as a museum curator may pick a handful of artifacts or art objects from among thousands, so must a stakeholder curate the full scientific corpus down to a small collection of relevant facts, findings, and conclusions. *Interpretive* SST involves emphasizing one argument over another in a case of inconsistent or inconclusive findings. Stakeholders can highlight one finding while ignoring or downplaying another, or describe findings in a way that reveals different preferences regarding statistical certainty, case study relevance, or precautionary policy making. The third category, *inaccurate* SST, involves incorrect communication. A stakeholder may falsely attribute findings or conclusions to a scientific study, or incorrectly characterize the strength or quality of the evidence supporting their positions.

Though I present them separately, the boundaries between these three categories are certainly fluid. For example, a stakeholder may simultaneously engage in selective and interpretive SST when deciding how to summarize the existing literature, including some articles because they are judged to be particularly relevant and excluding others a priori because of a disagreement with the research findings.

Selective SST

Selective SST involves the selective use of scientific evidence, and is common among all science policy stakeholders, including practicing scientists. An EPA researcher explained that for topics rich with prior research, it is "easy" to find papers that support any preconceived position, because "every study is different" and a "snapshot of reality." It's like the blind man feeling the elephant, she explained: everyone picks up a different piece of "the Truth with a capital T."

Crafting a scientific literature review requires selective SST, as it is impossible to summarize or refer to all potentially relevant prior studies. Instead, authors choose to cite one study and not another to best set up their research project, emphasize the research questions of interest, and demonstrate their understanding of their field.[68] I myself engage in extensive selective SST throughout this book to summarize the literature on flame retardants, environmental health risks, and the sociology of science.

The need to summarize the literature means stakeholders can also strategically select papers that are particularly well-suited to their goals. In this way, selective SST could be thought of as an extreme version of cognitive bias, the tendency to "cling to" information that confirms previously held biases "and ignore or reject as unreliable everything else."[69] However, SST involves the strategic, not just unconscious, translation of science for a certain purpose to target audiences. It is not just that certain findings are more subtly appealing, but that these findings are strategically presented in order to make one's case or lay the foundation for one's argument. A regulatory scientist described this process:

> If you really want to get into it and really understand it, you would have to order a hundred articles and really sit down and study them all and look at the weight of evidence . . . I mean, it takes a lot of time to really weigh evidence, and a lot of the dogs in the fight don't bother . . . There's these polarized groups that don't listen to each other's science.

The "weight of evidence" around any issue, she argued, "should change every time a new research article comes out." Instead, most stakeholders engage in selective SST by summarizing scientific evidence in a way that supports their argument.

Selective SST can be described as one's "spin" on an issue. An EPA toxicologist described how she evaluated research from different stakeholders: "You have to be careful to think about the lens they use. Their stake, their angle, why they're giving you that information, because it might be framed in a way that ignores certain issues or avoids certain issues." This is especially common when stakeholders summarize the state of the science for nonscientific audiences. For example, environmental advocacy groups rarely include references to studies that do not find risks in their releases, fact sheets, and reports summarizing research, while industry websites and blogs rapidly report those same studies.

Selective SST may be connected to structural features of the scientific field beyond the particular stakeholder's desire to make their argument. For example, several characteristics of the scientific field may explain why some stakeholders pay more attention to studies that find connections between exposures and health effects, but tend to ignore industry-funded studies that conclude there is no cause for concern. Methodological and statistical limitations and a widespread scientific preference for false-negative results over false-positive results mean that identifying positive effects is particularly difficult.[70] The bar for statistical significance in environmental health research is so difficult to achieve that any study finding significant effects suggests cause for concern. Additionally, research on the funding effect has shown that industry-funded studies are less likely than academic or government-funded research to identify serious health risks, suggesting that industry-funded findings of chemical safety may warrant additional scrutiny.[71] These structural characteristics of the scientific field contribute to selective SST by stakeholders.

Interpretive SST

Interpretive SST occurs when stakeholders present implications or conclusions of scientific evidence to support their goals. A toxicologist explained that this is common in assembled review panels:

> I've seen several times where . . . some stakeholder group, for example, has put together an expert panel, but it's an expert panel that is completely or highly composed of people with the same biases. And they come out with . . . a consensus document [that says], "Hey, we don't have any problem here." And then you get a different stakeholder group that will get a group of scientists around the table that have a different set of

biases and say, "Hey, we really need funding here, this is a big deal. We've got to take action."

Interpretive SST is common in environmental policy arenas because moving from scientific findings to policy recommendations requires stepping away from the data, or, as one EPA chemist described it, "building a scaffolding" between data and conclusions. Categorizing findings for policy utility requires qualitative and value judgments about the scale of an issue and needed responses.[72] Even in environmental health research areas with established research protocols and well-understood toxicity mechanisms, applying results for decision making requires interpretation. As an environmental scientist at EPA asked rhetorically, "Do we classify this dose as a high, moderate, or low concern? That's policy, that's not really science. It's science policy. So it's based on science, but it's a policy decision."

Stakeholders engage in interpretive SST when they summarize or draw conclusions from scientific evidence to support their interest-based goals. An expert in chemicals regulation used alcohol as an example:

> It's definitely a neurotox, but does it . . . meet your levels of high concern or not? . . . When you have chemicals that are on that threshold, you're going to get scientists outside of politics disagreeing . . . and then when you put them into a political environment, then they're going to sort of tilt in a certain direction.

Interpretive SST is particularly prevalent when stakeholders discuss the weight of the evidence, reducing a large research area to one overarching interpretation.[73] An advocacy representative of the flame retardant industry practiced interpretive SST when he described risk assessments for a widely used chemical:

> If you look at the risk assessments . . . what their job is to do is to take all that science—good science, counter science, all this stuff—put it in a pot, look at the chemical in use, and weigh out this issue of hazard . . . And I think what most of these reports are saying is that the scale is like this [holds out hands like a balance]. It's not tilted this way or that way.

Similarly, an industry scientist explained that "if there is debatable science . . . I think we can argue what is good science and bad science and you can go on both sides of the thing." By presenting the issue as though the scientific community is divided on the risks of flame retardants, both industry representatives are engaging in interpretive SST that is designed to convince an audience that the chemicals present limited environmental health risks.

Sometimes science-policy interpretations are guided by transparent, established protocols. Government programs often identify the criteria used to assign qualitative evaluations to quantitative results. For example, the EPA's DfE program developed a set of standard hazard assessment criteria that delineate different hazard evaluations and explain that the program will rely on the finding of greater hazard in the case of multiple and differing results.[74]

Often, however, interpretive schemas are more opaque, based on disciplinary training, risk tolerance, or unnamed interests. Scientists themselves may object that strict evaluative criteria are too black and white. An EPA toxicologist noted risk assessment toxicologists dislike overly stringent guidance documents because they "expect" some level of variability in chemical assessments. Additionally, formal guidance documents are rarely able to routinize all interpretive decisions, and some would say this is not a desirable objective. For example, even with the official criteria document, some DfE hazard determinations rely on "expert judgment" because exposure metrics for particular hazard end points are not specified. Although some interpretations are thus clearly delineated, other interpretations remain squarely in the hands of individual evaluators.

Interpretive SST is particularly visible when stakeholders describe their evaluation protocols. For example, the flame retardant industry engages in interpretive SST by redefining hazard so that chemical persistence—a common characteristic of flame retardants—is not seen as concerning. As I described in chapter 3, an assessment methodology developed by the flame retardant industry does not include chemical persistence as a distinguishing characteristic because "persistence is part of flame retardant design." However, hazard assessment methodologies developed by governments and environmental organizations do include persistence, thus reinforcing a belief that high levels of persistence are concerning.[75]

Stakeholders also engage in interpretive SST when they argue that research findings must be compared with other studies or completed risk assessments that arrived at preferred conclusions. For example, an employee of a flame retardant manufacturer wrote that the authors of a study on polybrominated diphenyl ether (PBDE) levels in food had "missed the opportunities to put their results in contrast" by comparing the levels found in the study with earlier reference doses.[76] Similarly, industry representatives strategically claim that data gaps exist, ignoring potentially relevant data to accentuate existing uncertainty. Paralleling the arguments of agnotology researchers, a public health scientist said that in her experience, industry stakeholders "are going to use the information that supports their cause and, and scream it from the highest mountain, you know . . . either ignore or downplay the science that does not suit their cause." This involves a strategic interpretation of the data to support a certain goal: keeping a controversial chemical in commerce.

Inaccurate SST

A final form of SST involves incorrect translations or the identifiable manipulation of science in support of stakeholders' goals. When translations move knowledge between different discourses or sites, they can be erroneous or strategically altered to give off a certain impression.[77] Inaccurate SST is similar to the manufacture of doubt described by agnotology researchers, but conceiving of these activities as SST highlights both similarities and differences across the range of stakeholders also using science to make an argument.

Inaccurate SST is often directly linked to economic goals such as delaying regulation of a chemical of concern and protecting market share. For example, after wooden storage pallets began to lose market share to lighter, more durable plastic pallets, a "public relations war" began between the wood and plastic pallet manufacturers, with each side using scientific arguments to claim that their product was more sustainable and safer.[78] As a flame retardant industry representative described the actions by the Wood Pallet Association, "The wood [pallet] guys go, 'I've got a competitive disadvantage. How do I deal with that?'... Write an article about ... 'this is the worst chemical ever and it's in your food.'" Engaging in inaccurate SST allowed wood pallet advocates to use science as a weapon in policy debates. (Furthermore, this industry representative engaged in interpretive SST himself by reframing the pallet safety debate as one of competitive disadvantage, rather than health and safety.)

Stakeholders also engage in inaccurate SST when they misrepresent research findings by other scientists. For example, a public health researcher who had published a paper on PBDEs described how the chemical industry had published a critical letter to the editor under the byline of one European researcher and five industry employees:

> It's total bullshit ... They're in the business of manufacturing doubt, you know? They didn't really have, like, one valid scientific argument. And really, they weren't even necessarily critiquing our science. But they were just like poking mini-holes ... But you know you first read this and you're like, "they're critiquing me," you know? Like, "I'm under assault!"

An epidemiologist shared a similar story:

> The couple papers I've written in [environmental journal], I think both of them have gotten responses from industry-related groups. And I sort of expect it. And actually, maybe it's a good thing ... That's how you know you're successful, you know. You've written something that industry cares about [*laughing*].

The public health researcher described feeling like she was being attacked when industry representatives wrote a critical response to her work, but the epidemiologist saw it partially as a mark of doing relevant work. This is a common pattern in the flame retardant scientific literature, and these critiques go far beyond the organized skepticism that is a defining characteristic of the practice of science, or the public skepticism of science that is widespread in American society today.[79] Instead, they are frequently examples of inaccurate SST.

I identified over two dozen published letters to the editor in journals in which industry-funded authors critiqued published studies on flame retardants. Though these industry-backed letters are occasionally sole-authored, they usually have many authors, including flame retardant company employees, paid consultants who acknowledge industry funding, and sometimes academic or independent researchers. These letters share common themes: they compare the study to industry-funded studies, accuse the analysis of ignoring potential confounders, claim that the study's authors inaccurately summarized the relevant scientific literature, and critique the study's scientific and analytical methods. For example, an industry-supported letter critiquing an academic paper on PBDEs and pregnancy outcomes inaccurately summarized the paper's methods, statistical models, and treatment of confounders in order to discredit the paper's conclusions.[80]

I most often observed or heard inaccurate SST by industry actors, perhaps because they had the most identifiable economic interests and frequently needed to defend controversial products. However, inaccurate SST is not limited to the industry sector. Other stakeholders engage in this form of SST as well, albeit with potentially less harmful repercussions. For example, activists, politicians, and media representatives sometimes summarize a study by speaking broadly about "flame retardants" as a class of chemicals, rather than speaking precisely about the particular chemical that was studied. A public health scientist was critical that people speak about studies on PentaBDE or OctaBDE as though they implicated all flame retardants: "So you get . . . 'flame retardants are known to cause bla bla bla bla bla.' When in reality, the data that they have been using to make those cases comes from studies regarding Penta or Octa." This form of inaccurate SST allows advocacy stakeholders in particular to make broader and more precautionary arguments about flame retardant risks. Although it is inaccurate to jump from PBDEs to all flame retardants, in some ways it is in fact appropriate to group flame retardants together by chemistry for scientific and regulatory reasons. For example, a consensus statement signed in 2010 by over a hundred and fifty scientists suggested that all brominated and chlorinated flame retardants, regardless of specific chemistry, are of significant concern.[81]

POWER ASYMMETRIES AND STRATEGIC
SCIENCE TRANSLATION

This analysis offers important insights about the structure and role of power in science policy fields, insights that are unavailable if the contested and pliable nature of science is assumed rather than investigated, or if only part of the field of stakeholders is studied. Because all groups of stakeholders engage in SST when they participate in contested environmental fields, their actions in those fields and the resulting policy outcomes often reduce not to the settling of scientific truths but to types of SST put forward and the resources deployed in service of a preferred scientific interpretation. Profitable companies can rally significant resources to defend products and markets from potential restrictions, resources typically unavailable to the general public or advocates who might support those greater restrictions.

The recent DfE alternatives assessment of functional alternatives to the flame retardant DecaBDE provides a strong example of the asymmetric nature of SST.[82] The organizations or individuals representing the environmental health advocacy field did not submit a single set of formal comments in response to a draft assessment; the flame retardant manufacturers, supply chain users, and consulting firms submitted fifteen sets of comments.[83] After this comment period, numerous small changes were made to the final DecaBDE assessment, and several hazard end points were downgraded between the draft and final assessments from high to low. Thus, it appears that the chemical industry's extensive SST efforts were partially successful in achieving a lower assessed hazard for key products.

Agnotology would point to this as a further example of how industry manufactures doubt and manipulates science to protect and expand markets. SST shares much intellectual territory with agnotology research, but differs in its attention to multiple types of scientific communication and to the full range of stakeholders participating in science policy dialogues. In this way, SST offers a step back from normative science studies that attempt to identify legitimate or preferred uses of science. It also allows for a more holistic understanding of how stakeholders select, interpret, and—sometimes, but not always—misuse research findings for policy goals. Although all stakeholders engage in SST when they write policy briefs, craft literature reviews, or summarize the state of the science, they do it in different ways. Thus SST is a concept that can be used to evaluate different stakeholders' scientific claims and make sense of environmental controversies without assuming that industry actors are manufacturing doubt or that activists lack the expertise to accurately communicate research findings.

STANDING STRONG ON A SHIFTING SCIENTIFIC FOUNDATION

In contested areas of environmental research and policy, all stakeholders can be expected to claim that their position is grounded in science, even though they are likely to disagree about the appropriate conclusions to be drawn and courses of action to be taken based on the scientific evidence. Around such issues, conflicts can be traced back to methodological, data-driven, and conceptual uncertainties which open the door for selective, interpretive, and inaccurate SST. Stakeholders engage in SST in the context of decision making about chemical risk and safety, using science to justify arguments that advance their political and economic interests. Furthermore, these interests explain much of the wide divergence in different stakeholders' interpretations of the same or similar pieces of science. The malleable and uncertain nature of scientific evidence means that it has to be interpreted and applied in order for it to be useful for policy. Thus flexibility in science policy is needed because the scientific corpus is always incomplete, and relevant cutting-edge research on chemicals of concern is not yet replicated or validated.

Yet waiting for definitive proof of harm is not an option that protects public health. Part of activists' concerns about scientific uncertainty is that effects that are unknown at the present moment may prove to be irreversible. Once persistent and bioaccumulative chemicals are in use, it becomes nearly impossible to prevent exposure. In the words of an environmental health leader, "It's like jumping off a cliff. You can't un-jump off the cliff." Activists argue we must act on the information we have now, while recognizing that the information is incomplete: "[We] can only work on the knowledge we have now." Risk assessments, then, should be treated as "living documents," able to incorporate new pieces of evidence as they emerge, and should be flexible enough to respond to and incorporate these changes.

How should environmental policy proceed in the inevitable absence of scientific certainty, especially when scientific practice is imbued with challenges over conflict of interest and data ownership? Industry is frequently accused of nontransparency or data manipulation when it comes to research findings, and academic scientists are often hesitant to share raw data out of fear that their data will be inappropriately mined and interpreted. Furthermore, the scientific field is uneven in terms of power, resources, and reward systems. Industry stakeholders likely have greater resources to hire consultants to write review articles or commission original research projects to fill specific data gaps or respond to emerging research. Simultaneously, academic researchers may be discouraged from review articles because the scientific community rewards original research over secondary and review analysis. Furthermore, standards for scientific proof can vary across contexts. A person may evaluate chemical risk quite differently in his or her professional activities as a scientist than in personal activities as a parent and consumer. And a company may evaluate chemical risk differently based on whether the chemical is a newly conceptualized molecule with low infrastructure investments, as opposed to a long-term "workhorse" chemical with production facilities, complex supply chains, and a dedicated sales force.

Together, the multifaceted nature of scientific uncertainty on the one hand, and the requirement that science be translated out of the scientific field into other fields of decision making on the other, suggest that science on its own can never definitively justify policy or regulation.[84] In the words of a toxicologist, "you just have to learn to live with uncertainty." Furthermore, the ubiquity of SST ensures that findings can never stand on their own, and instead will be translated, praised, or critiqued by stakeholders from all sides. The decision then becomes, how should a course of action be decided upon? While science must play a role, it cannot be expected to provide definitive proof. As I was told by an environmental health expert, unacceptable risk "is a policy judgment. It has nothing to do with science." And yet our current political tendency of waiting until nearly definitive proof emerges before taking action may actually promote unacceptable risk, because this involves waiting for the unacceptable outcomes of "bodies in the street," documented illness, or neurodevelopmental disorders in children.

If regulatory inaction is not a preferred solution, several different science-based options emerge as possibilities. A "weight of evidence" assessment looks holistically at multiple sources and types of data to arrive at a "best guess" of whether certain risks are likely to be seen.[85] For many, a weight of evidence approach allows evaluations and decisions to be made. For example, a toxicologist told me that it was a "no brainer" that DecaBDE was dangerous based on a weight of evidence assessment: "It just wasn't equivocal. When you look at the environmental fate, when you look at the biomonitoring data, the fact that they're increasing exponentially in wildlife and human body, and then you look at the toxicity, it all comes together."

But a weight of evidence approach has limitations: by seeking a continually increasing number of studies, it can delay action on research-poor topics, it demands commensurability of different end points, and typically it requires a documented exposure and disease pathway, something that is very difficult to prove.[86] That is, demanding a weight of evidence analysis can be used strategically to produce doubt and inaction. As an environmental health advocate told me, "Industry always likes to say, you know, you need weight of evidence. You need to have a great weight of evidence . . . They would like you not to be able to stand up under the weight." And because most areas of environmental health research, including research on flame retardants, have numerous inconsistent findings, it is possible to pick papers that support a preconceived position or preferred policy outcome.

A related but slightly different approach would take a protective stance within a weight of evidence analysis, privileging the most sensitive end point or lowest observed effect from a comprehensive analysis of scientific findings. If multiple high-quality studies arrive at different conclusions, the evaluating scientific team can go with the more conservative estimate: the tie goes to the greater hazard, so to speak. Whatever process is used, risk management activities need to be nimble, able to incorporate new science and emerging concerns. This is particularly the case in areas of environmental health research where what is "known" in the current moment might later be confirmed, or might be proven to be incorrect or incomplete.

■ ■ ■

Uncertain science thus acts as a shaky foundation for environmental health regulation. Because of the caution ingrained in scientific practices, scientists cannot be expected to speculate too far beyond their own research. It then becomes the role of policy makers to move forward with the scientific evidence at hand and make decisions that minimize risk. In the next chapter, I focus on this decision-making process and describe how scientists and regulators at the EPA balance competing interests to develop science policy, partially negotiating what counts as science and how data should be utilized.

5

NEGOTIATING SCIENCE, POLITICIZING SCIENCE

D r. Deborah Rice is a toxicologist who spent decades as a scientist and risk assessor at the EPA's National Center for Environmental Assessment studying the developmental effects of exposure to lead, methylmercury, and polychlorinated biphenyls (PCBs) before moving on to a second career with the Maine Center for Disease Control and Prevention.[1] In 2004, Maine passed the country's first presumptive ban on the flame retardant decabromodiphenyl ether (DecaBDE), outlawing it once a "safer, nationally available alternative" could be identified by state experts.[2] Working in her capacity as a regulatory toxicologist, Dr. Rice contributed to annual assessments on DecaBDE, and with academic collaborators she studied the developmental impacts of DecaBDE exposure in mice.[3]

As part of her work, Dr. Rice occasionally testified in legislative hearings to update state lawmakers. By the third DecaBDE report in 2007, she felt confident in her scientific assessment that DecaBDE represented too great a risk to remain in commerce. A rapidly accumulating amount of toxicological and exposure evidence had been developed by European researchers who had first identified neurotoxicity following exposure to DecaBDE; EPA scientists replicated these studies, as did Dr. Rice's own research collaborations in Maine. Exposure scientists had found DecaBDE in people's bodies, in wildlife, and in environmental media. Additionally, alternative chemicals were available and looked promising, though like most chemicals they were inadequately or unevenly studied. Based on all this information, Dr. Rice and scientists with Maine's Department of Environmental Protection told a legislative working group on February 15, 2007, that DecaBDE exposure and toxicity were concerning.[4] In her opinion, she said, there was "no question" that DecaBDE should be removed from commerce based on the scientific evidence.[5]

Also throughout the mid-2000s, the EPA became increasingly interested in polybrominated diphenyl ethers (PBDEs), motivated by growing scientific concern about PBDE toxicity and exposure, the industry's voluntary phase-out of two commercial formulations, and growing regulatory pressure in some U.S. states and in Europe. As part of the agency's regulatory and scientific efforts, the EPA conducted a formal PBDE risk assessment through the Integrated Risk Information System (IRIS), which was housed in EPA's Office of Research and Development (ORD). The goal of the assessment was to conduct a thorough toxicological and dose–response review to identify a *reference dose*, a level of daily exposure that would be safe for the average person for a lifetime. In late 2006, this risk assessment was drafted and released for public comment; following common agency practice, an EPA contractor was charged with assembling an external panel of experts for peer review.[6] Dr. Rice was asked by that contractor to chair the five-member committee, which also consisted of three academics, in pathology, biostatistics, and neurotoxicology, and one toxicologist from a consulting firm.

On February 22, 2007, a week after Dr. Rice testified in front of a Maine legislative work group regarding the ban on DecaBDE, she joined the EPA's external review panel for a one-day review session in Washington, D.C. According to her accounts of the panel, her role as chair was largely procedural, and she did not influence the comments of the other reviewers. As she would testify in front of a congressional panel the next year, "My function as chair was to ensure that all scientific issues were discussed, and all reviewers had an opportunity to express their views."[7] The EPA compiled the review panel's final written comments and posted them online in March of that year.

A few months later, however, Dr. Rice saw her scientific credibility and her role on the review panel called into question. The American Chemistry Council (ACC), representing the interests of brominated flame retardant manufacturers, wrote a ten-page letter to Dr. George Gray, then the EPA's assistant administrator for research and development. The ACC's letter conveyed "serious concerns" about the expert review panel, notably that "the peer review panel's leadership might lack the impartiality and objectivity necessary to conduct a fair and impartial review of the data."[8] Based on Dr. Rice's testimony to the Maine legislature in February and her peer-reviewed publications on DecaBDE's toxicity, the ACC argued, "there is no doubt that she [Dr. Rice] has taken a very public position" about DecaBDE's toxicity and "has been a fervent advocate of banning deca-BDE—the very sort of policy predisposition that has no place in an independent, objective peer review."[9] The letter urged the EPA to reject the recommendations of the review panel and to "base its final Toxicological Review on data, opinions, and conclusions *other* than the Chairperson's."[10] They closed by asking for a meeting with Dr. Gray.

On June 15, 2007, ACC representatives met with EPA regulators in a closed-door meeting. No public records from this meeting exist, but a follow-up letter from Dr. Gray and the EPA to the ACC summarized the meeting this way: "EPA acknowledged your concerns and indicated that we would take actions to respond to some of these concerns."[11] In August, the EPA followed through, removing Dr. Rice from the panel and erasing her name and comments from the online peer review summary, though an ACC representative later said the trade association had not requested these specific changes.[12] This August document retained the original February date and made no note of the editorial change.

The ACC soon spoke out against this action as well, demanding formal acknowledgement that Dr. Rice had been removed from the panel because of a perceived conflict of interest. Again the EPA complied, and a third version of the peer reviewers' comments was released in November 2007, still bearing the February 2007 dateline, with the following statement: "EPA modified this report in August, 2007 to include only four of the five reviewers' comments. One reviewer's comments were excluded from the report and were not considered by EPA due to the perception of a potential conflict of interest."[13] Explaining this decision in a January 2008 letter to the ACC, Dr. Gray said that "EPA listened to your concerns," and that "the Chairperson's comments would not be considered in the EPA's response to the external peer reviewers."[14] The letter also attempted to comfort the ACC that Dr. Rice had not "influenced the views of the other PBDE peer review panelists."

The redaction of Dr. Rice's comments from the peer review panel inspired significant criticism from a broad range of interested parties. Maine's Governor John Baldacci wrote a letter to the EPA protesting Dr. Rice's removal from the panel.[15] The Environmental Working Group (EWG), a large environmental nonprofit, submitted a formal Freedom of Information Act (FOIA) request for all communications surrounding the decision to remove Dr. Rice from the peer review panel, and published a detailed exposé on the timeline of Dr. Rice's removal from the panel.[16] The Congressional Committee on Energy and Commerce's Subcommittee on Oversight and Investigations invited representatives of the EPA, the ACC, and several environmental and scientific nongovernmental organizations (NGOs) to testify at a September hearing provocatively titled "Science under Siege: Scientific Integrity at the Environmental Protection Agency."[17] At the hearing, Dr. Jennifer Sass of the National Resources Defense Council called Dr. Rice's removal "another example of politics trumping science," while Sharon Kneiss of the ACC defended her organization's actions as pursuing "transparency" in EPA reviews. The EPA's Dr. Gray argued that his agency's decision was procedural, not politically motivated.[18]

At the heart of the numerous critiques of EPA's actions in this case were questions regarding the politicization of EPA science and the chemical industry's influence on scientific practices. During this time, concerns were mounting about scientific censorship in President George W. Bush's administration. A growing chorus of voices argued that the EPA was ignoring science, delaying needed environmental regulations, and catering to industry at the expense of public and environmental health. In 2008, for example, the Union of Concerned Scientists conducted a survey of 1,600 EPA scientists and found widespread evidence of "political interference in their work, significant barriers to the free communication of scientific results, and concerns about the agency's effectiveness."[19] Many in the environmental and scientific communities believed that the EPA at this time was not following through on its primary mission to protect human health and the environment.

The removal of Dr. Rice's comments from the peer review document, seemingly at the direct behest of a powerful industry trade association, smacked of blatant corporate manipulation of the EPA's scientific work. Critics noted that this was the first time a panel member's presence and comments on a panel had been completely erased after the panel's work was completed. Dr. Rice was a widely respected career scientist with no industry ties. In contrast, industry scientists and consultants regularly served as external reviewers on all types of EPA panels, in spite of their more identifiable conflicts of interest. On this same IRIS panel, for example, was Dr. Richard Bull, a toxicologist and consultant who had been asked to resign from a previous National Academies committee after he failed to disclose industry connections.[20] As part of its investigation into the Dr. Rice case, the EWG identified seventeen cases of conflict of interest on seven other IRIS review panels convened in 2007, including academic scientists who made public statements or conference presentations suggesting chemicals presented little or no risk, consultants and scientists employed by the companies making chemicals they were charged with reviewing, and scientists at industry-funded research institutes.[21] None of these seventeen panel members were removed from their peer review panels, perhaps, EWG suggested, because in each case they sided with industry instead of siding with public health.

Scientific credibility and bias were also central to the debate. Dr. Rice did have an opinion on DecaBDE, but it was an opinion based on her analysis of relevant scientific evidence and her significant scientific expertise. As she said in her testimony at the congressional "Science under Siege" hearing, "I had no opinion on the hazard posed by deca BDE before I started my review. I believe that having an informed scientific opinion constitutes neither bias nor conflict of interest."[22] Commentators agreed: Representatives John Dingell (D-Mich.) and Bart Stupak (D-Mich.) wrote that the ACC "seems to argue that scientific expertise with regard to a particular chemical and its human health effects is a basis for disqualification from a peer review board. This does not seem sensible on its face."[23]

In essence, this was a case where a science policy process fell victim to extreme politicization and manipulation. The ACC likely went after Dr. Rice with so much intensity not because the organization objected to her comments on the IRIS panel, which numerous observers agreed were in line with other reviewers' comments, but as part of the trade association's broad campaign to discredit the EPA's entire assessment of DecaBDE. Facing recent DecaBDE restrictions in Europe, challenges in several states, and mounting scientific evidence of toxicity and exposure concerns, the industry's defense of this profitable workhorse chemical was slipping away. Based on suspected neurotoxicity effects, the EPA's reference dose was three times lower than previous systematic evaluations of the chemical, sufficiently low that additional regulatory action might be expected.[24] By critiquing Dr. Rice, the ACC may have hoped to discredit not only her reputation as a scientist but all regulatory activities targeting DecaBDE. This in fact was Dr. Rice's interpretation of the events: in a media interview, she speculated that the ACC's motivation was "to say, she's biased, she has a conflict, she's discredited. These other states shouldn't pay attention to what Maine has done."[25] Critiquing Dr. Rice's work on the EPA review panel became "a good opportunity" to fight the EPA's and the states' broader actions to restrict DecaBDE.

In short, Dr. Rice was a recognized expert and respected scientist, but she became caught in the middle of a science policy controversy over chemical safety. In this chapter, I show that whenever science is used for policy at an agency like the EPA, it becomes in some ways a social act of contestation and negotiation. This is especially the case in the numerous EPA activities requiring stakeholder participation, including the type of external peer review panel on which Dr. Rice was asked to serve as well as voluntary programs that evaluate safer alternatives to chemicals already in commerce.

These assessments are presented as scientific documents, but the data included, the evaluations made, and the conclusions presented reflect stakeholder input, both informal and codified evaluator schemas, and an often deliberate balancing of interests and perspectives. Further complicating the supposed neutrality of government assessments is the revolving door that moves industry representatives and EPA officials back and forth between the two institutions. Thus, the decisions presented as being purely or primarily scientific in nature are in fact heavily influenced by social, political, and economic considerations.

ASSESSING SCIENCE FOR POLICY'S SAKE

Interpreting science for chemicals assessment at the EPA is a social act of negotiation, visible in several ways: the stakeholder participation invited or even required by some assessments; the multiple iterations of data development and analysis that support most chemical evaluations; and the publication of regulations and assessments in draft format, inviting widespread public comment, before the documents are finalized.[26] In this chapter, I focus on two programs in which the EPA engages directly with stakeholders to produce scientific assessments: evaluations of newly developed chemicals conducted by the New Chemicals Program, and alternatives assessments conducted by the Design for the Environment (DfE) program. Both are part of the EPA's Office of Pollution Prevention and Toxics (OPPT), headquartered in grand offices on Constitution Avenue in Washington, D.C., across the street from the Smithsonian Museum of American History.[27]

Examining these two assessment processes brings to life the concepts developed in the previous two chapters: conceptual risk formulas and strategic science translation (SST). When stakeholders help to define the scope of an alternatives assessment, their preferred risk formula guides the emphasis they place on toxicity data, exposure assessment, and weight of evidence evaluation. And when they submit formal and informal comments on draft documents, they engage in SST to argue that a report should be revised in certain ways, supporting certain conclusions or policy actions.

Power comes into play from the start because it delineates who partici- pates and who is excluded from the process of evaluating and developing a scientific assessment. Scientific authority is privileged to such an extent that even stakeholders who do not typically occupy scientific roles are compelled to frame their arguments in scientific terms. Additionally, not all voices are equally loud in the process. Profitable corporations, intent on defending their products from proposed restrictions or negative as- sessments, may dedicate significant resources and staff time to these EPA processes or may commission additional sets of comments or reanalysis from hired consultants. Environmental and public health activists, on the other hand, are often overextended and underpaid, with little time to devote to complicated regulatory reviews as their attention typically is divided among a dozen or more projects, with little or no ability to hire outside help.

The two programs I describe differ in significant ways: DfE is volun- tary, not regulatory, while the New Chemicals Program is a core part of how the EPA regulates chemicals through the Toxic Substances Control Act (TSCA). But both serve the agency's core mission of protecting human health and the environment, and both involve interacting with stakeholders including industry actors, environmental and public health activists, other government agencies, and scientists.

Toxic Substances Control Act

The TSCA inventory of industrial chemicals currently includes around 84,000 chemicals, though not all of these are actively produced or used today.[28] Today, TSCA (pronounced *tosca*) is the only major federal environmental law that has not been significantly updated since its enactment in 1976. As this book went to press, bills to dramatically reform TSCA were advancing through both chambers of Congress. Observers in industry, government, and the advocacy world were optimistic that passage of a major piece of TSCA legislation was possible, even if many held strong reservations about some of the proposed reforms.

TSCA provides the EPA with some authority to regulate chemicals in order to protect people and the environment from "unreasonable risk of injury," though this emphasis on unreasonable risk has been interpreted as a nearly impossibly high level of proof before actions can be taken.[29] When TSCA was enacted, the EPA's Administrator Russell Train called it "a major step toward an increasingly effective preventive approach" to global environmental problems. He said that TSCA would "balance risks against benefits as well as costs against benefits" and "draw upon as much outside expertise and advice as we can."[30] Despite these optimistic predictions about TSCA's potential, today the statute is widely seen as insufficiently protective of public health and the environment.[31] Certainly the strengths and weaknesses of TSCA are nuanced, but there are several widely agreed upon shortcomings of TSCA.[32]

First, the EPA has limited authority to impose regulations on the roughly 62,000 chemicals that were "grandfathered in" when TSCA was enacted, called "existing chemicals" in EPA parlance. As one EPA scientist explained to me during an informal meeting, "TSCA is ineffective against old, nasty shit." Caught in a "data Catch-22," the agency cannot require further testing on a chemical or enact restrictions without data demonstrating that the chemical presents an unreasonable risk. As a result, it has enacted only 211 Test Rules to require the development of new data on existing chemicals, far short of the thousands of chemicals of potential concern for toxicity and exposure.[33] The EPA also must justify all risk-based decisions as the "least burdensome" using cost–benefit analysis, a high bar when dollars of economic benefit are easily quantified while environmental and health harms are difficult to accurately estimate. Because of these limitations, the EPA has succeeded in placing restrictions on only five chemicals in commerce before 1976.[34] Since its proposed ban on asbestos was struck down in 1991, the EPA has pursued no other restrictions through Section 6, the portion of TSCA that allows for bans of existing chemicals.[35]

TSCA is generally seen as being more effective with the regulation of new chemicals, and has processed over 36,000 new chemical applications through the New Chemicals Program since 1979.[36] As part of the review process, the EPA can require the development of additional data or impose restrictions on new chemicals through "Consent Orders" with a much lower required burden of proof, but this happens relatively rarely. Out of those 36,000 applications, roughly 4 percent have been subject to a Consent Order.[37] Manufacturers are not required to submit any toxicity or exposure data when requesting approval for new chemicals. According to EPA statistics, two-thirds of new chemicals are submitted with no measured data whatsoever.[38] Only 15 percent are submitted with any data on health or toxicity, 10 percent with information on environmental fate or exposures, and only 4 percent with "repeated dose" studies, which track toxicity over a twenty-eight- to ninety-day period.[39] Multigenerational, neurodevelopmental, or cancer studies in animals are almost nonexistent, despite the fact that these health outcomes are of significant concern for human health.[40]

Third, many chemicals are exempt from full reporting. For newly developed chemicals planned for production below 10,000 kilograms per year, companies can request a "low volume exemption" with a significantly abbreviated review process. Additionally, nearly 29,000 chemicals currently listed on the TSCA inventory are exempt from regular reporting requirements because of their chemical structure or other factors. Over 1,000 chemicals have qualified for "polymer exemptions" since

1995.[41] These chemicals are exempt from full review because of the assumption that a chemical with a large enough molecular weight is unable to be absorbed by living organisms and therefore has insubstantial risk from exposure.

A final significant concern with TSCA is the amount of data that manufacturers can claim as confidential business information (CBI). Manufacturers are understandably protective of the structure of a new chemical or the details of their manufacturing processes, but CBI claims regularly extend to how the chemical will be used, its chemical identity, and even the volume produced. Companies also routinely claim health and safety data as CBI although this is not allowed under TSCA.[42] There are no penalties for falsely claiming CBI status, so there is little disincentive against claiming CBI as a default in all submissions of data to the EPA.[43] Over 16,000 chemicals on the TSCA inventory do not even include a publicly available name.[44] These chemicals are identified only with a generic description like "polyurethane" or "metal salt," insufficient information to allow interested parties outside the EPA to come to any conclusions about risk.

The EPA is exploring ways to reduce the number of CBI claims and has released some formerly confidential information, but in interviews EPA scientists expressed frustrations with the amount of CBI information involved in chemical evaluations.[45] They also identified common-sense mechanisms that could be easily implemented to decrease the amount of CBI data, such as requiring extra fees for each piece of CBI in an application or requiring periodic renewal of CBI claims.

The EPA recognizes the significant shortcomings of TSCA, and in 2009 EPA Administrator Lisa Jackson published "Essential Principles for Reform of Chemicals Management Legislation" to improve chemical evaluation, improve protection of sensitive populations, prioritize chemicals of concern, and enhance green chemistry solutions.[46] The ACC's "10 Principles for Modernizing TSCA" makes similar points about chemical prioritization and assessment, but it is more focused on risk-based evaluation, streamlined EPA processes, and expected-use scenarios.[47] Despite widespread and growing support for updating TSCA, reform efforts at the federal level languished for years due to lack of political will and intense lobbying, and recent efforts to pass competing pieces of legislation have been alternately praised as going too far and critiqued for ignoring needed reforms by different stakeholders.

Working within this context of political paralysis and great environmental need, the EPA has pursued various avenues to motivate, prod, or require action by industry, including a proposed but never implemented "chemicals of concern list," a handful of chemical Action Plans, "low-hanging fruit" risk-management targets such as stopping the use of lead in tire weights, and an increasingly public presence in support of chemicals reform.[48] Additionally, the EPA has pursued voluntary efforts such as the now-inactive Voluntary Children's Chemical Evaluation Program (VCCEP), which gave industry significant control over the timeline and composition of needed testing, and the DfE alternatives assessment program.

Evaluating New Chemicals: The EPA's New Chemicals Program

The EPA evaluates new chemicals through Section 5 of TSCA, which requires manufacturers to formally submit "new" chemicals (chemicals that have not previously been produced or registered with TSCA) at least ninety days before production is scheduled to begin. The EPA uses a complex, multistage process for reviewing new chemical submissions, also called Pre-Manufacture Notices (PMN) chemicals. This process aligns with the exposure-proxy risk formula I described in chapter 3: it identifies possible exposure and toxicity concerns, evaluates persistence and bioaccumulation potential, and synthesizes this information to determine whether the chemical represents an "unreasonable risk." I conducted interviews with EPA scientists all along this review process, as outlined in Figure 5.1, which was shared with me by an individual with the New Chemicals Program.

When a chemical is submitted to the EPA, it first goes through an administrative check to make sure that the application is complete. It then moves to the Industrial Chemistry branch, where chemists and EPA contractors develop a profile of the compound's physical and chemical properties and identify appropriate "analogs," existing compounds with similar chemical structures that will inform the analysis.[49] For example, a

Day 1	Day 8–12	Day 9–20	Day 21–23	Day 24–78	Day 79–82	End of review
New Chemicals Chemicals notification receipt	Chemical review meeting (CRSS)	Hazard/fate (SAT) meeting and exposure reviews	**Initial risk management decision meeting (FOCUS)** 80% of PMNs are dropped from further review based on no unreasonable risk determination	20% of PMNs go through further review and regulation	**Final decision meeting** 8% of PMNs are dropped with no regulatory action	12% of PMNs require regulatory action: 5(f) actions 5(e) consent order 5(e) SNUR Non–5(e) SNUR Withdrawn in the face of regulation

FIGURE 5.1. New Chemicals Review Process for Pre-Manufacture Notices

Source: Personal communication, EPA New Chemicals Program representative

researcher told me that a new brominated flame retardant chemical might be assigned a PBDE analog that would then suggest "PBDE-like concerns." The EPA uses over fifty chemical categories that allow for comparison using chemical structure, and can use multiple analogs to "bracket" a wider range of properties a chemical is predicted to have or multiple analogs for different end points to provide the most conservative estimate.[50] Approximately 5 percent of PMN chemicals do not have a clear analog, and these automatically go through a more detailed review.[51] After developing a chemical profile, the chemists meet at the Chemical Review and Search Strategy meeting to decide on one of three options: they can "drop" the chemical if it is judged as exempt from TSCA reporting (e.g., if it qualifies for the polymer exemption); they can request more data from the manufacturer because the application lacks all the information needed or because their professional judgment does not align with the submitted data; or they can send it on to the next step, an analysis by the Structure Activity Team.

In the Structure Activity Team meeting, toxicologists, exposure scientists, chemists, and environmental fate experts assess the health and ecological toxicity by reviewing data on the chemical analogs identified by the chemistry team (or, rarely, actual data provided with the PMN). Scientists at this stage will identify toxicity concerns and determine the exposure analyses needs. They then pass this information to the Chemical Engineering branch, which assesses occupational exposures, engineering controls, and environmental releases, and to the Exposure Assessment branch, which assesses environmental fate, consumer and general population exposures, and environmental exposures. All these pieces of data—toxicity concerns, environmental exposure levels, and estimates of exposure for occupational groups, consumers, and the general population—are synthesized at a culminating Focus Meeting, when scientists, risk assessors, and program managers compare the predicted exposure levels to predicted toxicity levels of concern. If exposure is predicted to exceed these toxicity levels, there is a finding of potential risk.

Approximately 80 percent of PMNs are dropped from further review at this stage because there is a determination of "no unreasonable risk" based on an assessment of toxicity, exposure levels, and exposure potential.[52] Some of these dropped chemicals will also include a "concern letter" in which the EPA informs the submitter about a potential hazard or risk: for example, the agency can encourage but not require the company to amend required safety sheets to include greater personal protective equipment requirements. The remaining 20 percent of chemicals are sent for a Standard Review, a more detailed investigation that often leads to some regulatory requirement. At this point, the ninety-day timeline is typically extended to allow for a more detailed investigation or the development or communication of more data.

Several options exist for chemicals that go through Standard Review. The EPA can decide to ban a chemical pending upfront testing. The agency can implement a Consent Order requiring more testing in coordination with production volumes, greater exposure, or release restrictions, or higher levels of transparency and communication. The agency can also implement a Significant New Use Rule (SNUR), which states that the use proposed in the submission is allowed but any other uses must be reviewed by the EPA. At this point, 12 percent of all PMNs (or roughly three-fifths of the chemicals that underwent Standard Review) will face some sort of regulatory action: for 7 percent of the chemicals, the manufacturers fulfill the regulatory requirements; for 5 percent of the chemicals, the manufacturers withdraw them from the review process, so the chemicals cannot be used in commerce.[53]

Throughout this process, the EPA and the chemical manufacturer often communicate significantly. A chemical engineer told me that his group frequently goes back and forth with the manufacturers (or with the managers coordinating the chemical submission), who supply additional data and agree to exposure controls in order to avoid additional regulatory activity. Though the company has little power to negotiate the EPA's actual review of their PMN chemical, they have significant control over the information they release to the agency and at what point in the review process, because they are only required to share data produced during the chemical development process if it documents "substantial" risk.

This review process is designed to be protective of public health and the environment, and EPA scientists frequently told me that the New Chemicals Program does a good job of protecting the public. As one EPA manager told me confidently, "in all cases, we think we've controlled the risks that might be unreasonable." EPA representatives told me that the new chemicals being submitted today are generally "improvements" over existing products, meaning their chemical structure suggests a lower risk concern. This aligns with the chemical industry claims that while historically toxicity concerns were not a focus of new product development, today companies place much greater emphasis on evaluating the risks of chemicals before commercializing them.

Additionally, EPA scientists gave many concrete examples of how they restrict chemicals, impose exposure controls, or essentially force companies to provide data during reviews of new submissions, all without taking formal regulatory action. Because the EPA relies on expert judgment in their assessments, especially in the absence of data, scientists have some flexibility to disagree with industry statements and identify potential concerns. For example, a chemist described how they might require companies to resubmit a low-volume exemption (an "L case") as a regular

PMN (a "P case") if they have concerns about a chemical's use and exposure patterns: "With the L case, we're more likely to deny that, because it has to meet certain conditions . . . that it's low volume, lower exposure. So if we disagree, we just say, 'no, you have to submit a P case.' And then they have to do that." Most of this activity is invisible to those outside the EPA, because chemicals are submitted with so many CBI claims that data requirements, exposure controls, or challenges to the submission are not publicly shared.

However, these efforts are hampered by data gaps, time and resource limits, and the quality of models and predictions based on prior cases. Uncertainty is a constant condition when it comes to assessing the risks of new, understudied chemicals, and determinations about the safety of new chemicals require significant interpretations beyond the available data. One chemist told me that decisions about things such as which analog to use were "not always black and white. I think sometimes it's a matter of discussion." In the face of uncertainty, EPA scientists told me they regularly make "conservative" (protective) assessments, following features of the exposure-proxy risk formula detailed in chapter 3. These conservative estimates—such as "worst-case scenarios" and 95th percentile values—also act as negotiating tools to incentivize manufacturers to develop and share more data on their products.

EPA representatives supported these practices as allowing the agency to be protective of public health in the absence of data. An EPA contractor praised the use of conservative assessments but qualified the practice: "Now, that's not actually part of the scientific process, mind you. But that is also another part of what I think the assessments try to accomplish." However, this reliance on conservative assumptions is connected to two significant weaknesses in the PMN review process.

First, these conservative assumptions are needed because the vast majority of chemicals are submitted to the New Chemicals Program with no hard data. As I have noted, only one-third of chemicals are submitted with any data whatsoever, and sometimes the submitted data are modeled, not experimental. TSCA does not require PMN submissions to include any toxicity or exposure data, though any data that show a "substantial risk" are supposed to be shared with EPA under TSCA Section 8(e). Some worry that this disincentivizes toxicity or exposure research by manufacturers: if no data are required and any data provided could potentially derail the chemical's review, companies may choose to develop little or no data unless and until data are requested by the EPA. Numerous EPA representatives told me that companies sometimes do not provide all existing data with their PMN application. A program manager said that companies often do not provide all the data they have developed with their initial application, but after the EPA evaluates the chemical, "we tell them what the concerns are, and then if they have specific information, like, oh, all of a sudden they found data." This shows that the data included in the PMN review can involve significant exchanges between the EPA and companies.

In other cases, companies may choose to develop significant amounts of data, especially for high-profile classes of chemicals such as flame retardants, to protect their products from surprises and critiques down the road. FlameCorp representatives told me the company proactively conducts research because providing the data up front preempts conservative risk assessments. This description of industry's process was confirmed by an EPA scientist who has worked in the PMN program for much of his career. He told me that the EPA actively shares its evaluation models with industry, and companies today often do toxicity testing ahead of time anticipating conservative assessments by the EPA based on those models. This person said that twenty-five years ago the thought of giving industry the models used by the EPA would have been "like letting the fox into the henhouse," but now the idea is that it is better to share the models with companies so that they can anticipate the agency's evaluation and conduct more initial testing.

I attempted to learn whether flame retardants typically are submitted with more data than the average chemical or face greater than average regulatory review during the PMN process, as I had been told by industry representatives. Because of the amount of PMN information claimed as CBI, this is impossible for anyone outside the EPA to determine, and I was given conflicting information in interviews with EPA employees. One senior researcher within the New Chemicals Program told me that "most" flame retardants developed since the 1990s have been approved with Consent Orders or Test Rules, meaning that at least some data were required before the chemical was approved for unrestricted use. In contrast, another New Chemicals Program manager told me that flame retardants are "often" submitted with no data, though the amounts of data can "vary widely": some had "next to nothing" while others were submitted with "lots" of data. This person told me that when data on human health toxicity are provided, it is usually an *in vitro* test for carcinogenicity or skin sensitization, even though most flame retardants are not a concern for sensitization.

A second and related limitation of the New Chemicals Program is its reliance on models and assumptions. In the absence of measured data, EPA reviewers evaluate chemical hazard and exposure potential with *structure–activity evaluation*, which evaluates the chemical's structure using quantitative computer models or qualitative evaluations by EPA scientists to predict its toxicological and exposure properties. The EPA also relies on over a dozen quantitative computer models that estimate physical-chemical properties, environmental fate, exposure scenarios, and different types of toxicity for chemicals submitted with insufficient data. But these structure–activity models are largely based on confidential data from chemicals that the agency has reviewed in the past, information that is not transparently available to outsiders.

Although the agency uses conservative assumptions in the absence of data, in the end they are just that—assumptions, which are sometimes not accurate or are limited in relevant ways. For example, to predict occupational exposure, the EPA makes assumptions about chemical use and behavior. Some of these assumptions are based on technical factors, such as the assumption that particles are respirable only if they are less than 10 microns in diameter. Other assumptions are more social in nature, such as the assumptions that there are 250 worker days per year (a Monday to Friday work week) or that if a chemical is corrosive the workers will be wearing protective gloves. It is not hard to imagine scenarios where these assumptions would inaccurately estimate exposure.

If a PMN chemical is approved for use as a result of its review, the chemical is added to the TSCA inventory, and the company must submit a notice of "Commencement" within 30 days of starting production. Once the chemical is listed, any future restrictions or data requirements must meet the much higher standard of scientific proof and cost–benefit analysis required in the Existing Chemicals program. But in spite of the protective goals of the New Chemicals Program, several potential problems remain and can impact chemical risk after a new product is approved for use.

First, the potential risk of a chemical is evaluated based on the specific use scenarios described in the PMN application as well as any other uses identified by the scientists reviewing the chemical. But unless the agency places a SNUR on other uses of that chemical, all other uses are considered allowable once it is listed on the inventory. As a chemist told me, "If a company says they're going to do X, Y and Z, the common practice is they probably are. But that substance goes on the chemical inventory, and Company 2 may eventually produce that chemical and not know anything about the toxicological properties of that chemical." These possible future uses might lead to exposure scenarios that are different from those listed in the PMN application, potentially altering the risk calculation in an important way.

A second concern is that testing requirements or limitations are often linked to production volume, based on the assumption that low production volume chemicals generally should be less of an exposure concern, and allowing companies to make enough money from the chemical to afford the requested tests. For example, a company may be required to complete a certain toxicological study once they have produced more than 100,000 kilograms of a chemical per year. If they never reach this volume of production, they never have to do the testing, even if they produce the chemical year after year. But the potential exposures from annual production of 99,000 kilograms of the chemical may be significant.

Another problem is that TSCA only applies to industrial uses of chemicals in the United States. A chemical on the TSCA inventory might also be widely used in a cosmetic and thus be subject to Food and Drug Administration (FDA) regulation, but the EPA's assessment only evaluates the quantity of the chemical used for industrial applications. It ignores the amount used for cosmetic applications, even if the same manufacturing company produces the chemical for both sets of products. Thus TSCA assessments may underestimate potential total exposures.

Finally, TSCA also only applies to chemicals that are produced for domestic use. As an EPA chemist told me, "You can make two million pounds, for example, of nasty [chemical], and dump it into the environment. But as long as you're exporting all of that [chemical], you don't fall under the regime of TSCA [because] you're making it for export only." All these limitations potentially lead to underestimations of exposure, which may translate into an underestimation of the total risk to humans and the environment.

In short, the New Chemicals Program reviews chemicals for their potential risks using a combination of hazard and exposure data and predictions. The program aligns with the exposure-proxy risk formula, relying on conservative assumptions in the (frequent) absence of hard data on a chemical's properties, exposure potential, and toxicity. Now I describe a very different way of making decisions about chemical risk and safety, the DfE alternatives assessment program.

Assessing Alternatives to Existing Chemicals: Design for the Environment

DfE conducts chemical alternatives assessments for chemicals identified by the EPA as a priority for other risk management activities, especially restriction.[54] This comparative chemical hazard assessment methodology allows stakeholders to "place chemicals on a continuum of relative hazard to inform decision making on chemical use."[55] Interest in chemical alternatives assessment has grown substantially on the part of industry, regulatory, and activist stakeholders, who recognize the method's potential to provide useful information to guide substitution away from identified chemicals of concern and avoid regrettable substitution.[56] Alternatives assessments at DfE are voluntary programs used to complement regulatory activities such as bans; participation by companies or stakeholders is voluntary, though once a chemical is included in the TSCA inventory, it can be included in DfE assessment without the support of the manufacturer.

The alternatives assessment process begins by clarifying the scope of the project: what is the chemical that may need to be replaced, and for what uses? Then DfE invites all interested participants to join a Partnership. For example, the DecaBDE assessment had over 170 participants from academia, consulting companies, NGOs, the flame retardant industry, compounders and resin manufacturers, the automobile industry, the aerospace industry, the electronics industry, the shipping pallet industry, the textile industry, recyclers, and state, federal, and international governments. Through in-person meetings and conference calls, the Partnership refines the scope of the project and identifies the functional alternatives to be evaluated. These include "drop-in" substitutes, chemicals that require little to no reformulation of the rest of the product, and chemicals (and sometimes alternative manufacturing techniques) that would require more significant reformulation.

Then DfE conducts a hazard assessment on each chemical included in the project. For example, each of the thirty-four chemicals in the DecaBDE assessment was assessed for fifteen hazard end points: acute mammalian toxicity, carcinogenicity, genotoxicity, reproductive toxicity, developmental toxicity, neurological toxicity, repeated dose toxicity, skin sensitization, respiratory sensitization, eye irritation, dermal irritation, acute aquatic toxicity, chronic aquatic toxicity, environmental persistence, and bioaccumulation.[57] EPA scientists and hired contractors gather the peer-reviewed literature on each chemical, paying special attention to reviews and assessments completed by other government agencies or authoritative bodies, especially prior EPA reviews of the chemicals. In addition to the peer-reviewed literature, hazard evaluations are derived from computer models, structure–activity relationships, and expert judgment. In the final report, these "hazard calls" range from Very Low to Very High when they are based on actual data, or from Low to High when they are based on models and predicted values.

The assessment's organizer then collaborates with other DfE and EPA scientists to write a narrative report describing the scope of the project, the chemicals assessed and their uses, and specifics about their functions. The reports also contain information on exposure and life-cycle considerations, and a concluding chapter summarizing the report and discussing additional considerations for selecting a replacement product, including social, economic, and production factors. Drafts of the reports or selections of the text are shared with stakeholders, who have the opportunity to comment, suggest changes and additions, and, in the case of manufacturers, offer more data on their chemical.

To establish each hazard call, DfE identifies adverse effect levels for all relevant exposure routes, and assigns hazard calls based on predetermined criteria that DfE developed through a stakeholder process in 2011.[58] The program follows an established weight of evidence process in the case of multiple studies with inconsistent results. Because the program takes place within the EPA, DfE has access not only to the peer-reviewed publications but also to all information submitted for the chemical through TSCA or other reporting mechanisms, even if that information is CBI. DfE navigates this by summarizing data when appropriate and omitting CBI information. For example, in the DecaBDE review, four chemicals had "confidential" structures, and many chemicals were evaluated using confidential studies.[59]

Because of the data gaps for most industrial chemicals, DfE assessments, like TSCA assessments generally, rely greatly on modeled or predicted data, structure–activity relationship evaluation, and expert judgment. In the written alternatives assessment reports, hazard calls based on modeled or predicted outcomes are differentiated from calls based on experimental data. As with reviews in the New Chemicals Program, default assumptions and hazard calls based on expert judgment are common. Looking over a chemical's hazard assessment with a DfE scientist, I noticed that there were a few values (e.g., 10^{-10}) that were frequently repeated. When I asked about this, the researcher explained that these were default values assigned based on expert judgment. Additionally, researchers can use conservative benchmarks for chemicals with missing data. For example, several chemicals in the DecaBDE assessment were predicted to have "Moderate" concern for bioaccumulation, and the researcher explained to me that "Moderate" is a conservative estimate because no data existed. The researcher explained that it is hard to make predictions about bioaccumulation because scientists only understand a small number of chemicals (like PCBs) in a small number of species, and therefore scientists assign a conservative estimate based on expert judgment.

A high level of stakeholder involvement is a key feature of DfE's work. In part this is an inevitable consequence of DfE's identity as a voluntary program: the assessments are only useful if decision makers trust the information and use it to make future chemical-selection decisions. Behind-the-scenes communication between the leaders of the alternatives assessments and participants is common and encouraged. A scientist who has directed several DfE alternatives assessments explained how she reached out to stakeholders:

> Every communication is "Please call me or email me if you want to discuss anything" so that they can have the one-on-one with me and get a reaction or understand or share information that's not appropriate to share, or . . . is confusing or difficult to share in the more public area. Or they can call me first, and then they can feel more comfortable in a conference call . . . It's important to make everybody feel that they are important and they have something to offer and they shouldn't hesitate to speak.

This high level of stakeholder engagement requires a significant amount of time. Most reviews in the New Chemicals Program are wrapped up in under ninety days, but DfE assessments can take several years to complete.

DfE representatives have a lot of experience "balancing" stakeholder involvement from different sectors, as an alternatives assessment leader told me: "More often than not, there's a small group who are very vocal, so you may have one or two NGOs who are particularly vocal, you may have several trade associations that are particularly vocal, and then one or two academics. So you might have a core group of, I'd say, ten organizations that really seek to drive their interests." Part of balancing these different perspectives means recognizing the motivations of different stakeholders. The scientist continued:

> We're talking about often businesses who are advocating for what they do. So they may overlook an area that is an issue, because they don't want, if they raised it, they are compromising themselves. They don't mean badly. And the same way an NGO has their position and they want to stick to that. And so they're going to present their stake. So you have to consider why this stakeholder is saying what they're saying.

The overall process is a balance of trust and triangulation. DfE actively seeks input from stakeholders with identifiable positions and biases, including environmental NGOs that would like the chemicals removed from commerce completely and companies that manufacture the chemicals under review. DfE managers recognize that these stakes impact the type of information shared. As a result, a DfE leader told me that she is "more trusting of technical information, the chemistry and the engineering, because that's difficult to spin."

As with the review of new chemicals by the New Chemicals Program, industry will often provide additional data in response to conservative calls made by the EPA. A DfE scientist explained that after DfE completes its initial evaluation of each chemical, the assessment leader will communicate the findings to members of the assessment Partnership for review: "And then sometimes, there are like, 'oh, we don't agree with this call.' And then they'll supply the data. So it's like a game almost, where, you know, there's no data, and then you go ahead and make these calls, and then they don't like the calls that you're making." Protective assessments provide significant leverage for companies to disclose more data on their chemicals. These early reviews also allow stakeholders to fill data gaps about how chemicals are used, or to correct misinformation in the reports. But, DfE representatives emphasized, they will not change hazard assessments just because a company objects: "If they don't have the data to support it, there's nothing they can do about those estimations."

As I showed in chapter 3, DfE's assessment process clearly reflects the hazard-centric risk formula, with its emphasis on reducing risk by reducing hazard and its use of a complex understanding of toxicity and exposure potential as part of hazard identification. In a 2010 feature publication in *Environmental Science & Technology* by DfE representatives, the authors emphasize the ability of the alternatives assessment to reduce risk and "minimize the potential for unintended consequences."[60] Their model emphasizes a broad and multifaceted assessment of toxicity end points and exposure routes, including persistence and bioaccumulation. Additionally, the authors follow the hazard-centric risk formula when they argue that reducing hazard is an effective way to reduce risk, even as they maintain the multiplicative function of the classic risk formula: "risk (defined as hazard multiplied by exposure) can be reduced through a reduction in chemical hazard."[61]

DfE is not the only entity to conduct chemical alternatives assessments. An industry coalition of trade associations, including the ACC, released a statement of "Principles of Alternatives Assessment" in 2012 that focuses on performance and economic factors in addition to environmental health and safety.[62] The eleven-page document shows what the exposure-centric risk formula could look like in practice. The report specifies a multifaceted description of exposure that includes exposure pathways, use patterns, exposure levels, and exposure potential. Alternatives assessment should be guided by "best practices from the R&D [research and development] process," including comparable or superior cost and performance profiles, consumer acceptance, and protection of confidential business information. When describing the process of assessing human and environmental safety, the industry coalition notes that chemical hazard should be assessed alongside of "product use and exposure" as part of a "comprehensive risk-based safety assessment of alternatives." They specifically reject the hazard-centric risk formula, stating "any comparative assessment methodology that relies solely on hazard can be grossly misleading." They also explicitly reject the either-or risk formula while embracing a multiplicative approach to risk, stating, "The presence of a chemical in biomonitoring studies does not necessarily indicate there is a likelihood of harm." Thus, the industry's favored model of alternatives assessment directly reflects the exposure-centric risk formula, with its focus on multiple elements of exposure and its rejection of decision making based primarily on chemical toxicity.

The chemical alternatives assessments favored by DfE and the ACC differ dramatically because they draw on different risk formulas, which lay the foundation for risk assessment and management activities. The exposure-centric alternatives assessment does not involve assessment of persistence and bioaccumulation as proxies for exposure potential, but would require documentation of exposure pathway, use profiles, and human exposure before exposure is considered significant enough to influence decision making. This product use and exposure information would also be used in a multiplicative formula and could thus potentially conclude that the ongoing use of toxic chemicals was preferable given certain use patterns. All these attributes make it less likely that an exposure-centric alternatives assessment will arrive at a determination of significant hazard.

In contrast, a hazard-centric risk formula emphasizes that reducing hazard is likely to reduce overall risk, and that hazard-based decision making should proceed even without a multifaceted picture of exposure.

This approach has the potential to align with precautionary action, green chemistry, or a preference for false-positive results rather than false-negative results.[63] Although the hazard-centric risk formula maintains the multiplicative risk formula, it downplays the classic assumption—an assumption built into many risk-based regulatory frameworks—that controlling exposure is possible and likely, and instead focuses on hazard reduction. As an EPA representative explained, risk assessment as it is typically practiced assumes that systems will work "perfectly well" and are "perfectly safe," until a lack of safety is documented. Instead, the hazard-centric risk formula assumes that exposure concerns are a strong possibility for widely used chemicals, and it advocates risk reduction through the use of less hazardous chemical substances. The hazard-centric risk formula also facilitates long-term industry decision making by giving companies greater confidence in chemical substitution choices, because no matter what future exposure patterns exist, chemicals with low expected hazard are less likely to face regulatory action in the future.

PROMISE AND PERIL OF NEGOTIATING SCIENCE POLICY

Interpreting science for chemicals assessment at the EPA often involves social acts of negotiation that go beyond supposedly objective applications of scientific data. This can be readily seen in the high levels of stakeholder participation in DfE assessments, and in the ways that the EPA publishes regulations and chemical assessments in draft format and invites widespread public comment. For both programs, initial or draft assessments partially function as a leveraging tool to motivate companies to share or develop additional data on the products being evaluated or engage more actively with EPA assessment processes. A scientist in the New Chemicals Program praised stakeholder involvement at EPA: "The great thing about being here at EPA is we can sit down at the table with anyone and everyone and collect good ideas, regardless of the source."

DfE in particular cultivates productive working relationships with stakeholders, and program representatives argue that this participation increases the breadth of available data and quality of their assessments. Because the group has completed several assessments on flame retardants produced by the same companies, studied by the same scientists, and protested by the same environmental advocacy organizations, DfE scientists and staffers have worked with the same key stakeholders across multiple assessments. As a senior DfE representative told me, "We figured out how to bring all the stakeholders into the conversation, educate all the stakeholders about the process and the challenges, and then display the information in a transparent fashion . . . By bringing everybody together, I think we found a way to educate the full range of stakeholders about each other's issues, and to find . . . a path forward." From this perspective, stakeholder participation increases mutual understanding.

Additionally, broad participation often improves the quality of the information developed. An EPA researcher expressed this sentiment: bringing "a whole lot of people together that have different bits and pieces of information . . . , you might have insights that you'd otherwise not have." I saw numerous instances of this during my work in their office. For example, on a DfE planning call, a chemical manufacturer clarified that a chemical had been listed twice in relevant documents: the chemical had two separate identification numbers because there are two ways to manufacture the substance, but for evaluation these chemicals should be combined. A DfE scientist explained these technical benefits this way: "You can't have a good report or a good product without those points of view, because you may miss an issue, or you may miss a technical piece of information that explains why something is the way it is." This is particularly the case when it comes to identifying functional alternatives for a chemical of concern. Flame retardant chemicals are very application specific, and relatively few drop-in substitutes exist for any given use. Typically only the manufacturers and their supply chain customers have comprehensive knowledge of which chemicals really work for which uses.

However, this type of stakeholder process also has drawbacks, precisely because it involves balancing different interests and biases. DfE works closely with industry representatives in their alternatives assessment process and relies on them to identify possible alternatives, provide data on chemicals, and provide feedback on drafts. Because it is a voluntary program, they need the buy-in of the very companies whose chemicals they are evaluating for risk, as well as the participation of NGOs, academics, consultants, and regulators. As one DfE employee told me, "We fend some off [*making a blocking motion with her left arm*] and try to pull others in [*waving with her right arm*]." People told me repeatedly that DfE aims to be the neutral, in-the-middle voice, and they expect NGOs to provide a counterweight to the industry perspective, especially in their human health focus. Thus, they actively encourage the NGOs to participate and comment, but this is challenging because nonprofit groups typically have fewer resources and direct stakes in the process than industry. As a DfE scientist told me, "We want to have a report or product that people, everybody is generally comfortable with. So we don't want to have something that would be seen as industry-centric or very government-oriented or too NGO-centric. It needs to be balanced by everybody's viewpoint, everybody needs to feel that they were able to be heard."

This crafting of participation in science policy development is not unique to the EPA. Although one model involves balancing participation from industry and NGO actors, others have suggested that arriving at consensus means managing or delimiting participation to some degree, to prevent processes from being overtaken by unruly or uncompromising actors. For example, a state-level regulator described recent efforts to update the state's fish consumption advisories:

> REGULATOR: I actually put together a science advisory committee made of [state residents]. There were people from the university, from Fish and Wildlife Service, from [state health and environment offices], had a couple of MDs on there from different places, and—to try to get this balanced, you know, kind of consensus. We tried to reach consensus, and we actually got it! It was amazing. But, I must admit we left the tails [off], or we never would have reached a consensus, you know what I mean?
>
> AC: What do you mean?
>
> REGULATOR: Like, we didn't have anybody from the seafood industry, and we didn't have [environmental health organization] (*laughing*).

She felt that consensus in a formal regulatory process was only achievable by controlling participation and limiting the involvement of "extreme" groups.

I noted earlier that companies have significantly more resources than most other stakeholders to dedicate to EPA reviews. This means that companies participating in DfE reviews or communicating with scientists reviewing TSCA chemicals frequently submit extensive comments, suggestions, and often additional data. Sometimes responding to industry feedback results in relatively minor edits. For example, the DfE flame retardant assessment for DecaBDE changed the title of a brief section in the report from "Pallets" to "Storage and Distribution Products" at the request of the pallet industry.[64] But in other instances DfE scientists expressed concern about industry collaborations. A scientist told me that a flame retardant company had approached DfE to discuss possible collaborations related to shifting the market away from DecaBDE and toward a more expensive replacement chemical with a preferable hazard profile. I said that it seemed like the company's interest in getting customers to switch to this new chemical aligned with DfE's goals of getting manufacturers to move to significantly safer chemistry (as opposed to the only slightly preferable drop-in replacements for DecaBDE). The DfE scientist agreed, but said that the concern was in being too closely aligned with this company, and that bringing in an NGO who could play a significant role in the initiative would be a good thing.

Other times, however, the relationship appeared to be hostile. Someone had shared with DfE researchers a PowerPoint presentation by an industry participant in an ongoing alternatives assessment that outlined the industry "strategy" to challenge DfE assessments. The slides talked about "battles," "battlefields," and "winning," and concluded by encouraging the industry to spend money to fight DfE "to avoid losing 'the big one' or sustaining 'a thousand cuts.'"[65] As a DfE researcher said to me, stakeholders can be one's best friend on the phone, but "then you realize you're the target of their war."

THE POWER OF THE REVOLVING DOOR

Science is both the foundation of and the explanation for the EPA's risk assessment practices. That is, science is always the explicit justification for the EPA's decisions, even when it is not the only factor motivating decisions and even when the state of the science around the issue at hand is highly contested. This reliance on science begins with establishing which types of science are seen as credible and how scientific evidence is evaluated in decision-making processes. The EPA can only use peer-reviewed findings for its assessments—even data produced within the agency or funded by an EPA grant cannot be used until it has been published in a peer-reviewed document. For example, an EPA researcher in ORD replicated and confirmed the findings of a controversial study on a flame retardant, but because the results were never published they could not be used for EPA assessments.

Furthermore, EPA assessment processes have clearly defined criteria for data identification and assessment. This generally involves a hierarchy of data preference, ranging from well-conducted empirical studies on the chemical itself to models and expert judgment. For example, the DfE alternatives assessment Criteria document states that "DfE will use data in the following order of preference: 1) measured data on the chemical being evaluated, 2) measured data from a suitable analog, and 3) estimated data from appropriate models."[66] Similarly, the New Chemicals Program prioritizes exposure data in the following way: exposure monitoring data for the chemical itself, monitoring data for a chemical analog, established exposure scenarios for an identified industrial process, mathematical models, and professional judgment.[67] Expert judgment is widely used in the absence of data, and experts are identified by scientific credentials and professional experience.

But there are cases where even completed studies provide insufficient data. For example, a hazard call in the DfE DecaBDE assessment was based on predicted values, even though a study existed, because the study did not provide enough data. As an EPA scientist told me, "One study is never enough, never . . . You have to take all the pieces together." In the case of multiple sources of evidence with conflicting conclusions, EPA relies on a weight of evidence approach, though this is a slippery term that is not easy to concretely define. For example, weight of evidence has been defined by industry panels in ways that privilege industry-funded Good Laboratory Practices (GLP) studies over academic research, because the GLP studies often involve a larger number of animals and more standardized measures.[68]

The EPA's explicit reliance on science, however, does not mean that science is the only force at play. As one longtime ORD researcher told me, sometimes "politics can trump science" or science can be dismissed because of "the realities of cost–benefit analysis or financial issues," and sometimes the issues are so political that science will never be the "primary indicator." In short, the EPA always describes their regulations as science-based, but they're not always purely—or even primarily—scientific.

Because stakeholder participation is a frequent, often required, component of EPA assessments, the data included, the evaluations made, and the conclusions presented often reflect political considerations and a careful balancing of stakeholder interests. Further complicating the supposed neutrality of governmental assessments is the revolving door that moves industry representatives and EPA officials between the two institutions. These career pathways were particularly prevalent under President George W. Bush, who recruited from corporate ranks more heavily than did President Bill Clinton.[69] Prior research on the revolving door phenomenon has shown that the strengthened connections between governments and corporate groups enable lobbying and make it more influential over policy.[70]

As an example, consider the career of Dr. George Gray, the EPA's assistant administrator for research and development who oversaw the removal of Dr. Rice from the EPA's review of DecaBDE in 2007. Before coming to the EPA in 2005, he was the executive director of the Harvard Center for Risk Analysis (HCRA), a multidisciplinary research group at Harvard University. The HCRA was founded by Dr. John Graham in 1989, and throughout its existence it has operated with a mix of university, foundation, government, and industry funding. Dr. Graham was particularly interested in cost–benefit analysis and risk/risk trade-offs in risk assessment, which made his work attractive to industry funders. Described by colleagues as a strong fundraiser, he solicited and received funding for the HCRA from agricultural, pharmaceutical, chemical, and transportation industries, among others.[71] Dr. Gray was at the helm of the HCRA from 1999 to 2005. During that time, he received funding from the Society of the Plastics Industry to investigate low-dose effects of bisphenol-A (BPA), the National Food Processors Association to investigate health and fish consumption, and the U.S. Department of Agriculture and the FDA to investigate mad cow disease and risk communication.[72] Today the HCRA receives funding from a mix of grants and private foundations, though many environmental health scientists and advocates continue to view the HCRA as a corporate-driven enterprise hostile to regulation and community concerns.

Dr. Gray left the EPA in 2009, and is now a professor of environmental and occupational health at George Washington University. He remains connected to industry, and often speaks and writes publicly about risk assessment and science policy. In 2012, he topped the program at Croplife America's Science Forum, and he coauthored a critical *Nature* commentary accusing the EPA of slow, inefficient, and biased risk assessment practices.[73] He is also on a steering committee at the International Life Sciences Institute's Health and Environmental Sciences Institute (HESI), a research organization funded by chemical, food, and pharmaceutical companies. At the organization's 2013 annual meeting, he gave a presentation on behalf of HESI's "Frameworks for Alternative Chemical Assessment and Selection of Safer, Sustainable Alternatives" committee, emphasizing the need to move "Beyond Hazard" in alternatives assessment.[74]

Another example of the revolving door between industry and the EPA is even more striking. Dr. Todd Stedeford is a lawyer and toxicologist. After receiving his advanced degrees, he spent three years working as an EPA scientist. He then moved to industry, and from 2008 to 2012 he worked as an in-house counsel and toxicology advisor to the flame retardant manufacturer Albemarle.[75] During that time, he coauthored over forty-five articles, letters, and commentaries in the scientific literature, many of them defending flame retardant chemicals from critiques or proposed regulations. He was a coauthor on the industry-funded study concluding that DecaBDE caused no neurodevelopmental toxicity in rats, wrote numerous letters to the editors critiquing the research methods and analytical strategies of studies finding health effects from exposure to flame retardants, defended tris(1,3-dichloro-2-propyl)-phosphate (TDCPP) as it was being listed by California's Proposition 65 list of chemicals known to cause cancer and reproductive toxicity, and accused other scientists of dangerous and irresponsible practice for not providing their raw data for reanalysis.[76]

In 2012, Dr. Stedeford left Albemarle and joined the EPA as the chief of the Existing Chemicals Branch, the EPA department charged with regulating chemicals already on the TSCA inventory, including many of the flame retardants he defended in his work at Albemarle. According to EPA officials, Stedeford has "recused himself from any direct involvement in matters related to Albemarle and . . . steps have been taken to ensure his staff does not report to him about any work regarding flame retardants."[77] Since 2012, the EPA has announced ongoing review of several clusters of flame retardants, and several formal risk assessments are ongoing, but as this book went to press, no further regulatory actions have been announced.[78] The ACC spoke favorably on his move from Albemarle to the EPA, praising Stedeford as an example of how "qualified individuals with private sector experience . . . have real-world understanding of how policy and regulatory issues can affect American businesses." But the appointment raised critical eyebrows in the environmental science and advocacy world. According to Julie Herbstman, an epidemiologist who studies the links between PBDE exposure and children's development, "It's hard to imagine going from one job where you are a hired gun to another where you are supposed to be protecting the public."[79]

PRODUCT DEFENSE AND MERCENARY SCIENCE

As I showed in chapter 4, stakeholders will use SST to debate scientific proof and certainty, discuss or invent data gaps, or criticize scientists rather than talking directly advocating for policy changes. The flame retardant industry often engages in inaccurate SST in its product defense work. As epidemiologist and Assistant Secretary of Labor for occupational safety and health David Michaels writes, "Industry has learned that debating the *science* is much easier and more effective than debating the *policy*."[80]

Michaels writes about the "mercenary science" conducted and disseminated by product-defense consultant groups, companies that "profit by helping corporations minimize public health and environmental protection and fight claims of injury and illness."[81] This type of activity is prevalent in the flame retardant world, and flame retardant companies regularly hire consultants to develop original research or review existing studies. For example, Gradient is a consulting firm specializing in environmental sciences and risk assessment. Their online self-description hints at their proindustry reputation in a subtle way: the text below a beautiful color photo of a river weaving between mountains reads, "Since 1985, Gradient has employed sound science to assist national and global clients with resolving their complex environmental and human health challenges."[82] This description appears innocuous at first glance, but "sound science" refers to science that justifies a preferred position. This catchphrase was created as a public relations strategy in the 1980s by the tobacco industry to perpetuate doubt about the dangers of cigarettes: big tobacco claimed there was no sound science that cigarettes caused cancer.[83] Today, sound science remains a code phrase for science that generally supports industry positions. Sound science is frequently contrasted with "junk science," a critique typically invoked by corporate spokespeople against science that challenges their business interests.[84]

Gradient's industry ties are less subtle when it comes to articles written by staff scientists because contemporary publication standards typically require that authors disclose conflicts of interest, including funding sources. As an example, Dr. Julie E. Goodman is a principal toxicologist with Gradient whose work on flame retardants has been funded by Albemarle. In one toxicology publication, she reviewed the studies used by EPA scientists to develop a reference dose for DecaBDE, and concluded that four studies that found developmental neurobehavioral effects after in utero exposure to the flame retardant did not support the development of the reference dose.[85] These types of articles can then support industry arguments directed at the scientific community. Dr. Goodman's article, for example, was cited in the industry-funded Biesemeier toxicity study of DecaBDE as evidence that previous developmental neurotoxicity studies were inconsistent and of questionable quality.[86]

Articles of this type have benefits that surpass their initial publication because once a paper appears in the peer-reviewed literature, it can be used in other types of product-defense work, can be referred to in regulatory activities, and can be cited and disseminated by industry stakeholders. With the recent DfE review of DecaBDE, the EPA received 499 pages of public comments (and additional unreleased comments containing CBI information) from fifteen entities representing chemical, automotive, aerospace, electronics, and shipping pallet industries, and in the end released fifty-seven pages of responses to these comments.[87] Albemarle hired three outside consultant groups to write detailed scientific rebuttals and critiques of the EPA's assessment of chemicals that Albemarle manufactured. Albemarle may have hired these consultants because they lacked sufficient in-house staffing or expertise to develop such lengthy and detailed reviews on a short timeline, they may have funded additional analyses to increase the quantity of comments supporting their position, or they may have chosen to outsource these scientific assessments for the appearance of greater impartiality.

Some of the comments were of strong merit or presented new information, improving the final project. For example, companies shared several additional studies on several flame retardants, resulting in some hazard calls staying the same but others being changed to incorporate additional studies or to reflect measured values instead of expert judgments.[88] Other times, however, the critiques involved extreme acts of interpretive or inaccurate SST. Some comments misstated the hazard calls given to certain chemicals, asked the DfE to move from hazard assessment to risk assessment, and ignored the role of expert judgment and EPA review in the absence of data. Additionally, overarching arguments were reinforced through their very repetition across the scores of pages of comments. The Albemarle-funded comments cited Dr. Goodman's study nine separate times.[89] These comments on draft assessments and reports require that EPA scientists spend a significant amount of time and resources to respond to each point.

ECONOMICS, POLITICS, AND SCIENCE

Science policy activity at the EPA is not only influenced by economically motivated industry arguments but by direct economic and political considerations. For example, many risk assessment activities at EPA must include an assessment of potential economic impacts. This emphasis on cost–benefit analysis is simultaneously evidence that economic or political considerations matter greatly and that science is used to justify EPA decisions, because cost–benefit analysis is always framed as a scientific process even though it is far from neutral.[90]

The Economics of Science

Research in all institutions is heavily influenced by economic considerations. Academic researchers make choices about research projects, methods, and equipment procurement based in part on the availability of different funding streams. Federal funding is often tied to political pet projects or national security goals.[91] A shift in recent years in the availability of federal funding has significantly contributed to the formation of research centers and institutes at universities around the country dedicated to the exploration of interdisciplinary research topics instead of individual academics investigating disparate research questions.[92] And academic scientists are increasingly subject to industry and corporate funding sources and institutional logics.[93]

The source of funding for a given project is also closely tied to the publication of findings. Work on the "funding effect" has demonstrated that industry-sponsored research publications support the safety and/or effectiveness of the product being evaluated, while academic- and government-sponsored articles are much more likely to call safety and effectiveness into question.[94] The funding that supports a research project is generally seen as a potential conflict of interest, broadly defined as "potential sources of bias, including affiliations, funding sources, and financial or management relationships."[95] Industry funding and control over the dissemination of research findings are clear examples of a conflict of interest. Yet in interviews several industry representatives expanded the boundaries of financial conflicts of interest from owning stock in or receiving a paycheck from a company, to academic researchers who are dependent on grant funding for their work. One industry representative argued that so much attention was devoted to flame retardants not because they were risky chemicals but because they were "a hot area to get grants." This person maintained that anyone whose salary was reliant on grant funding—so-called soft money positions in academia—was necessarily biased: "You can't tell me that somebody's not biased if they stand to lose their job if they don't publish, if they don't get grants. That's a strong incentive to find or interpret data in a manner that's going to continue your funding." However, grants are funded based on research questions and programs, not on a search for specific positive or negative findings.

Industry representatives critiqued other components of scientific publishing as well. One industry representative asserted that the identification and pursuit of additional research questions—a central tenet of scientific practice—is in fact a ploy to get more grant money. Another critiqued the concept of peer review: "You've got a group of people all peer reviewing each other's stuff, right? . . . This is not objective. The peer review process in academia has become non-objective." An industry toxicologist asserted that academics should be required to make their raw data available: "You can reanalyze it and do whatever you want . . . Academics, they'll publish a summary of the data in a journal. That's it. And if you try to get the data . . . you're a bad guy for even asking about it." A high level of data transparency has become the standard for some journals. The influential journal *Toxicology*, for example, requires researchers "to share the original data and materials if so requested."[96] However, the chemical industry rarely makes any of their own raw data available, as evidenced by the fact that a tremendous volume of data submitted to the EPA is CBI and the frequency with which these submissions include only data summaries instead of the actual raw data.

As would be expected, academic scientists maintain that the peer review process is appropriately rigorous. While I was observing an environmental chemist in his laboratory, I asked for his thoughts on the peer review process and bias in academia. He responded that academics are competitive and critical of each other, and even friends and close colleagues will "tear apart" papers during the peer review process. In a small subfield, everyone is after the same grant money and working on similar research questions. This aligns with economic historian and public intellectual Karl Polanyi's description of the scientific community as a self-policing community.[97]

Economic requirements are also influential for chemical companies developing new products. Indeed, when FlameCorp representatives described their product development process, they told me that market forces are just as significant a hurdle in product development as performance and potential toxicity. When a new chemical substance is initially synthesized in their R&D laboratory, the researchers' first test is to see if it does actually slow combustion. The next step at FlameCorp is a cost-efficiency analysis of the chemical components, manufacturing process, and potential to "scale up" from laboratory to commercial production, because new products cannot be significantly more expensive to produce than current-use compounds. This seriously limits the scope of chemistries that can be developed and commercialized. I was also told that with few exceptions, the company will only develop products that can fit within existing plant operations and machinery because it would be prohibitively expensive to build new plants or to completely retool existing plants for new chemicals.

Market pulls also influence the commercialization of new products. As one FlameCorp representative told me, product development of flame retardants "is really market-driven. It's a market-pulled process." Because of the time and expense involved in developing new chemicals, new products will almost inevitably be more expensive than existing chemicals, because a chemical that has been on the market for thirty years has long ago recouped its initial investment costs. In contrast, a new chemical may have a fresh $5 million price tag that needs to be paid off within a few years. But regulations and supply industry purchasing policies can create space for newly developed, more expensive products.

A new product development manager at FlameCorp described how the company was able to introduce a new nonhalogenated flame retardant in response to current market demands, even though it cost much more than the commodity chemical it was replacing:

> Right now, a company like IKEA has made a stance that they don't want the old product in any of their products, any of their foams or furniture or anything like that. That has allowed us to introduce this new molecule and new product. Without them saying that they're banning the old one and they want something new, that never would have happened. It would just sit on our shelf, because the IKEAs and the Walmarts and the people making the products don't want to pay more unless there's someone who is demanding it. And we can go out there and advertise our new replacements all we want. If someone's not willing to buy it, it's not going to go anywhere.

This represents the overlap of economic and risk concerns. To develop a new flame retardant chemical and bring it to market typically takes at least five years and several million dollars, according to FlameCorp representatives. Because of this high level of initial investment, the company does not want to invest a lot of money in a chemical that will be found to have significant hazard or risk problems later on. In the words of a toxicologist who has worked in both government and industry, companies "don't want any bad news, so they usually would much rather do the work and understand the molecule . . . Bad press destroys your product line and your company brand name."

Similarly, the industry is attentive to potential liability issues. One former industry scientist argued that the industry would never move forward with known toxic chemicals: "There's no incentive out there for an industry to cover something up, because the penalties for doing so are so extreme." Such penalties are rare, but they can be severe.[98] According to industry representatives, chemical companies reject most new molecules they develop early on in the R&D process, many for performance reasons but some because of concerns about toxicity or exposure potential. Even early in product development, however, the company's position that higher toxicity chemicals may be developed if they are not expected to have exposure concerns is rooted in the exposure-centric risk formula.

The Politics of Science

In addition to the economic considerations that impact the development of new products, the production of new scientific data, and the dissemination and uses of scientific evidence, political considerations regularly influence science and science policy at the EPA. Sometimes this involves direct political oversight of science policy activities. For example, most EPA regulations must pass political and economic muster with the Office of Information and Regulatory Affairs (OIRA), part of the Office of Management and Budget within the executive branch. Proposals can die a "silent death" through OIRA review, as appears to have happened with a proposed Chemicals of Concern List that was withdrawn after being stuck in OIRA review for over two years.[99] OIRA has been called the "stumbling block when it comes to transparency" around many aspects of environmental regulation.[100] Rena Steinzor, an environmental law expert, has described OIRA as "a politicized place where rules go to die," and elsewhere has written that the office has shown a "relentless drive to quell controversial rules going back forty years."[101]

Although political influence on scientific decision making at the EPA is rarely that blatant, policy needs or political concerns frequently play an identifiable role in the conduct of scientific research and the application of science to decisions. As an EPA leader explained, "Sometimes you have to put the cart before the horse because of political or scientific reasons, but most of the time you like the horse to go before the cart." That is, EPA representatives prefer to have science—"the horse"—lead political decisions—"the cart"—but they reluctantly recognize this will not always be the case.

Policy requirements also impact the type of research conducted in the EPA's ORD. This office contributes to the agency's priorities, filling data gaps and trying to reduce scientific uncertainty. As one researcher explained, "My job is to help the Agency solve problems, and that's mostly because they have a lack of information." Many ORD researchers described their work as "very applied." In addition to driving the development of certain types of data, policy needs also motivate what analyses are done and what articles are written for publication. One researcher in ORD described a multisite, multiyear project he had directed, and said, "It's a huge database, and we published a lot on it, but probably there's a lot of more stuff that we could publish on. But so we chose what to analyze, how to analyze it . . . [based on] what we thought would be the best way to . . . be used in policy." Thus, data analysis choices were made to maximize policy utility.

The FOIA process provides another example of visible overlaps between science and politics. Industries regularly request documents on government-funded research through FOIA.[102] In the late 1990s, the tobacco industry drove the passage of two data quality acts that allow interested parties to use FOIA to request most types of federally funded research.[103] This strategy was confirmed by industry respondents, who said they submitted data requests via FOIA when they did not agree with the analysis conducted by the original researchers. A regulatory toxicologist at the EPA told me that her work had been requested via FOIA two dozen times in her career. In one instance, her research on flame retardants had been requested via FOIA by one of the flame retardant companies: "After we did our studies, after publishing our studies, [the company] FOIAed all of my data. Oh, yeah. It's a standard thing. They go around FOIAing as many people as they possibly can. And of course, because we're the government, we have to turn it all over." But, this researcher continued, she had expected this: "I kind of knew it was going to be coming, so it wasn't a big deal. And actually I've never seen anything after we sent them that, so they probably went through everything with a fine-toothed comb, like they do, and couldn't find anything to write a letter to the editor about. So, it just kind of died." Her work had been requested via FOIA so many times in her career that it no longer bothered her.

Another regulatory toxicologist said that her work was often targeted by FOIA requests at strategic times. Because research findings must be peer reviewed before they can be used for regulatory activities and EPA reviews of chemicals, the industry producing the chemicals she studied used FOIA to demand her data just as the relevant regulatory deadlines approached so that she would not have enough time to publish her findings. She told me that in one case she had spent eight months responding to industry FOIA requests when she should have been publishing her findings.

The industry-sponsored letters to the editor in journals, which I discussed in chapter 4 as an example of inaccurate SST, are another way that the industry pushes back against government or academic science. These letters to the editor critique research design and analytical methods, misrepresent findings, and, in the words of a public health researcher, poke "mini-holes" in the article's argument. These letters also frequently assert that future research on a given end point is not needed once a single GLP study has found no associations. For example, a 2010 study by two Albemarle employees and two laboratory employees tested the prenatal developmental toxicity of decabromodiphenyl ethane (DBDPEthane) in rats and rabbits. The researchers administered different doses of the flame retardant to adult female animals during pregnancy. They then tested for acute toxicity and reproductive failure in the adults and gross developmental deformities in the pups, and found no evidence of toxicity or teratogenicity.[104] The last paragraph concluded, "Therefore, *no further experimentation is warranted* to test the reproductive/developmental toxicity of DBDPEthane."[105] Similarly, an industry letter to the editor stated that because an earlier industry study had found negative results, a subsequent academic study that found positive results amounted to "unnecessary duplication and use of laboratory animals."[106]

It is unquestionably common practice for scientific articles to conclude with a call for more research. This fact was frequently discussed in interviews, as some respondents bemoaned scientists' constant desire for more research and others reiterated the need for replication of preliminary findings. In contrast, industry-funded studies sometimes make the argument that after a single Guideline-compliant study, no further studies on the topic should be conducted. Industry-sponsored publications also tend to contain fewer details, are less likely to be replicated by other researchers, and are less frequently duplicated, often because of the ambiguity in their methods.[107]

As I pointed out in chapter 4, SST is frequently unbalanced. I found over thirty instances of critical letters to the editor from the chemical industry about academic studies; I found only one instance of a critical letter to the editor from an academic team critiquing the findings of an industry study.[108] The greatest barriers to academics or independent researchers writing this type of critique may be time, resources, and disciplinary reward structures, but in other ways the imbalance is even more substantial. It is impossible to use FOIA to demand access to raw data from industry-funded studies because only government-funded research is subject to the data-disclosure rules.

NEGOTIATING SCIENCE FOR POLICY

The inevitable uncertainty of science means that research findings can be strategically used by different stakeholders for competing policy claims. All types of science-based decisions, whether science policy and chemical assessments at the EPA or new chemical products by chemical manufacturers, involve social acts of negotiation—negotiation of multiple interpretations of scientific evidence, of participation and input, and of political and economic concerns and constraints. Additionally, the various efforts to assess and regulate chemical risks are political, not just scientific. This can clearly be seen in the patchwork of regulations on individual chemicals in individual states across the country. If chemical regulation was purely scientific, all states seeking legislation at the same time would arrive at the same legislative outcome because the body of science would point to a single answer. Instead, Maine passes a chemical ban, Connecticut fails to do so, and Kentucky does not even try, because of their different political, social, and economic contexts.

Even with uneven regulations across different geographic units, laws and regulations clearly exert influence beyond their own borders. DfE based its criteria for hazard calls partly on international standards such as the United Nation's Globally Harmonized System for the Classification and Labeling of Chemicals.[109] California's flammability standard for upholstered furniture has had national influence on the use of flame retardants in furniture. Regulations in Europe have impacted regulations in the United States, with the chemical companies agreeing to phase out chemicals only after they were restricted in the European Union. The impact of Europe's Registration, Evaluation, Authorisation, and Restriction of Chemicals (REACH) may continue to grow, as the reporting requirements for companies doing business in Europe increase and the REACH database of hazard and exposure studies becomes more broadly available.

Stakeholder participation is a common, often required, component of EPA assessments. In this chapter, I focused on EPA reviews of chemicals through the New and Existing Chemicals Programs and the DfE alternatives assessment program. For these and other EPA offices, stakeholder and engagement matters greatly. But as the removal of Dr. Rice from an EPA review of DecaBDE demonstrates, stakeholder engagement does not just involve expert evaluation of EPA documents. Science policy at the EPA—like science-based decisions in other institutions—is deeply influenced by political and economic considerations. When these considerations are taken too far by stakeholders and given excessive weight by decision makers, or when the revolving door between industry and government gives individuals influence over the very regulations they previously fought against, science-based processes lose their impartiality, and, potentially, their ability to make the best possible decisions based on the best available evidence.

Instead, the asymmetrical nature of the science policy field means that some groups of stakeholders are better equipped and better resourced to use SST effectively. Industry scientists and hired consulting groups can criticize academic publications, submit lengthy comments on draft regulatory documents, and even conduct entirely new research projects in the service of product defense, activities largely beyond the time and monetary budgets of NGOs or academic scientists. The examples in this chapter have also highlighted the importance of risk formulas in how stakeholders engage in science policy debates. A chemical alternatives assessment guided by the exposure-centric risk formula, for example, could use measured exposure levels or product use patterns to defend highly hazardous chemicals, a line of defense unavailable to those following the hazard-centric risk formula.

A high level of stakeholder engagement at the EPA is unlikely to disappear, nor should it. In many cases, stakeholder participation improves the processes and outcomes of important environmental regulations and assessments, and it is particularly important when it involves community groups, creates opportunities for broad public involvement, and recognizes the values that inform environmental decision making.[110] DfE alternatives assessments take engagement further than most EPA programs, partly because it is a voluntary program with no regulatory teeth. Indeed, the fact that it is voluntary incentivizes buy-in from stakeholders, which increases the amount of information, and provides new data sources. This improves the likelihood that DfE reports and products will be useful for stakeholders in their intended ways.

There are potential drawbacks to the close involvement of stakeholders in these programs. Some policy options—for example, that flame retardants should not be used at all—will necessarily be off the table in order to bring together all the needed participants. Additionally, DfE alternatives assessments take a long time, at least two years from project scoping to final publication. However, even though regulations are typically seen as being more definitive, several EPA representatives told me that, given the layers of bureaucratic review required of any regulation, the voluntary programs may in fact get more done in some instances. As one DfE scientist noted, although the voluntary programs "can't force changes down people's throats, they can still be effective when regulation takes so long."

Other voluntary programs, however, have clearly fallen short. The now-defunct VCCEP program failed to achieve any identifiable reductions in risks to children's health. An Office of the Inspector General review of this program expressed significant concerns not just about VCCEP's shortcomings but about voluntary programs in general, stating that the "EPA has not demonstrated that it can achieve children's health goals with a voluntary program."[111] An additional limit of voluntary measures related to chemicals is that sometimes the parameters are defined by the participating industries themselves. For example, the flame retardant industry developed the Voluntary Emissions Control Action Program (VECAP) to reduce exposure to its chemicals in response to a European Union request for greater monitoring of flame retardants in the environment.[112] But rather than identifying exposure sources or conducting environmental monitoring, VECAP is focused exclusively on manufacturing processes, even though research shows that most exposures for the average person come from the home environment. This suggests that regulatory efforts are needed and that individual exposure-reduction efforts are insufficient, the conclusion arrived at by other scholars of chemicals exposure and risk.[113] As I discuss in greater detail in the next chapter, flame retardants have become a poster child for the limits of TSCA to protect public health partly because of these extensive regulatory delays and conflicts of interest.

When science policy is guided too heavily by economic considerations, public health is likely to suffer. In developing new flame retardants, the industry pays significant—and understandable—attention to the cost of the new product and how it will fit into existing manufacturing processes. Unfortunately, this means that new product R&D is likely to maintain the status quo rather than significantly transform dirty manufacturing processes.[114] But there is a way out of this pattern: companies are more able to develop new, less hazardous, typically more expensive products that require new production facilities when regulations and supply chain decisions create markets for safer chemistry.[115] Regulations and progressive policies that protect public health and the environment can be good for business and good for innovation, incentivizing the development and production of safer products and forcing their adoption along the supply chain. Corporations that are required to use flame retardants in their products as well as distributors selling those manufactured products have increasingly played a major role in changing flammability regulations, both in response to consumer and social movement pressure and in light of their own concerns for health.

Unfortunately, the flame retardant industry has resisted regulations that would move the market, at least until they have developed, registered, and commercialized replacement chemicals. The flame retardant story has too often been one of regrettable substitution, delayed action, and perversion of the scientific process, a story that becomes head-to-head conflict when we examine the interactions and debates between environmental and health advocates on the one hand and chemical corporate advocates on the other. The next chapter turns from science policy debates at the EPA to activism and controversy in the public sphere. In the past decade, a broad and unexpected coalition of environmental, health, consumer, and fire safety advocates has rallied around campaigns to restrict the use of flame retardants in consumer products, but this coalition has faced dramatic resistance from industry front groups and corporate advocates. These two sets of stakeholders favor different risk formulas, engage in different forms of SST, and offer dramatically different paths forward, one characterized by precautionary action and the other by product defense.

6

SCIENCE FOR ADVOCACY

I n the spring of 2012, I attended a meeting in New York City with three dozen advocacy leaders from around the country representing traditional environmental groups and antitoxics nongovernmental organizations (NGOs), environmental justice and urban health organizations, firefighters and burn survivors, the medical and green building professions, and several scientific disciplines. We gathered in a long, windowless conference room with butcher paper hanging on the walls and laptops and notes covering the tables. We spent the first day going over the past year's successes, challenges, and defeats. Scientists called in over a speaker phone and reviewed emerging science on flame retardant exposure and toxicity. Activists from around the country debriefed their encounters with the "opposition," including the chemical companies, trade associations and lobbyists, and conservative business groups, and they discussed market engagement strategies for collaborations with product manufacturers, suppliers, and retailers.

At the end of the long day, about half the group went to a soul food restaurant a few blocks away from our meeting site. I sat across the table from an environmental health organizer and told her about my research. She responded, "I try to never use the word 'risk' because risk assessment is so deeply flawed. I talk about hazard and about exposure. There should not be toxic chemicals in children's products, period." She continued by distinguishing between "chemicals" and "toxics": "We shouldn't have toxics. Of course we're going to have chemicals, but it is possible to have a toxic-free world."

This organizer's description of risk and hazard epitomizes the environmental advocacy community's perspective on chemicals. In the last

decade, environmental and health activists and organizations have
played central roles in building public awareness and pushing for legislative, regulatory, and market-based changes regarding flame retardants. A broad coalition led by environmental organizations, public interest nonprofits, and firefighters has used science in multiple ways by
translating it for their audiences, collaborating with scientists, and conducting their own research. In their work, they reinforce and personalize the either-or risk formula I described in chapter 3, explicitly rejecting
the classic definition of risk as a function of hazard and exposure, and
arguing instead that either hazard or exposure can provide sufficient evidence of risk for action to be taken.

In opposition, a small number of well-funded industry front groups
have fought against restrictions on flame retardants and changes to flammability standards, acting as the public face of the chemical industry and
working to institutionalize a very different risk definition. These *corporate advocates*—direct representatives of the industry or organizations
funded by industry to lobby decision makers or build public support—
share the exposure-centric risk formula and understanding of fire science
favored by the flame retardant industry.

The debates between these flame retardant supporters and environmental and public health advocates demonstrate that struggles over risk
definition and strategic science translation play out not just in the pages
of scientific journals or the offices of the EPA, but in the halls of Congress,
in NGO conference rooms, and on the front pages of newspapers. For
these groups of stakeholders, scientific arguments are required but not
sufficient. Advocates on both sides of the issue produce, disseminate, and
translate scientific research, and also marshal economic, social, and moral
arguments to make their case. For all involved, flame retardants have
become a case study of chemicals regulation. Environmental and public health advocates see flame retardants as the poster child of a broken
system of chemicals regulation, while corporate advocates see the
chemicals as an example of misguided public concern and overly precautionary regulation. Each side, however, lobs similar critiques at the
other: they reject the opposing side's interpretation of the science as
flawed and lacking rigor; they accuse their opposition of "getting away"

with whatever they want; and they point out their opponent's strategic science translation (SST) while maintaining that their position is the scientifically grounded one.

Most of the book until now has focused on scientific conflicts and behind-closed-doors regulatory activities, but I end with a discussion of activism and public sphere debate because that is where these controversies are often the most visible, and because environmental health advocates offer a dramatically different path forward: one characterized by precautionary action and a different valuation of scientific uncertainty. The world of flame retardant activism is vast, covering years of contention, dozens of involved nongovernmental organization (NGOs), hundreds of individual actors, numerous legislative challenges in over a dozen states, and various strategies used by different activist groups. Rather than attempt to outline everything that has happened, mention every group involved in advocacy, and investigate every strategy and response, I draw on a few empirical examples and on interview data and observations to reveal the general characteristics of the flame retardant movement, how these advocates think about risk and hazard, and how science and risk inform activist decision making.[1]

A BROAD COALITION: "NOT JUST ENVIROS ANYMORE"

The coalition that has coalesced to campaign against flame retardant chemicals is situated within the environmental health movement and is led by many of the expected players: nonprofit organizations focused on environmental, health, antitoxics, and public interest issues. But some of the key allies are more surprising: ecologists working in state regulatory offices, environmentally minded legislators, breastfeeding advocates, furniture makers, firefighters, advocates for burn victims, and fire scientists.[2] With the partial exception of the Green Science Policy Institute, a small but influential science-advocacy NGO founded in 2008 by chemist and environmental advocate Arlene Blum, no organizations focus primarily or exclusively on flame retardants.[3] Rather, organizations and activists work on flame retardants as one campaign among many that contribute to their larger goals.

The wide range of organizations involved demonstrates the breadth of topics of concern in the flame retardant arena. Whether you are a breast-feeding advocate concerned about breast milk purity, a conservationist angry that wildlife is contaminated by industrial pollutants, an antitox-ics activist wanting to reform federal regulations, a firefighter troubled by the high rates of occupational cancers, or a learning disabilities advocate interested in the rise in neurodevelopmental problems in children, flame retardants are an issue you can rally around. Activists credit the breadth of the coalition with much of their regulatory success: as one leader told me, "When you assemble enough unlikely allies together in one place, in front of a legislature, you can help to overcome the noise and the doubt-sowing."

Throughout the past decade, organizations have chosen flame retar-dants as campaign targets for several reasons. First, scientific research has shown that halogenated flame retardants as a class are often hazard-ous to humans and the environment.[4] Exposure from consumer products is common and largely unavoidable, and the actual sources of exposure are highly visible. As one environmental health organizer said, "One of the reasons that we chose to work on flame retardants was because of the widespread use in consumer products. Flame retardants were in every-body's homes." Additionally, activists see polybrominated diphenyl ethers (PBDEs) specifically as "a good case study because there are very clear safer alternatives."

Restricting flame retardants is also seen as a good campaign because it is possible to target a relatively small, identifiable industry. As an orga-nizer said, "These kinds of pollutants, whether it's PBDEs or whether it's triclosan, another chemical that we work on, or bisphenol-A, these are pollutants that are manufactured by a relatively small number of compa-nies."[5] Having an identifiable target clarifies messaging and improves the likelihood of campaign victories. Finally, flame retardant bans are seen as more winnable than other chemical bans because they have an identi-fiable constituency in mothers who are concerned that flame retardants accumulate in breast milk, are measured in cord blood, and can be trans-mitted to children, who are more vulnerable to their health effects. The availability of funding from environmental philanthropists and foundations has also encouraged other organizations to join this well-networked campaign.

Despite widespread concern about environmental health issues and growing awareness of chemicals such as flame retardants, many *activists*— that is, people who are actively campaigning for social change— intentionally avoid the "activist label," at least in some contexts. Respondents told me that it could be a "pejorative label," and that activists were often seen as "short-sighted" in their work. Some tried not to use the label at all. For example, one scientist said she used the term "communication or outreach specialists" instead of activists. Several respondents who were actively working for social and policy change specifically requested that I identify them as "advocates" and not as "activists." One regulator went so far as to say that conservative legislators, whose support for chemical bans is often critical, see environmental activists as "enemies" who are "not trustworthy."

In light of these critiques of and concerns about the activist identity, it is no surprise that organizers used various strategies to bolster their legitimacy. Their use of science is one such strategy because scientific evidence is highly valued in public and regulatory spheres. They also strategically reference information from non-advocacy sources. One organizer said that when reaching out to legislators, she tried to find news sources that were not seen as left leaning, and she showed me a recent monthly health bulletin with clippings from the *Washington Post* and *Chemical and Engineering News*.

Coalition building also allowed activists to benefit from the credibility of other professions. One activist said that the environmental community "has a credibility problem," so "when a legislator's working on this bill, they'd rather have the nurse there as opposed to the Sierra Club lobbyist, or they'd rather have the firefighter there or the doctor there testifying, as opposed to the college graduate who's running PIRG [the Public Interest Research Group]." Another activist echoed this position: "It's easy to be perceived as 'oh those tree-huggers, they always want something.' It's helpful to have people who are the most affected, like with learning disabilities, or childhood cancer, or breast cancer, a diverse set of voices. People that are very credible to the project, like nurses and teachers." By bringing in coalition members, activists hope to avoid sounding like a "broken record."

Advocates for restrictions of flame retardants often intentionally framed their work around the issue of health, especially children's health. As one activist explained it, "Part of the whole strategy was to change the frame from an environmental frame to a health frame. And within that health frame, children's health was held up first and foremost: both threats to the developing baby in the womb as well as infants, toddlers, and children." In sociology, *framing* refers to the ways that social movements actively construct meanings around contested issues, offering specific interpretations to increase support.[6] One policy expert put it bluntly when I asked why chemical exposure is compelling to the public: "Kids and body burden. Body burden and kids." This "children's health frame," as activists call it, is effective for advocacy.

The children's health frame is also firmly rooted in scientific evidence from the last decade. Research shows that children have higher body burdens of flame retardants than their parents.[7] There is strong and growing evidence that some flame retardants are particularly dangerous for children, especially the developing fetus and young children, and that prenatal exposures lead to developmental and behavioral problems in childhood.[8] Additionally, children are not just small adults: their bodies metabolize chemicals differently, they eat more food and breathe more air for their body weight, they spend more time on the floor, they engage in more hand-to-mouth behaviors, and their developing bodies are more sensitive to environmental exposures.[9] As a scientist working in an environmental advocacy organization said, "Even though they are breathing the same air, eating the same diet, generally, and living with the same products . . . kids have more exposures."

This children's health frame, as one advocate told me, is both "accurate" in its scientific foundations and "successful" in its appeal to the public and decision makers. A speaker at a flame retardant conference called children "charismatic mega-fauna": focusing on children, she explained, "personalizes" the issue and appeals to the public's "sense of wanting to protect children," a universal societal value. The fact that breast milk is contaminated by industrial chemicals is especially resonant for parents. This invokes the either-or risk formula, because the presence of a chemical in a child's body becomes grounds for action, regardless of the dose.

Another key strategy by the flame retardant coalition has been to develop coalitions with the occupational groups, including firefighters, transportation workers, flight attendants, and pilots. There are also on-going collaborations with the medical professions, especially through the "Healthy Hospitals Initiative," a partnership of several NGOs and over five-hundred hospitals with more than $20 billion in purchasing power, which has sponsored informational calls on flame retardants, urged man-ufacturers to develop products that do not contain halogenated flame retardants, and educated hospitals to avoid building materials and furni-ture with flame retardants.[10] Recently, for example, Kaiser Permanente committed to only purchasing furniture without flame retardants for its network of hospitals and medical offices.[11]

Firefighters have played a particularly important role in successful campaigns against flame retardants. According to interviews and obser-vations, this is the first example of firefighters directly participating in environmentally oriented activism.[12] Previous *blue-green alliances*, or collaborations between labor and environment organizations around a common policy goal, have been facilitated through the use of health frames, and this has certainly been the case in the flame retardant story.[13] Health is a significant concern for firefighters because they and other first responders have elevated rates of many types of cancers and other ill-nesses.[14] This evidence has compelled thirty-three states to pass "pre-sumptive disabilities laws" establishing that certain illnesses and types of cancer will automatically be judged to be service related.[15] When a prod-uct containing halogenated flame retardants burns, the smoke released contains dioxins and furans, which are known carcinogens; the resulting smoke is more toxic than if the product had no flame retardants, with more carbon monoxide, hydrogen chloride or bromide, and partially burned hydrocarbons.[16]

Several firefighter organizations in the United States have been in-volved with chemical regulation in some way. The International Associa-tion of Fire Fighters (IAFF) is a union representing more than 300,000 professional firefighters and emergency medicine personnel from 3,200 fire departments around the country.[17] The IAFF has supported multiple

PBDE restrictions, as have local and state IAFF chapters.[18] Also influential in the fire safety world are organizations representing burn survivors and burn doctors. The National Association of State Fire Marshals (NASFM) has often opposed flame retardant bans. As I described in chapter 2, tobacco companies and flame retardant manufacturers provided funding to NASFM in the past, and at times they have even shared lobbyists.[19] According to an active firefighter, NASFM is part "real" fire service but part "astro-turf."

Firefighters have been involved in state regulations of flame retardants in different ways, most actively supporting bans of specific chemicals only after safer alternatives have been identified. Firefighters sometimes testify at public hearings wearing their uniforms, and some firefighting organizations have written letters to legislatures in support of new regulations. For example, firefighters from four professional associations wrote to the American Chemistry Council (ACC) in 2012, urging the trade association to expel the flame retardant manufacturers.[20] Representatives of the Associated Fire Fighters of Illinois and Illinois Fire Fighters Association submitted a memorandum to the Illinois legislature that stated, "The elimination of deca-BDE [decabromodiphenyl ether] will not compromise fire safety, but we believe it will be a step in the right direction for improving the health and safety of our firefighters."[21] Firefighting organizations are also interested in the science linking their occupational exposures to health outcomes, and they have collaborated with environmental health activists and researchers on biomonitoring projects to understand firefighters' exposure to chemicals.

Firefighters recognize that chemical exposure is not a typical issue for them. One day I visited an urban fire station and spoke with several firefighters, including the union's health and safety representative who had been involved with an unsuccessful attempt to ban PBDEs in his state. As we walked around the station, he showed me how the white walls of the engine bays darken toward the ceiling, discolored by exhaust from the fire trucks. He explained that today's firefighters are more health conscious than earlier generations in the profession, and he said that firefighters are "a really simple blue collar group" with their own priority issues, including improving medical disability coverage and protecting their pensions from political attacks. But his union decided to support a

PBDE ban in his state because "solid research shows it's a harmful chemical," and they know this from their lived experience: after exiting a fire, he said, "You're blowing soot out of your mouth, your nose." This health frame aligns with both the firefighters' goal of preventing fires and illness and the environmentalists' goal of preventing consumer exposure, serving as a *frame alignment* between the interpretations and explanations for an issue between these different groups.[22] Firefighters who work on chemicals regulation are highly aware that their profession suffers high rates of illness as a result of occupational exposures. In the powerful words of a career firefighter, "We are the canaries. I think we are the ones that are being put into the mine shaft with this stuff and hopefully come out alive."

Other firefighters had their resolve strengthened by what they described as dirty or strong-arm tactics used by the flame retardant manufacturing companies and their lobbyists. Many firefighters were motivated to get involved after the chemical industry falsely invoked firefighters' authority. As one firefighter explained, industry lobbyists "used our image and our reputation . . . To be honest with you, I was OK with flame retardants being all jacked up and being stupid and hurting the environment and people, but then when they used our reputation and our name, that's when it pissed me off [*laughing*]." A firefighter in another state got involved after a lobbying organization came into his state and "tried to disrupt the firefighter community" with a "robo-call" that challenged the credibility of the fire safety community. Thus, in some cases firefighters got involved in flame retardant legislation because they felt the chemical industry was making inaccurate claims about the fire service.

Activists and legislators give significant credit to firefighters for the success of chemicals reform. In the words of a firefighter, "There isn't a government or legislative body that's going to pass a ban on flame retardants unless the fire service weighs in and has some significant involvement." A state regulator agreed: the participation of firefighters "carried a lot of weight because the major argument that industry was trying to make is that by banning the substance, we were compromising fire safety . . . And so [firefighter support] really counteracted that very well." Indeed, no state level bans have passed without some support from the fire service, though sometimes the fire service was divided, with some groups in support and some in opposition.

Because of the needed authority that firefighters bring to the issue, activists work hard to cultivate these relationships. Environmental groups recognize that the coalition is fragile because it is relatively new and because questions exist about how it might extend beyond relatively narrow flame retardant campaigns. They therefore devote significant energy to maintaining these alliances. They bend over backward to involve the fire service, even when it means downplaying their own short-term goals. As an example, environmental organizers deferred to the firefighters' desires to not release the results of a biomonitoring study. Activists also talk about engaging in "reciprocal work" that supports fire departments around issues such as spending cuts to fire departments, protecting union pensions and bargaining rights, and supporting presumptive cancer laws.

CORPORATE ADVOCACY

The opposition to flame retardant bans and restrictions has its own advocacy movement.[23] This corporate advocacy is organized by the flame retardant companies and the lobbying organizations they fund.[24] Corporate activism is not unique to the flame retardant world; for example, it is a powerful force in climate change denial.[25] This is often called *elite* or *astroturf activism*, in contrast to the *grassroots activism* that emerges from concerned citizens.[26] Astroturf, a term coined in 1985 by Senator Lloyd Bentsen (D-Tex.), describes "the artificial grassroots campaigns created by public relations (PR) firms."[27] Corporate activists use strategies similar to those of traditional social movements, including direct lobbying, writing letters to elected officials, publishing op-eds or letters to the editor in newspapers, and gathering petition signatures.

Corporate advocacy on behalf of the flame retardant industry has been extremely well funded. In California, in many ways the epicenter of flame retardant contests, the flame retardant industry spent at least $11.3 million on lobbying between 2007 and 2011.[28] At a minimum, lobbying expenditures and campaign contributions improve access to decision makers and legislators. California State Senator Mark Leno stated, "Almost without exception, as I'm leaving my colleague's office, there's a lobbyist for the chemical industry in the waiting room to go in to get the last word. And of course, there's a dozen of them and one of me."

Three corporate activist groups, all funded by flame retardant manufacturers, have been active in the United States.[29] The Bromine Science and Environmental Forum (BSEF) is the industry's global organization headquartered in Brussels, Belgium. BSEF mostly focuses on European initiatives, especially REACH (Registration, Evaluation, Authorisation, and Restriction of Chemicals) and Europe's electronics regulations. BSEF is funded by corporate members—Albemarle, Chemtura, Israeli Chemicals Limited Industrial Products (ICL-IP), and the Japanese company Tosoh Corporation. It commissions the development and dissemination of "innovative research" on brominated chemicals, and "represent[s] the bromine industry on issues of environment and human health."[30] The group's activities are coordinated by the lobbying firm Burson-Marsteller, known for creating the industry front-group the "National Smokers Alliance" in 1993 with funding from tobacco companies.[31] BSEF is less active in the United States than in Europe but has lobbied in California, Hawaii, and Montana, spending over $6 million dollars on lobbying "re: flame retardants" in the third quarter of 2007 alone.[32] (A 2007 California bill, AB 513, would have been the first in the United States to ban DecaBDE, but it failed in the California Assembly.) According to observers, BSEF engages in direct and aggressive lobbying, commissions pro-bromine research, and "uses 'science' as a political tool."[33] The group is also active in disseminating information about the Voluntary Emissions Control Action Program (VECAP), the industry's voluntary exposure reduction program for flame retardant manufacturers and users.

Citizens for Fire Safety (CfFS) was a small but active trade association funded by Albemarle, Chemtura, and ICL-IP from 2007 to 2012. CfFS was originally coordinated by Burson-Marsteller; more recently, BSEF and CfFS shared lobbying firms in California.[34] Though CfFS described itself as "a coalition of fire professionals, educators, community activists, burn centers, doctors, fire departments and industry leaders," according to tax filings the organization's expressed purpose was to "promote common business interests of members involved with the chemical manufacturing industry."[35] With the exception of a paid executive director, all officers were employees of the flame retardant industry.[36] CfFS presented itself as a group primarily focused on education and research, but in 2010 they spent $241,000 on education and $38,000 on research out of nearly $4 million in "functional expenses" (including an itemized $1.4 million in lobbying fees).[37]

In interviews, respondents described outreach efforts to the fire service by CfFS including funding scholarship efforts, sending children to camps for burn survivors, or buying equipment, efforts that were perceived by the fire service and environmental activists as attempts to buy the support of fire departments and fire safety experts. An advocate for burn survivors said, "I've heard firsthand . . . there's been offers to various local employee unions from the Citizens for Fire Safety: 'just give us an invoice for $10,000 a month and no questions asked.'" After months of criticism following an investigative journalism series in the *Chicago Tribune* detailing their lobbying efforts, CfFS closed its doors, and the industry announced that all advocacy efforts would henceforth be coordinated through the ACC, though a representative of the flame retardant industry told me that the timing was coincidental.[38]

The ACC is the trade association for the American chemical industry, headquartered in Washington, D.C. In 2014, the ACC spent $11.4 million on lobbying and gave $562,256 in campaign contributions, making it the twenty-sixth biggest lobbying operation in the United States.[39] Much of the ACC's current lobbying focus is on efforts to reform the Toxic Substances Control Act (TSCA). The North American Flame Retardant Alliance (NAFRA) is the ACC's flame retardant panel; Albemarle, Chemtura, and ICL-IP are its only members. NAFRA describes itself as "the lead advocacy organization in North America for flame retardant producers and users."[40] Since August 2012 when NAFRA assumed all advocacy work for the flame retardant companies, NAFRA has written press releases in response to emerging science and regulation on flame retardants. No lobbying or funding information is available for NAFRA separately from ACC.

The lobbyists and scientists who speak on behalf of the flame retardant industry rely on two sets of arguments. The first is that flame retardants are a proven and important tool for fire safety that saves lives, and that "every second counts" in a fire. The second is that the health and environmental risks of flame retardants are insignificant, nonexistent, or applicable only to "legacy chemicals" (i.e., PBDEs), not today's "greener" flame retardants. To buttress both arguments, they engage in extensive SST.

As an example, since mid-2012 the industry has repeatedly cited a recent white paper by Dr. Matthew Blais, a chemist and fire scientist at the Southwest Research Institute.[41] Dr. Blais reanalyzed the data from a fire modeling study funded by the Department of Justice, and concluded that flame retardants in residential furniture provide a substantial safety benefit. Industry representatives have pointed to this study as "new" evidence that their products are effective. At a federal hearing on flame retardants in June 2012, Marshall Moore of Chemtura discussed the Blais findings in his testimony, saying the research shows that flame retardant foam provides "up to 10 additional minutes for an individual or family to escape to safety."[42] Industry representatives, trade associations, and lobbyists have emphasized the "independent" institutional location and the federal funding source of this research. At another hearing, chemist Dr. Gordon Nelson testified on behalf of the flame retardant industry and incorrectly said that the supplementary analysis by Dr. Blais was funded by the Department of Justice.[43]

However, the methodology of the Blais analysis appears to be seriously flawed. It extrapolated from the results of a single burn test—one piece of furniture that had originally been excluded from the Institute's full analysis as an outlier—and it described the theater-grade furnishings used in the study as common household furniture.[44] The foam treated with flame retardants was of a higher density than the flame retardant–free foam, meaning it would likely burn more slowly irrespective of added retardants. Furthermore, Dr. Blais is not fully independent—he is a "technical advisor" to NAFRA, and was paid honorariums and reimbursed for travel expenses to present his findings at fire science meetings, though he told reporters that he invested the money back into his research institute.[45] In response to an exposé published in the *Chicago Tribune*, the ACC accused the reporters of "attacking the credibility, character, and work of individual, highly-respected scientists who are now engaged in flame retardant research, rather than looking at how and why flame retardants have saved lives since they were introduced more than 30 years ago."[46] The ACC press release concluded, "Every minute counts in a fire, and extra time for first responders can help save lives." It did not address the substantive criticisms of the *Tribune*'s analysis, instead portraying flame retardants as an irrefutable tool for fire safety.

The main product defense strategy for the flame retardant industry is to engage in interpretive SST around fire safety and environmental health science. Lobbyists and corporate advocates routinely argue that the science conclusively shows that flame retardants save lives but is inconclusive or absent about flame retardants causing harm; therefore, flame retardants are a needed and effective tool for fire safety. Flame retardant representatives typically concede that the legacy chemical PentaBDE was problematic, but generally insist that other chemicals, including DecaBDE, tris(1,3-dichloro-2-propyl)-phosphate (TDCPP), and newly developed replacements, do not pose significant risks to human health. To buttress these assertions, they cite risk assessments conducted by government bodies, though they often cite them incompletely or incorrectly. They also paint the environmental health science as inconclusive or lacking, and argue that regulators should err on the side of adding "layers" of fire safety.

This strategy appears to be at least somewhat effective. California State Senator Lou Correa told a reporter about his decision to vote against changes to California's Technical Bulletin 117 (TB117):

> Correa said he voted against the bills partly because the science is inconclusive—"you've got papers on all sides"—and partly because he worries about young burn victims. "You take the totality of the testimony and the danger of ever-growing concerns of environmental risk versus the real existing danger of children being burned. Seeing some of the horrible, horrible cases of children being burned in their cribs, that stuff is very powerful," Correa said.[47]

In short, corporate activists engage in interpretive SST when they argue that flame retardant use should continue because there is scientific certainty about fire safety benefits but scientific uncertainty about health effects.

Additionally, industry supporters argue that opposition to flame retardants comes from well-heeled environmental groups, self-interested occupational groups, and manufacturing sectors capitulating to pressure from uninformed consumers. A representative of a flame retardant trade association called antitoxics activism "an area of opportunity for the green side." Industry representatives also argued that firefighters supported bans on flame retardants for self-interested reasons and were misguided in their efforts. A flame retardant industry representative said that firefighters engaged in "horse trading" to become involved in flame retardant bans: the same legislators who support bills increasing benefits for firefighters (such as presumptive cancer laws) want their support for environmental legislation. Yet such a charge assumes that it is a problem to support multiple types of health and environmental legislation.

Though flame retardants are a multi-billion-dollar sector, the industry portrays itself as an underdog that is unfairly attacked because their chemicals are used in high-profile consumer products and because fire protection is intangible. As a trade association representative said, "We're talking about your couch, or baby pajamas. That's newsy. That's going to get people's attention. The flipside is that the benefits messaging is complex, because fire is something that happens to other people. And there's this feeling that you can control fire, you know, that if I'm smart, I can manage that. It's not going to happen to me." Echoing conservative dialogue, industry respondents regularly expressed concerns about "chemophobia," which they see as an unjustified, unscientific, and ignorant fear of chemicals. An industry scientist said that chemophobia "is just general ignorance . . . People just hear 'chemical' and . . . they think, 'oh chemistry, chemicals are bad.'" From this perspective, then, flame retardants face a double barrier of being a chemical—and thus automatically dangerous—and providing an invisible, future-oriented benefit. As an industry scientist said, "Unless you've been in that situation [a fire], it's not really a tangible problem for you." In this way, the industry reiterates the benefits of flame retardants while arguing these benefits are poorly understood by the public.

In addition to direct lobbying of decision makers and testifying at public hearings, the flame retardant lobbying groups have paid for mass mailings that feature burning buildings and children, and "robo-calls" to residents about upcoming state bans. In some states, these tactics have backfired. An organizer in New England said that aggressive lobbying tactics actually resulted in legislators strengthening the bill against flame retardants:

> The bromine industry started to pay for robo-calls across the state . . . They . . . would call you, have a firefighter introduce themselves on the call, invite you into a conference call to learn more about these flame retardants, and then the conference call was like, full of people asking pre-screened questions that led to answers like, "there are no safe alternatives," "if we don't have Deca your children will die in fires" . . . Luckily it was so over the top and outrageous that when legislators started getting these calls they just got even more upset, and we ended up with an even stronger bill.

These tactics were seen as especially out of touch in smaller states. In the words of one East Coast legislator, "It was totally disproportionate to the size and scale and the way that politics gets done [here], it actually helped us because it was so over the top and extreme." A regulator in the same state told me a similar story, and concluded, "The chemical industry couldn't just come pour a lot of money in and win. They got smoked." Similarly, in a western state, an activist said that at least four representatives in the House had voted for the bill specifically because of the industry's aggressive lobbying: one "held up their flier and said, 'I was going to vote 'no' until I got this flier in my mailbox, and now I'm voting yes. And it has nothing to do with the bill. It has everything to do with making a statement against this kind of lobbying." Thus, aggressive lobbying tactics by the flame retardant industry backfired in multiple ways—by inspiring legislators to vote in favor of chemical bans and by inspiring the fire service to support those bans.

Despite obvious differences in their resources and messages, environmental and corporate activists sometimes use similar tactics. In addition to citing the scientific literature and encouraging supporters to write letters, call decision makers, and testify at hearings, they use blogs and e-mail lists to communicate their positions to a wider audience. The "American Council on Science and Health," a research and educational organization funded by corporate interests, publishes a daily "Dispatch" that includes their responses to scientific findings, and the ACC has a similar blog, "American Chemistry Matters."[48] The Safer Chemicals, Healthy Families group similarly has a network of dozens of bloggers who post or repost findings about chemicals.[49] This confirms social science research findings that industry-driven activism adopts supposedly grassroots tactics.[50]

TARGETS BEYOND THE STATE

Social movements are "collective challenges to systems of authority."[51] Though traditionally scholarship on social movements has focused on nongovernmental campaigns to influence the behavior of central governments, contemporary social movements have many targets, including formal political structures, corporations and market institutions, the media, and other social movements.[52] This is particularly true of environmental health movements and *alternate pathway* social movements, which target industry and technology directly.[53] Environmental health movements also challenge the *dominant epidemiological paradigm*—the codification of science, government, and private sector beliefs about disease identification, causation, and treatment.[54]

Flame retardant campaigns direct their efforts against all these targets, showing how these multiple strategies complement each other. A typical campaign includes lobbying for regulatory changes at the state and federal levels, conducting public education and media outreach, participating in processes to develop or reform codes and standards, and engaging with supply chain manufacturers.

Market Campaigns and Corporate Challenges

Flame retardant activists frequently engage in "market campaigns" and relationship-building with industries. Market campaigns, in the words of one advocate, work to "change products through markets, versus the old strategy of government policy." They focus on the full flame retardant supply chain, from raw material and chemical inputs to consumer sales. These are similar to the *product- and technology-oriented movements* that sociologist David Hess has described, in which social movements aim to change products and technologies by targeting consumers, manufacturers, and researchers.[55]

Some market campaigns involve public collaborations between NGOs and businesses; for example, the Green Science Policy Institute regularly works with the furniture industry. But oftentimes market campaigns target specific companies or industry sectors, trying to convince or incentivize them to change their practices. Indeed, some activists are hesitant to engage in formal industry collaborations or are skeptical of positive engagement with companies, which is seen by some as "jumping into bed" with corporations.

But market campaigns have played prominent roles in the flame retardant story because corporations' reputations may be vulnerable to protest tactics.[56] Though these market campaigns are sometimes controversial, this type of activism has often been able to extract meaningful commitments from individual companies more quickly than governmental action. In the mid-2000s, for example, Greenpeace targeted Hewlett-Packard (HP) with the campaign, "HP=Harmful Products." An industry respondent said this motivated HP to reformulate their products: "Like when Greenpeace goes in and storms Hewlett-Packard, its headquarters, and climbs ropes up to the roof . . . Companies don't want to have that sort of a PR problem. So basically that drives their decisions."

Recently, environmental health activists have targeted the baby products sector, building on academic and advocacy studies detecting flame retardants in the foam of nursing pillows, car seats, and strollers.[57] According to activists, baby products are a "target-rich" sector that is particularly vulnerable to attacks on its reputation. Additionally, most baby products, with the exception of car seats, are not required to contain flame retardants.[58] In 2012, several environmental health activists and parenting bloggers targeted the baby products company Graco after studies found TDCPP (commonly called chlorinated Tris) in its strollers. Within one week, an online petition asking the company to remove flame retardants from their products, disclose all materials, and turn to safer alternative chemicals had gathered thousands of signatures. Soon after, Graco contacted the campaign leaders to say they would stop using two controversial chemicals, and wrote a collaborative press release with NGOs announcing their reformulation plans.[59]

Advocates talked about this type of market pressure being an important piece of the social change puzzle; in the words of an environmental health activist, "I see it as, you've got scientific pressure, you've got market pressures, you've got regulatory pressures, you've got the pressure of the general public and public perception, and all these things work together to help change behavior at the corporate level." Modifying corporate behavior creates immediate changes in the chemicals used in consumer products. As one policy expert said, "You know, you can reform TSCA all you want, but if Walmart says, 'you can't sell to users unless your chemicals do *XYZ*,' you're going to switch, it doesn't matter if there's regulation or not."[60] Even advocates working primarily on regulatory and legislative change will support market activities, and several told me that this "retail regulation" leads to quicker and more effective change.

Beyond potentially quick changes by flame retardant users, market campaigns can also create or enhance ruptures between different industrial sectors. The chemical industry typically opposes retail regulation, which they also call "black lists" or "product de-selection," for two primary reasons: it is outside of their control, and it usually is more protective than regulatory standards. Although retail regulation and supply chain hazard assessment practices are increasingly common, most companies are less open about their internal decision-making processes regarding potentially toxic chemicals. Liability is a concern: if a company admits they stopped using a certain chemical because of hazard concerns, then it effectively is admitting past use of a dangerous chemical. This could make them vulnerable to lawsuits.

Product manufacturers and retailers thus occupy an interesting position when it comes to antitoxics activism. The complexity of supply chains can make product manufacturers more vulnerable to public and NGO pressure than their chemical industry counterparts. An environmental health organizer explained that activists generally have little leverage over the chemical manufacturers and more leverage over "high profile companies" whose products contain flame retardants.

Simultaneously, however, the long supply chain can shield retailers and product manufacturers from responsibility, since they do not make the flame retardants and may have little choice about including chemicals to meet flammability standards. Product manufacturers may not even be aware that the chemicals are in their products if they buy already-formulated materials based on performance characteristics. Indeed, the complexity of most supply chains means that chemical-related information "can be conflicting, protected by trade secrets, lost in supply chains, or nonexistent."[61] Though flame retardants have been found in many baby products, an industry representative told me that he believed that the product manufacturers did not request flame retardant foam but ended up with it inadvertently when they ordered foam based on certain performance characteristics: "The [manufacturer] never says, 'I need flame retardant foam.' . . . There's no step or screening in that process to say, we don't want flame retardant in this foam . . . They just ask for a foam that meets their performance needs."[62] This can make supply chain companies appear eager to get out of flame retardants just to avoid the public scrutiny.

Some stakeholders expressed concern about market campaigns against product manufacturers. One environmental health scientist who regularly works with environmental advocates said, "The evil are the people who are making these chemicals, not the couch makers. I'm worried about targeting the wrong people." Some are cautious about pushing too hard against product manufacturers out of fear that they will, in the words of another activist, "unwittingly drive the furniture industry into the arms of the chemical industry." But they also see benefits in pitting one segment of industry against each other and in "isolating" chemical manufacturers until they agree to innovate.

Codes and Standards

Beyond market campaigns, environmental, health, and corporate advocates target a variety of regulatory mechanisms, especially performance-based manufacturing standards and fire codes. Some of these codes are developed through state and federal governments, especially the U.S. Consumer Product Safety Commission (CPSC). Other performance-based standards or fire codes are developed and revised through fire safety organizations and international regulatory bodies. Some of these standards involve lengthy consensus-based processes involving numerous stakeholders; others are developed, revised, and approved through largely behind-the-scenes proposals and votes. Activists participate in these standards procedures relatively rarely, but they have achieved noteworthy successes. For example, a broad coalition has successfully organized opposition to a proposed international candle standard for electronics three times in the last decade.[63]

These flammability standards are very consequential. The performance and development of flame retardants is based on flammability standards and codes. Indeed, a FlameCorp scientist told me that regulations drive the industry: "People wouldn't have flame retardants in their products unless someone told them they had to be there." Another fire scientist explained that flame retardant development is "a unique field of science in that it's regulation-driven, not necessarily driven by what are the newest scientific breakthroughs . . . The regulations dictate the fire performance of a product, and then you design to that product." Furniture flammability standards, including California's TB117 and the more stringent British standard, are the most well-known and widely discussed. Recent revisions to TB117 were the result of years of organizing by the flame retardant coalition, and they happened in spite of extensive work by corporate activists defending the open flame standard. Beyond these high-profile examples, hundreds of other building codes and product-specific standards touch on flame retardants or flammability in some way.

Although industry representatives told me in interviews that their work was constrained by these standards, the industry also devotes significant resources to influencing and changing them through technical assistance and direct advocacy. In high-profile cases, this can take the form of direct lobbying. For example, NAFRA protested changes to TB117 that replaced the open flame standard for foam with a smolder standard for fabric.[64] More often, however, industry scientists and advocacy representatives participate in the standards-setting process in a technical or behind-the-scenes capacity. At a meeting at FlameCorp, one of the company's research scientists described his participation on an influential committee about the environmental impact of electronics products. He serves on the committee, and said that FlameCorp needed to be an official member of the standards-setting organization in order to more fully influence the standard: "We need to be members of [the organization], in order to kill [the standard] later on." The meeting leader explained that the company participated because FlameCorp products were often part of the discussion: the proposed purchasing standards involved "de-selection" based on certain chemistries, not a full evaluation of chemicals, and halogenated flame retardants had been proposed for de-selection. Thus, participation and membership allowed them to directly influence the proposed standards.

Later that day, I met individually with this FlameCorp scientist. A Ph.D. chemist by training, he spends most of his time developing new products with a larger team of chemists but is also active on technical committees with the EPA, supply chain standards associations, and fire safety organizations that set fire codes. He told me that he tries to "stay away from the political type of things" when he sits on these committees. This scientist distinguished between the "technical" work that he did on committees and the "higher level" concerns about issues like corporate culture or sustainability: "I just give my vote . . . [To] try to argue with people at such a high level, it just doesn't make sense." He said he did not participate in advocacy and called his work technical because he was promoting flammability standards generally, not flame retardants in particular.

Revisions to product-specific codes offer the flame retardant industry additional opportunities to defend or expand markets. For example, a regulator told me the story of how the industry responded to a proposal that would have allowed television manufacturers to meet a European fire safety standard without flame retardant additives by separating the power source with a metal barrier: the flame retardant industry "did a video showing a flame retardant TV on fire smoldering and smoking . . . and then they showed the TV without flame retardants, just roaring . . . They accused the TV industry of making TVs that weren't safe for the public." In response, TV manufacturers in Europe chose to keep flame retardants in their products.

In other instances, the industry identifies opportunities to expand into new markets. In a meeting at FlameCorp, an advocacy representative said the company had been approached by the fire marshals to develop a standard for plastic garbage cans, and he had offered to write up the "technical issues" about how flame retardants could contribute to that market and what codes currently existed. The representative said that they needed to understand why the market was not doing anything on its own. Was it because companies were lazy and avoided acting, or was it that codes were not being enforced? These examples show how companies can use codes and standards to defend and expand markets.

State-Level Legislation: A "Patchwork Quilt" of Different Regulations

Most activism on flame retardant restrictions has focused on changing regulations at the state level. Following the first ban on PentaBDE in 2003, over a dozen states passed bans of specific flame retardants, and several developed more comprehensive systems to evaluate and regulate chemicals, motivated at least in part by the case of flame retardants. Activists have pursued this "patchwork quilt" of state-level chemical regulations to convince product manufacturers, the chemical industry, and state and federal legislators that sweeping federal change is needed.

State-level bans of flame retardants differ slightly, with small or sweeping changes in language, different dates, and different ways of dealing with complications like recycled materials. This is due both to intentional activist strategy to create a regulatory landscape that is hard for companies to manage, and to the inevitable changes made by legislators and committees as a bill works its way through the legislative process.[65] As the Washington Toxics Coalition wrote in a report on flame retardants in children's products, "States are proven laboratories for chemicals policy, showing what actions will succeed in protecting health and providing a model for federal action. State action also motivates industry to seek a federal solution, to avoid a patchwork of regulation across the country."[66] Another activist said that each state ban is a little different, "and that's what drives them [the industry] crazy." In general, industries greatly prefer "uniform and universal regulations."[67]

These environmental and public health activists' choice to target state legislators instead of the federal government reflects their knowledge of the *political opportunity structure*, in which social movements adjust their strategies and tactics to take account of constraints and possibilities posed by political and economic conditions.[68] Although federal legislation (particularly TSCA reform) is seen as the ultimate goal, activists believe success at the state level is more likely, especially in more progressive states. As one activist who had worked on flame retardant bans said, "chemicals management policy should come from the federal government," but until this happens, states will take action. Each year, dozens of states introduce bills related to chemical hazard and exposure.[69]

These state-level challenges are often coordinated through national networks of like-minded groups. For example, Safer Chemicals, Healthy Families is a national network of environmental, health, and family-focused nonprofits and activists. Launched in 2009, it now has over 450 participating organizations working on TSCA reform at the federal level.[70] These national networks are useful for activists across the country, who share information about successful tactics and opposition actors. One state organizer had worked with out-of-state counterparts in this manner:

> The other coordination piece was with Washington State, so they were going through the same bill . . . and we were talking with them all the time . . . They let us know about some of the strategies that industry was using to try and defeat the bill in Washington State, and having some of the heads up on that was useful for us to know. Our state Fire Marshal was quoted in an industry publication in Washington State in such a way that it looked like he in his role at that time . . . was against a ban on Deca . . . They made us aware of that before that was able to happen here, and so we were able to work with him.

These cross-state collaboration and communication strategies are also in place around other environmental health and chemicals issues.[71]

The Perfect Poster Child for Federal Reform

Though the flame retardant coalition has achieved the most identifiable success at the state level, federal chemicals reform is almost always a priority. One activist commented that flame retardants are not the only chemicals of concern, but they are "representative of the broken federal chemical safety system." Environmental organizers called the movement of one flame retardant to the next in the absence of sweeping chemicals reform the "whack-a-mole" or "toxic treadmill" problem. Although successes in environmental leadership states such as California, Washington, and Maine create helpful regulatory "precedent" and "momentum" for other states, activists remain focused on building momentum for TSCA reform.

Because bans of specific chemicals are "an imperfect approach" to chemical safety, NGOs and activists explicitly link flame retardant campaigns to TSCA reform. Single-chemical bans open the door to bans of all halogenated flame retardants. Bans of chemicals in children's products are seen as "a really good entry point" for broader uses of chemicals. Most talked about flame retardants as a segue to reforming how *all* chemicals are regulated, through TSCA. As one activist said,

> We can't go chemical-by-chemical, because it would take a lifetime. And so our goal has been . . . [to] use individual chemical campaigns to ban these harmful chemicals but to also elevate this issue on the radar screen for the legislators and the public, and while doing so help them understand that we need a comprehensive system for chemical reform in this state and in this country.

A policy expert described chemical-by-chemical regulation as a "jump from the frying pan into the fire," because alternative chemicals are rarely well studied. Most activists working on flame retardants also participate in campaigns to reform TSCA, through their organizations or through national coalitions. For example, many involved in the flame retardant coalition attended the 2012 Stroller Brigade, a march in Washington, D.C., that used emerging science on flame retardants in baby products as a "potent symbol" of the need for TSCA reform.[72]

SCIENCE AND RISK IN ACTIVIST DECISION MAKING

The broad range of stakeholders involved in contestations over flame retardants use science extensively to make their arguments, developing new data, contributing to ongoing collaborations, and frequently engaging in SST to justify their policy goals using science. In doing so, they draw on the authority of scientific knowledge and in many circumstances reify the value and necessity of formal expertise. Simultaneously, however, activist participation in scientific practices and use of science for advocacy challenges the *science boundary*, a symbolic and social boundary that monopolizes claims to scientific authority and controls access to resources associated with scientific work.[73] Sociologist Sabrina McCormick describes how *democratizing science movements* use science to "counter [industry-funded] studies, open governmental institutions, and reclaim power by generating new research or reframing scientifically codified objects, bodies, and livelihoods."[74] This is similar to the *doing, interpreting, and acting* framework outlined by sociologist Phil Brown and colleagues, in which science is conducted, interpreted, and used by advocates to challenge dominant interpretations of environmental and health issues.[75]

Blurring the Science-Advocacy Boundary

Stakeholders in the flame retardant story blur the boundaries between activist and scientific fields in several ways. First, science provides legitimacy for advocacy efforts. Environmental and corporate advocates alike draw on science to make their claims, and they argue that their position is the most scientifically justified.[76] Advocates' decisions about how to develop campaigns, which flame retardants to target, and how to communicate with the public, legislators, and regulators are based in no small part on toxicology and exposure research findings.

Activists also draw on the implicit authority of the scientific field.[77] For example, in 2010 advocates and scientists published a statement with nearly 150 signatories, the San Antonio Statement on Brominated and Chlorinated Flame Retardants, which summarized scientific concerns about flame retardants and problems with how chemicals are evaluated.[78] On an activist call, an organizer invoked the authoritative weight of the San Antonio Statement, saying there are "about 10 pages of references documenting this paper." Peer reviewed publications, in particular, lend credibility to advocacy efforts. In the words of one organizer, peer review "does a lot for the industry push-back that this is junk science."

Some trained or practicing scientists have taken a public advocacy role. Environmental organizations commonly employ scientists with master's degrees or doctorates who contribute scientific expertise to advocacy campaigns. For example, the Environmental Working Group employs at least eleven master's-level and three Ph.D.-level staff.[79] A toxicologist in an environmental health research organization noted, however, that these scientists are often stretched thin and act as a "general stakeholder on . . . five hundred issues." These boundary-crossing individuals occupy difficult roles, and often they engage in compartmentalized work.

Environmental health activists also have scientific allies in academia, environmental research organizations, and the government who sometimes testify at legislative and regulatory hearings, attend advocacy conferences, and contribute advice or laboratory work to advocacy science. Activists also occasionally contribute to academic research. For example, a 2012 publication identified the flame retardant chemicals found in 102 samples of polyurethane foam from couches purchased in the U.S. between 1985 and 2010.[80] Some of these samples had been submitted by members of the advocacy coalition, who learned about the requests for samples through e-mail mailing lists. Scientists, health professionals, and environmental health advocates paid close attention to the publication of the study, and they coordinated nationwide media outreach when it was finally published.

Though policy engagement and activism challenges the authority of practicing scientists, researchers working on flame retardants have contributed in meaningful ways to environmental health policy making without losing their scientific credibility. Scientists working on flame retardants have written highly cited papers in top journals, received millions of dollars in federal grants, and earned tenure and promotion at top research universities. They have also testified publicly at congressional hearings, written supportive letters, and spoken with the media about their work.

Finally, some NGOs carry out their own science, moving beyond scientific translation to scientific production by measuring chemicals in humans or their home environments. Numerous environmental organizations, often via projects developed through national coalitions, have conducted research known as *public* or *advocacy biomonitoring*, in which participants often speak out about their body burden and their personal exposures.[81] This work often tackles areas of undone science by focusing on newly identified chemicals of concern or specific subpopulations of interest to the public and NGOs. But advocacy biomonitoring is also seen by activists to have power and influence beyond the scientific arena because it can personalize or individualize the issue rather than merely documenting population-level exposures. Advocacy biomonitoring is seen as a particularly effective tool because it documents the toxic trespass, proving that exposure is happening and providing support for the either-or risk formula that often defines environmental health activists' understanding of chemical risk.

Thus, activists use and produce science and collaborate with scientists in many different ways. Sometimes they attribute policy successes directly to scientific findings. For example, an environmental health activist attributed a recent state ban of the flame retardant tris(2-chloroethyl) phosphate (TCEP) to scientific findings: "We passed a law to ban one of the chlorinated Tris chemicals that replaced Penta[BDE] in polyurethane foam in baby products. And we did this as a result of a study that came out of Duke University that [Heather] Stapleton and others produced . . . [that] found either toxic or untested flame retardants in 80 percent of these products." For these activists, science on exposure demonstrates that consumers cannot shop their way out of the problem.

This builds on sociologist Andrew Szasz's work on consumer activism and chemical exposure.[82] Szasz argues that individual efforts to protect the body from toxic chemicals often take the form of *inverted quarantine*, as people respond to external threats by trying to isolate themselves from exposure. Because they take personal action, Szasz suggests, they are less likely to take political or systemic action, making environmental reform less likely. The story of flame retardants shows that many activists recognize this paradox; in addition to taking steps to limit their own exposures, they pursue multifaceted campaigns that target corporations, educate consumers, and push regulators in a broad effort to decrease chemical exposure for all.

Critiques of Advocacy Science

Like other stakeholders, activists of all types engage in SST when they choose how to present scientific findings for their audiences. This varies from selective SST, such as highlighting one finding over another when summarizing a paper, to more significant examples of interpretive or inaccurate SST, such as not acknowledging toxicological research that finds no negative health effects. The use of science by those advocating for greater flame retardant regulation is critiqued by some, who argue that activists sometimes misrepresent scientific evidence. In the words of a supply chain consultant, "The NGOs are great at raising alarms, but they're not great at taking the science and actually using it properly." Industry supporters commonly describe "the public in general and environmentalists in particular as lay people who are ignorant of scientific facts and relationships," thereby excluding them from policy-making processes.[83] I found that these critiques of advocates' use of science came not only from industry (though they were most pronounced there), but also from some regulators and scientists.

Sometimes these critiques reify the boundary between science and policy by asserting that political activity and science are incompatible. One public health scientist distinguished his use of science from that of the activists: "It's a political battle that they are fighting, and in a political battle, they are not really clinging tightly through the science. I mean, they are not interested in science in the same way that I am." Others argued that activist representations are too simplified, focus on the wrong issues, or are overly emotional. A scientist with the National Institute of Environmental Health Sciences said that advocacy messaging could be too "black and white," especially when pieces written by scientists on staff were rewritten by press or outreach offices within organizations. Others accused activists of overreacting and relying too much on emotional arguments. A chemical industry scientist said, "When you see someone holding a baby and complaining about chemicals and stuff, it certainly tugs at the emotional heartstrings of people. You know, even though the science doesn't really support their opinion . . . it's a hard thing to fight." In these instances, advocates were critiqued for oversimplifying a complex scientific story or turning a scientific discussion into an emotional—and thus nonscientific—argument.

Others, however, accused activists of blatantly misusing the science and engaging in inaccurate SST. A flame retardant industry representative said that activists "have got an axe to grind" and that they say things that are "completely wrong." Some regulators expressed similar concerns.

A state environmental regulator said that she gets frustrated with activists: "They collaborate with people whose science I don't respect. They can be very shoddy with their science and they can interpret things and say things that are pretty inappropriate." Similarly, an EPA scientist said some activists are "very dangerous" because "they find a cause, they run with it, and . . . they can't be swayed." Some observers specifically critiqued flame retardant advocates for not understanding fire science or chemical substitution options.

Critique even comes from close scientific allies. One scientist who regularly collaborates with activists said that he does not trust them completely: "I don't fully trust advocates, because I think they're looking for information that supports their point of view, rather than trying to be objective about it." Another scientist who does consulting work for activists said he had seen "activists misuse data while claiming that the industry is misusing data, so I've seen them step too far on the science too." But he saw this as part of what he called the "activist role": "They're pushing the agenda to get it out there. It's just they're free to talk about things maybe a little more subjectively." This represents a difference in opinion by scientists about whether this SST is appropriate, not just expected, from activists.

While advocates regularly use science to convince constituencies that a given issue demands action, relying on science can also be limiting. First, advocates must confront scientific uncertainties and limitations related to the slow pace of science, the need to build on prior methods development and research projects, and the inevitable uncertainty when researchers attempt to document all steps of the exposure-to-disease pathway. Analytical methods must be developed before a chemical can be measured in the environment or in people, thus limiting the research questions that scientists can investigate. NGOs with tight budgets are even more affected by this limitation. Unless they have a scientific ally who will conduct the testing at low or no cost, they can only test for chemicals that have reasonably inexpensive analytical methods.

Activists also recognize the uncertainty of environmental health science and the limitations of a scientific method that demands high levels of proof. One activist said, "We're playing a massive game of catch up, and all of us are really serving as guinea pigs, waiting for the effects of these chemicals." I asked another organizer what she found most discouraging about her work, and she replied, "The limitations of science . . . You can never get definitive proof, and explaining that to non-scientists is very difficult." Additionally, activists note that relying solely on science to make an argument forces them to debate industry interpretations of the science, and, as one policy expert argued, "It's not where the NGOs should be. All of this Deca[BDE] stuff and Deca research and, you know, NGOs having to learn about Deca-debromination, getting into debates about it, that's exactly what the chemical industry wants." These science debates are very hard for activists to win. Furthermore, their status as nonexperts makes it difficult to cross the "expertise barrier" to fully participate in scientific debates.[84]

Advocates also acknowledged that, while many of their decisions were based on science, other times they made political decisions that had little or no scientific grounding. For example, an environmental health advocate reluctantly admitted that the decision to exempt transportation uses from a flame retardant ban was "a political call. If we felt like we could win, we would have done that, but we felt like the evidence was strong enough to say—you know, people don't want to think—see, I don't want to answer this question. It was a political call." Thus, even when advocates present their work as being scientifically motivated, many decisions come down to trade-offs, negotiations, and the political opportunity structure.

Advocacy regularly requires simplifying complex scientific evidence for nonscientific audiences. An antitoxics organizer said at a planning meeting that the names of chemicals are alienating to the lay public, so her organization likes to call them "toxic chemicals" without using their scientific names. Many activists argue that in order to reach the lay public, which lacks a nuanced understanding of science and chemistry, some simplification and lack of precision is inevitable. However, others critique this perspective. A regulatory scientist who had worked for several years at a national environmental health NGO described his frustrations with the lack of scientific specificity by some advocates:

> The scientists will do this fantastic job and towards the end . . . it would be rewritten by the senior management people, and you'd look at it and

you'd go, "that's not what I said. It's not really quite true. You're over-stating X, Y and Z." . . . I understand your purpose. You've got to say something dramatic to get it on the news. But, I really think it's strong enough as it is . . . You don't have to overstate your findings in order to have a poignant point.

He described this simplification as building "scaffolding" around your data: how high of a scaffolding are you willing to build, and how far off the scaffolding are you willing to lean?

MAKING SENSE OF PRECAUTIONARY AND CORPORATE ACTIVISM

For stakeholders involved in the broad field of contestation around flame retardants, these chemicals are seen as a case study for toxics reform more generally and have been crucial in driving national efforts aimed at federal chemicals regulation reform. The clashes between precautionary environmental activism and corporate advocacy in the media, activist conversations, and the public sphere of environmental policy making can be understood more fully by examining the conceptual risk formulas preferred by each group of stakeholders and the various SST forms in which they engage. While corporate advocates tend to echo the classic or exposure-centric risk formulas developed and favored by the chemical industry, the environmental advocacy community invokes the either-or risk formula. Most environmental health activists agree that chemicals that are detected in cord blood or breast milk should be restricted, as should chemicals that are found to be carcinogens or reproductive toxi-cants. There is no need to wait for definitive proof, and it is better to err on the side of safety. That is, decisions made regarding environmental health risks should utilize the precautionary principle.

For environmental health activists working to reform how chemicals are regulated, the flame retardant story demonstrates the benefits of the precautionary principle. When PentaBDE and OctaBDE were phased out in 2004, there was no definitive proof that they were harmful to humans, yet stakeholders were drawn to the chemicals because they were an example of toxic trespass: high levels were found in breast milk, house dust, and people of all ages. That exposure was enough for environmental health activists to take a stance, and combined with evidence that they were bioaccumulative and toxic to laboratory animals, industry and regulators agreed. Advocates applaud this as an example of the benefits of precautionary action (even if they disagree about how precautionary the removal of PentaBDE from commerce actually was) because evidence continues to accumulate that PentaBDE does in fact represent a significant risk to human populations at current exposure levels.

Advocates' work on flame retardants also highlights the importance of multilevel policy engagement. Campaigns against flame retardants have included education of the public, direct lobbying of decision makers, challenges to industry and state-level standards, supply chain engagement, and legislation at the state, federal, and even international levels. Science plays a pivotal, though not exclusive, role at each of these levels. Statistics about rates of cancer among firefighters reinforce the identity-based claims that firefighters make when they testify in uniform at the statehouse. The detection of chemicals in breast milk strengthens a mother's moral argument that her child should not be exposed to chemicals during gestation. And the support of retailers and product manufacturers contributes to economic claims that decreasing the use of flame retardants in some products will spur technological innovation.

A key feature of the flame retardant coalition has been collaborations with the fire service. At least in the short term, this alliance is expected to continue, as environmental activists move from banning PBDE flame retardants to other types of toxic flame retardants. The environmental community hopes to bolster this relationship by supporting the fire service more broadly, as the fire service faces budget cuts and is attacked for its pensions. Ongoing research projects documenting firefighters' exposure to chemicals may also strengthen linkages with the environmental health community.[85] More generally, some are optimistic about future alliances between firefighters and environmentalists. As a firefighter remarked,

> How un-green is a fire? . . . Where do you get the building materials
> from? You ship them from too far away, the transportation cost is

prohibitive. And then we want you to use organic materials . . . So at the end of the day . . . if the building burns . . . Have you seen all the smoke and all the crap that goes up into the sky? It's the most disgusting thing you could ever imagine. And so I think fires are very un-green.

This suggests that perhaps the environmental community could support fire prevention more broadly, because fires are polluting and wasteful.

However, the ability of this coalition to endure may depend on whether both sides can rally around a shared issue. For example, environmentalists may not feel particularly excited about a campaign to require sprinkler systems through local fire codes, one of the key priorities firefighters identify in reducing home fires, and firefighters are unlikely to see TSCA reform as their top policy priority. The flame retardant coalition has been cultivated and led by environmental activists, but firefighters may need to be the main drivers in future work on fire safety more generally, with environmentalists playing a supportive role. Environmentalists also cultivate other blue-green alliances with the medical community, transportation unions, and the aviation industry, occupational groups with higher than normal exposures to flame retardants because their work spaces—hospitals, buses, and airplanes—are governed by strict flammability standards.

The strength of the flame retardant industry lobby is unlikely to diminish. Although CfFS is no longer active, NAFRA now plays an active role in flame retardant advocacy. Corporate and traditional social movement activism may be characterized by the process of *asymmetrical convergence* that has been found between industry and academic science, in which industry research acquires some of the benefits of academia while academia is increasingly subject to corporate and market-based pressures.[86]

Both corporate and environmental activists engage in SST, selectively using science or presenting a simplified version of a complex story. These inconsistencies and different definitions of risk shed additional light on why these competing stakeholders are usually unable to see eye to eye. Indeed, both "extremes" in the field of stakeholders—the flame retardant industry on one side and environmental health activists on the other—use the same language when they accuse the other side of ignoring rigorous science and saying whatever they want without consequence. For example, a flame retardant industry scientist told me that NGOs "can go and say anything they want, almost, and people are willing to listen to it. But, and we will not have the same thing to go and say, 'I know everything about this chemical.' We wouldn't say that . . . if we don't have the science. Whereas . . . the NGOs can easily go and say, 'I scratched this, I found PBDEs.'" This same perspective was shared by activists on the other side of the debate. According to an environmental health organizer,

> One of the biggest challenges that we face is that we have to be perfect. We can never be wrong. We can certainly never lie. We have to be infallible and bullet-proof and unassailable all the time. Whereas the chemical industry and their trade associations can be ruthless and, you know, lack scruples or just be reprehensible, really, in their behavior, say anything they want without, they can just, they have an incredibly unfair advantage in that respect. They don't have to be accurate. They don't have to be truthful. They don't have to have proof.

Thus, both sets of actors engage in interpretive SST by claiming that their side is held to the more rigorous standard of scientific proof and argumentation.

Each side also accuses the other of being self-interested. Industry representatives charge environmentalists with cashing in on flame retardants as a "hot topic" that is fundable and attracts public attention. They also accuse firefighters, legislators, and environmentalists of engaging in a quid-pro-quo exchange of support between flame retardant bans and pro-firefighter legislation. Conversely, environmental and public health activists accuse the industry of prioritizing profits over safety, buying the support of legislators and the fire service, and playing off people's fears of fires and burns.

An undeniable difference, however, exists in resources. The lobbying expenditures of the flame retardant industry outstrip the entire budgets of most environmental health organizations. Revenues for CfFS ranged from $4.3 to $7 million per year from 2008 to 2010, with an official paid staff of one person.[87] Many environmental nonprofits have much smaller coffers. Although some nonprofits may have larger budgets, they also work on many issues. For example, the Environmental Working Group, one of the largest environmental health nonprofits in the United States, had total revenues of $7.4 million in 2013, but this organization works on dozens of issues in addition to flame retardants with a staff of over forty people.[88]

Thus the industry and activist stakeholders find themselves on a highly uneven playing field with unequal access to decision-making spaces and asymmetrical abilities to influence the structures and participants who dominate science policy fields. As an environmental activist described it, at legislative hearings the chemical industry will "have ten people in the room, and we'll have two. They're up there lobbying every day, and we can't be. So there's an issue of just resources, and being able to basically pick your battles . . . We can't be everywhere all the time." When comparing different types of activism, attention to these social and political structures is necessary. As another activist explained, "It's a David and Goliath [story]. Many of us are volunteers, or running on a shoestring. We aren't even certain about our funding sources and what the future holds . . . all those advantages that come with more resources, which equals more power." Although environmental groups also engage in lobbying, theirs is typically of a different scope and scale than that of large corporations.

Finally, a significant tension in contemporary antitoxics advocacy exists between the realistic and the ideal. Sweeping chemical reform is exceedingly difficult to achieve, while the chemical by chemical bans offer a greater chance of success but may be relatively ineffectual. As one activist said, "We can't continue to go chemical by chemical because it would take a lifetime." And yet activists do continue to go chemical by chemical as one strategy of a multiprong campaign to improve environmental and public health.

Part of the problem lies with the types of public interest and environmental campaigns that are fundable. Foundations, donors, and boards of directors demand results, so organizations often prioritize campaigns that are winnable in at most a two-year time frame. In her work on illness-oriented social movements, medical sociologist Rachel Best notes that, though short- and long-term changes are frequently viewed as an either-or trade-off, social movements have multiple—and sometimes

unexpected—outcomes in terms of direct benefits for constituents, sys-
temwide distributive benefits, and systemic and institutional changes.[89]
This is likely the case for environmental health movements as well.

Activists' attention to the political opportunity structure also means
they choose campaigns that hold a greater chance of success in the present
moment. Although sweeping TSCA reform looks more likely today than
it has in the past, it still must pass through a Congress currently hostile
to increased regulation. Any reforms will require support from many
legislators who have taken campaign contributions from the chemical
industry, raising questions about whether legislation with broad and bi-
partisan support will include needed and meaningful revisions that
might not be in the industry's interests. State-based action and restrictions of individual chemicals are seen as more realistic, though they may
achieve little real change in the long term. In response to the success of
this state-by-state patchwork approach to chemicals regulation, current
TSCA reform efforts supported by the chemical industry contain significant preemption provisions that would prevent individual states
from taking regulatory actions on chemicals that had been evaluated by
the EPA.

Advocates' participation in scientific discussions and the production
of science improves the quality and relevance of that scientific knowledge.
Proponents of community-based participatory research have long argued
that those most affected by an issue deserve a seat at the research table,
and this case shows that activist communities also have much to contribute.[90] As an environmental health researcher who regularly conducts
community-based research said, "Activists define important questions
that might not otherwise be asked." The importance of science in environmental activism is unlikely to diminish anytime soon.

CONCLUSION

The Pursuit of Chemical Justice

I n November 2011, I spent a morning with James, an engineer by training who now manages several programs within the EPA's Office of Pollution Prevention and Toxics. The phrase "Sound Science" was written
on his dry erase board, but "Sound" was crossed out and replaced with
"Best Available." I asked him why "Best Available Science" and not "Sound
Science" was a guide for his work. He said that sound science is code
language for "doing science *ad nauseam*." It is like risk assessment, he
explained: you just keep doing research, rather than acting on early information about risks and danger. Instead, he wanted to rely on the best
available science in his work for the EPA.

Toward the end of the interview, I asked James whether he was concerned about his exposure to common chemicals like flame retardants.
He said that he was: "I think that the people who say . . . 'it's an experiment
on human subjects,' it's true. That's really what it is, because we don't
understand the impacts of these chemicals." And, he added, he "abso-
lutely" makes changes to reduce his exposure: "I live my life to avoid, to
minimize exposure," he told me. "It's probably a more significant threat
than global warming to . . . the future of humans." The complex scientific
uncertainties associated with chemical exposure make science-based regulation of chemicals extremely difficult and contested, but the consequences of not taking regulatory action are potentially severe.

This state of affairs has translated into constrained regulatory efforts
by the EPA and state governments, public and secretive debates over how
chemicals should be regulated, and strategic translations of the scientific
evidence that allow interested parties to pursue decidedly nonscientific

aims. Relying on the best available science to achieve greater chemical justice is a worthwhile goal almost universally supported in principle. In practice, however, the malleable nature of scientific evidence, the incomplete scientific story promulgated by individual stakeholders, and the incentives to strategically present a certain cut of the science in the most flattering way to one's own cause all combine to make science-based decision making hotly contested.

In many ways, the idea that science is politicized is far from surprising. Many scholars have illuminated aspects of the politicization of science, the socially constructed nature of scientific facts, and the manipulation of regulatory processes. Yet they have rarely investigated the full range of stakeholders involved in environmental controversies; as a result, they have mostly developed concepts that describe how one group of stakeholders engages with or contributes to scientific uncertainty rather than overarching theories that more fully explain the role of risk and uncertainty in environmental debates. In contrast, my methodological focus was on the full range of stakeholders involved in a single environmental controversy, including industry representatives and environmental activists who are often described as being at opposite ends of the spectrum. Instead of taking the contested nature of science in policy-relevant fields for granted or assuming that it is limited to industry actors, I have shown that acts of conceptual risk definition and strategic scientific translation (SST) are themselves worthy objects of empirical investigation and analysis, and that the relations of power that underpin these strategic acts can offer useful analytical leverage for understanding the uneven topography of environmental regulation we have today.

In the case of flame retardant chemicals, controversies over environmental risk are framed by most stakeholders as debates over science, even as those decisions also are influenced by topics as varied as market preservation, institutional autonomy, expertise qualifications, and the sanctity of breast milk. Flame retardants have attracted a remarkable amount of regulatory, scientific, and activist attention in a relatively short period of time. Though ten years ago flame retardants were unknown or obscure to many exposure scientists, chemical regulators, and environmental health activists, today they are the subject of foundational scientific papers, award-winning investigative journalism series, and mainstream documentaries. Thousands of environmental and public health activists have signed petitions or written letters to policy makers about flame retardant toxicity. The scientific literature has grown exponentially, branching out from a handful of legacy chemicals to dozens of emerging contaminants, and from laboratory-based chemistry to epidemiological studies involving hundreds of research participants. The growth in public awareness has truly been remarkable, and it has been informed by a unique combination of activist, regulatory, scientific, and industry attention.

Sociological studies of risk often emphasize the social construction or tangibility of environmental risks, and they identify analytical or scholarly definitions of risk and how risk should be assessed. When these scholars do focus on individuals' risk perceptions, they typically concentrate on individual level perception of specific risks, ignoring how stakeholders conceive of risk as a broad topic. In chapter 3, I demonstrated that the very concept of risk is also the subject of competing social definitions. Before actors can disagree over whether flame retardants pose a risk to human health or the environment, they have to identify how risk should be calculated and what goes into the *conceptual risk formula*. These formulas function as rules for how chemical safety should be assessed.[1] Risk, therefore, is a concept whose definition travels, changes, and operates within and across institutions.[2]

These definitions of risk and understandings of what levels of risk are acceptable vary greatly across fields and between groups of stakeholders. Following the either-or risk formula, many environmental activists consider a chemical to pose an unacceptable risk if it meets certain toxicity or exposure concerns: if it is detected in breast milk or cord blood, if it is an endocrine disrupting chemical, or if it is found to cause cancer or reproductive problems. By establishing what can count as a risk, the risk formula motivates these activists' activities: detecting known hazardous chemicals in tissue samples is a common strategy of environmental health organizations, who recruit participants and send samples to accredited laboratories. In contrast, the flame retardant industry's more stringent idea of what is unacceptable follows the exposure-centric risk formula, requiring significant exposure evidence before a chemical can be considered a risk. This leads to a very different set of chemical evaluation tools that emphasize reducing point-source emissions and designing chemicals with low predicted exposure potential. These different definitions of risk are strategic and allow institutions to use the concept of risk to pursue dramatically different goals.

The malleable nature of scientific findings, the inherent uncertainty in environmental health research, and the widely different options for risk evaluation mean that competing flame retardant stakeholders can strategically use science in order to achieve their goals. In chapter 4, I described SST, the process of interpreting and communicating scientific evidence to an intended audience to advance certain goals and interests. Because all groups of stakeholders engage in SST in contested environmental policy fields, their actions in those fields and the resulting policy outcomes often reduce not to questions of scientific truth but to the resources deployed in support of their preferred scientific translation. The selective, interpretive, and inaccurate forms of SST can vary across institutions, and therefore looking at all types of scientific manipulation as SST allows for comparative analysis of different manipulations of the scientific corpus.

Furthermore, the production of scientific claims and documents involves social acts of negotiation. Chapter 5 described several sites of scientific negotiations at the EPA and showed how supposedly scientific decisions are heavily influenced by social, political, and economic considerations. Because all stakeholders are attempting to influence environmental policy processes in their favor, final actions and decisions can be evaluated not only based on the strength of the scientific evidence at hand but the various exercises of power behind competing stakeholders' claims.

These debates are not confined to the pages of scientific journals or the conference rooms at EPA headquarters. In chapter 6, I described heated debates over chemical risk and safety that have played out in the public sphere. From the pages of national newspapers to the halls of Congress, a broad coalition of environmental and health activists, public interest nonprofit organizations, and fire safety professionals has campaigned to restrict certain flame retardants, achieving significant and in many ways unexpected levels of success. In contrast, a small number of well-funded industry activists have fought these restrictions, acting as the public face of the chemical industry and working to institutionalize a very different definition and assessment of chemical risk.

Interestingly, environmental advocates and scientists on the one hand, and industry advocates and scientists on the other, are caught in similar scientific binds. Both would like to definitively link flame retardants to complex public health outcomes. Environmental researchers, regulators, and advocates struggle to link chemical exposures to specific health outcomes, while the flame retardant industry struggles to show that flame retardants offer protection in real-life fire scenarios. Both of these complex public health puzzles require an extraordinarily high bar of evidence. This is widely discussed with epidemiology; complex exposure pathways, synergistic and cumulative exposures, and a long lag-time from exposure to illness make searching for chemical hazard akin to finding the metaphorical needle in the haystack. Similar data problems plague fire scientists, and fire data are considered unreliable by many.[3] Environmental health activists, fire scientists, regulators, and chemical industry representatives alike point to the difficulties of teasing out the factors that contribute to declines in fires since the 1970s. A public health researcher described the puzzle this way:

> There has been a drop in fire deaths in the U.S. over the last twenty years that the brominated flame retardant industry claims is due to their project. But in fact, other things have been going on as well, like reduced smoking rates, sprinklers now in buildings, and fire alarms. And so, you know, it's actually a very interesting epidemiologic question that might not be very easy to answer.

Just as the flame retardant industry points to the uncertainty in environmental health epidemiological research, environmental health activists point to the uncertainty in fire-related epidemiology to argue that flame retardants have not contributed significantly to fire safety. In fact, in the past decade the epidemiological evidence documenting flame retardants' harms has accumulated, but there has been no parallel increase in peer-reviewed studies identifying the fire safety contributions of flame retardants at the population level. Several of the older studies long held up by industry advocates as evidence that the chemicals save lives have been methodologically critiqued by fire scientists and journalists.[4]

Flame retardants are used for the decidedly important purpose of reducing damage, injuries, and deaths from fires in our highly flammable modern world. The best fire protection measures are those that reflect real-life fire scenarios, but not all fire standards do this. Flame retardants can be very effective in the laboratory: a piece of flame-retarded foam exposed to constant flame will not ignite as quickly as a non–flame-retarded foam under identical conditions. However, this does not necessarily translate into effectiveness in real-life fire scenarios, which rarely involve exposed foam and propane torches. Significant debate exists around the appropriate way to regulate the flammability of consumer products, especially furniture. California's recent revisions to Technical Bulletin 117 represent an important step in this direction. Based on fire statistics, expert recommendations, and new research from the Consumer Products Safety Commission, California updated the state's flammability standard for upholstered furniture so that it now protects against smoldering materials such as cigarettes, the most common furniture fire scenario.[5] This standard is spreading to other locations around the country, and major companies have publicly committed to manufacturing and purchasing furniture that conforms to the new standard without the use of flame retardant chemicals.[6] A shopping trip for a couch today offers many more flame retardant-free options than only a year ago, from inexpensive furnishings to top-of-the-line products. This represents a win for fire safety and environmental health alike.

THE NEED FOR SCIENCE, AND THE UNCERTAINTY OF SCIENCE

Any health outcome involves a complex interplay of social, environmental, and individual level factors. People are exposed to hundreds of chemicals on a daily basis. These exposures interact with their individual genetics and embodied experiences, and are filtered through their social and material worlds to impact health in myriad ways. The associated scientific uncertainty matters greatly for risk-based decision making. As a former flame retardant industry executive told me, "if I knew everything about everything about every chemical, I would agree" with the classic risk formula that risk is a function of hazard and exposure.

But perfect knowledge is nonexistent and should not be seen as a necessary precondition for action. When there is much at stake—for markets,
regulatory capacity, or public health—institutions will use this uncertainty
to their advantage. However, they have different resources at their disposal.
The chemical industry has vastly superior economic resources compared
with public health advocates, and it uses these resources to guarantee
access to decision makers and to fund certain types of science. In contrast,
the activists often have greater "public sentiment" resources, and they cultivate these by seeking alliances with respected constituencies such as
breastfeeding mothers. In the words of a state-level regulator, "Who is
going to be on the side of chemical industries over children?"

Science is incomplete and fallible, requiring decision makers to choose
between multiple trade-offs and sets of potential consequences. Currently, economic trade-offs are implicitly privileged in regulatory processes because they are easily quantified and documented, while public
health and environmental trade-offs are often difficult to identify, measure, and convert into economic values.[7] Trade-offs would exist even under the (unattainable) condition of perfect scientific knowledge. An EPA
scientist explained, "Let's say we have perfect scientific knowledge. We
know the number of deaths from fire, the number of illnesses from the
flame retardants, or the number of deaths prevented . . . Deciding which
of those is better is not a scientific question." Thus, decision making needs
to both incorporate imperfect knowledge and explicitly identify social,
economic, and political priorities that influence how that knowledge is
turned into policy. Delaying policy decisions until more complete information is available about a current-use chemical will almost always benefit industries at the expense of public health. As environmental health
advocate Richard Denison recently testified at a workshop on weight of
evidence analysis, "Under our system where a pending assessment means
no action can be taken, all of the rewards of delay fall to one side—the
(un)regulated industry—and all of the risks fall on the public."[8] This is a
valuable status quo for the chemical industry, so it is not surprising that
they spend so many millions of dollars on lobbying and campaign efforts
every year.

The flame retardant industry has defended some products after other
stakeholders have said the proof of risk is sufficient. For example, manu-

facturers defended tris(1,3-dichloro-2-propyl)-phosphate (TDCPP) when California listed it on Proposition 65, in spite of research showing that the chemical caused multiple forms of tumors in rats.[9] However, companies are also sensitive to the dynamic nature of chemicals regulation, market pulls driven by consumer pressure, and liability concerns should they continue to manufacture hazardous products. For this reason, companies can be expected to respond to scientific evidence that their products pose a risk, although they will require a higher burden of scientific proof (in combination with other market and regulatory drivers) than other stakeholders. As a FlameCorp representative told me, the company will adjust their position accordingly based on accumulated scientific evidence: "At some point, if the science is overwhelming . . . what value is it to us to keep telling everyone it's not?"

Supply chain pressure has been important in the movement from scientific evidence to advocacy concern to industry action. Activists working to regulate flame retardants have continued a social movement trend of pressuring nonstate actors, targeting high-profile companies. In interviews, these activists pointed to notable "successes" with flame retardants, such as getting Graco to commit to selling TDCPP-free products, and with other chemicals, such as forcing baby bottle manufacturers to stop using the plasticizer bisphenol-A (BPA). This has worked well with high-profile companies and industries, especially in the baby products sector, but activists have debated targeting other industries. Thus far, for example, activists have mostly favored collaborative, not confrontational, work with the furniture industry.

Of course, market campaigns cannot create full conditions of environmental equity. Targeting high-profile companies will not lead to full market coverage, and bargain or store-brand products may be less likely to reformulate. As a result, consumers will benefit unequally, and exposure inequalities may be exacerbated, adding to the already significant environmental injustices that exist along class, racial, ethnicity, and geographic lines.[10] But activism against corporations is seen as part of a multiprong strategy, combining scientific, market, and regulatory pressures to build public awareness and change corporate behavior.

These questions are not limited to the chemicals arena. The strategic use of science and definition of risk can be seen in institutional contexts ranging from homeland security funding to climate change research, and more research is needed in these areas.[11] Questions of scientific uncertainty and manipulation are widespread in other pressing environmental policy arenas, especially regarding global climate change. In parallel to the climate change denial movement documented by sociologists and public observers, a pattern of environmental health denial is widespread in the United States today.[12] Just as climate change denial is organized by conservative think tanks and the fossil fuel industry, environmental health denial may be organized by a different (though potentially overlapping) set of organizations and corporations. In both cases, industry stakeholders present the science as balanced when in fact the overwhelming evidence suggests cause for concern, they malign individual scientists, and they claim that academic research is biased because of its reliance on government funding.

ENVIRONMENTAL HEALTH FUTURES

You don't put a bicycle helmet on your kid only when you know it is the day he is going to fall and get in a car accident. You don't put your seatbelt on only on the day you know you're going to crash . . . We do the best we can to protect health and safety with the information we have all the time. The only place where regulatorily we're not allowed to do that is with chemicals.

—Environmental Regulator

The future of environmental regulation will largely come down to questions of risk. Whether the issue is chemical exposure, hydraulic fracturing, or global climate change, will we continue to use a reactionary approach that allows us to proceed with business as usual until there is evidence of harm, or will we switch to a more precautionary approach that lets us act more confidently to decrease hazards, even in the absence of scientific proof? The science around these emerging environmental issues will rarely be complete and clear cut enough to allow for easy decision making until negative consequences are upon us, and

as the case of climate change clearly shows, policy disagreements remain paramount even in cases of remarkable scientific consensus.

Of particular concern for chemicals like flame retardants are the
health impacts of low-level exposure to chemicals used ubiquitously in
consumer products because exposure is long-term, typically far below the
levels examined in animal studies, and difficult or impossible to avoid.
But we need not wait for dramatic, catastrophic, place-specific instances
of contamination or incontrovertible proof accumulated over decades before decisive action can be taken to prevent further harm. The case of
polybrominated diphenyl ethers (PBDEs) in particular shows the ability
of regulatory and industry actors to take meaningful action based on a
strong but by no means complete or conclusive body of evidence. By the
time action was taken on pentabromodiphenyl ether (PentaBDE) and
octabromodiphenyl ether (OctaBDE), scientists knew that the chemicals
were persistent and bioaccumulative, with ubiquitous population-level
exposure and documented toxicity in animal studies, even if findings
from human studies were not yet available. A combination of data
points—on the chemicals' physical-chemical properties, exposure to
humans, and toxicity to animals—was enough for governments and
the chemical manufacturers to take action. In this case, "precautionary"
regulation meant that evidence of human health effects was not yet conclusive, though the chemicals were still used in products long enough
that we continue to identify and observe the health effects today. This is
far from the *strong program* of precautionary action feared by many
critics of environmental regulation which would demand absolute
proof of safety, a level of proof as difficult to obtain as absolute proof of
harm.[13]

Basing regulations on the best-available science requires flexibility because science is always incomplete and uncertain. Chemical hazard and
exposure research on chemicals of concern is often cutting edge and not
yet replicated or validated. Epidemiological data in particular takes years
or even decades to emerge. Though the case of PBDEs in many ways represents precautionary action to protect public health, the consequences
of three decades of PBDE use are significant and will be with us as long
as furniture, electronics, and recycled foam carpet padding containing
PBDEs remain in widespread use.

In short, waiting for definitive proof of harm is not an option that protects public health. Risk assessment practices need to be flexible enough to respond to and incorporate changes in the scientific corpus, and should not be treated as final, written-in-stone documents to be endlessly relied upon regardless of scientific advances. Decision makers across all institutions must also recognize that decisions can have unexpected and undesirable consequences. For example, the end of PentaBDE's use in furniture foam created a vacuum in the supply chain that has largely been filled with two chemicals, TDCPP and Firemaster 550, which have their own suspected risks.[14] Those who advocated for PentaBDE to be removed—including nongovernmental organizations (NGOs), state and federal environmental regulators, and some in industry—did not intend for one toxic chemical to be replaced with another, but this has been the unintended consequence of the PentaBDE phase out.

The frequency of these incidents of regrettable substitutions could be significantly reduced through the use of comparative chemical hazard assessments to drive regulatory and industry decision making. This approach is gaining traction and attention, expanding beyond its base in the federal government and into academia, the nonprofit sector, state governments, and the chemical industry. A hazard assessment methodology grounded in the hazard-centric risk formula begins with the assumption that reducing hazard is the most reliable way to reduce risk because exposure conditions are rarely well studied or fully understood. This would shift the question from "How much is safe?" to "What is the safest possible option?" Industry could incorporate hazard assessment into their new product development practices and use the methodology to voluntarily withdraw more hazardous products from the marketplace in anticipatory moves that protect public health—and corporate profits—in the long term.

Hazard assessment is equally important for new product development and supply chain decision making. The development of the next generation of fire prevention tools should rely on green chemistry approaches, in which chemicals and industrial processes are designed to eliminate or reduce hazardous substances throughout the life-cycle of the product.[15] Many critics of brominated flame retardants consider this class of chemicals to be prime examples of "lazy chemistry." In the words of a public health researcher, "PBDEs are essentially polybrominated biphenyls with an oxygen added. So it's no longer PBBs, they're not banned, you can make them! Like, what genius came up with that idea? It's insane." Scientists told me that developing better fire safety tools was well within the realm of possibilities, if adequate attention was paid. As a toxicologist said, "We can come up with better ways of, you know, preventing fires, than, like, dousing our mattresses with toxic chemicals. It's just a laziness or something factor." Unlike classic chemistry, which prioritizes performance and cost during chemical development, green chemistry makes hazard and exposure concerns central to chemical design.

In short, hazard should be designed out of the product. In addition to designing safer molecules, manufacturers can think about different ways of achieving fire safety without the use of toxic chemicals. An environmental health advocate who was a strong supporter of alternatives assessment presented several options: "[With] TVs, they can isolate the heat source so that the enclosures don't need the flame retardants . . . [With] carpets or wall coverings and things, you use inherently flame resistant materials . . . so you don't need the chemical." This may not be an option for all uses. As that same advocate acknowledged, "Let's say there's wire cable, and . . . you can't find a material that has that inherent flame resistance. Then you need to add flame retardant to it," and you seek out the most "environmentally friendly flame resistant material" you can find. Thus, achieving fire safety without toxic exposures is likely to require advances in green chemistry, creative thinking by product manufacturers, and comparative chemical hazard assessments of possible alternative chemicals. A similar perspective emphasizes functional substitution, which involves identifying a broad range of alternatives that perform the same function at the chemical, product, or service level, rather than a chemical-by-chemical assessment.[16] In the case of flame retardants, this might mean switching to inherently fire-resistant materials and removing the flame retardant all together.

Biomonitoring research has motivated regulations and likely will continue to do so, especially when it measures chemicals in vulnerable populations such as fence-line communities or children. Many advocates support a significant role for biomonitoring in motivating and improving policies around potentially toxic chemicals. As leading examples,

California's Environmental Contaminant Biomonitoring Program monitors levels and trends for priority chemicals in a representative sample of Californians, and Maine's Toxics in Children's Products Law uses biomonitoring data as one criterion to identify priority chemicals.[17] Biomonitoring data can also be used to evaluate the effects of public health policies. For example, recent work suggests that PBDE levels in U.S. women may be declining due to legislation, not personal or lifestyle changes.[18] However, these programs are limited by levels of detection and the availability of laboratory methods, and they typically test only for chemicals already of concern. Additionally, some stakeholders have noted that advocacy biomonitoring in particular may lose its "luster" and novelty now that numerous studies have been conducted and publicized.

Though state level regulations and monitoring programs are useful, they are not sufficient on their own. I agree with observers from industry, academia, the NGO world, and the EPA who say that the Toxic Substances Control Act (TSCA) is badly in need of updating. Though it was based on the "best available" science when it was enacted, knowledge of chemistry, toxicology, the exposure pathway, and epidemiology has advanced so significantly in the last thirty-five years that many of the assumptions built into TSCA are no longer accurate. The result is that risk management decisions are too often incomplete, tardy, and insufficiently protective. Conscious of the limitations of TSCA and a sometimes hostile political opportunity structure at the federal level, activists have pursued state-level regulation or pressured the market. These activities have raised the bar on some issues but have taken an exorbitant amount of time and resources to achieve relatively limited change. Changes through the markets or in environmentally progressive states tend to benefit privileged consumers who are able to pay more for safer products and those who live in certain progressive states, raising concerns about the environmental justice effects of these methods of reform.

Overhaul of TSCA is needed to give the EPA greater authority to evaluate the risks of existing chemicals, end the grandfathering of existing chemicals, fortify data requirements to submit new chemicals for review, and evaluate the cumulative risk of chemicals by end point. TSCA reform should shift the burden of proof from the public and EPA to document a risk onto the producer of the chemical, not to prove safety but to demonstrate that chemical products do not pose significant harm to the environment or humans. The age of chemicals being judged as "innocent until proven guilty" should come to an end in the United States, as it has in Europe. Data requirements for evaluating new and existing chemicals should be flexible enough to allow for new forms of data that are less expensive and more rapid than traditional animal toxicology, likely through the use of tiered screening, in vitro studies, and computer modeling. This would reduce the number of animals used in *in vivo* testing and also lessen the economic burden on manufacturers, while allowing evaluations to be made based on the best available science. TSCA reform should pursue harmonization and data sharing with the European Union's Registration, Evaluation, Authorisation, and Restriction of Chemicals (REACH) in many areas, while learning from the European experience and incorporating changes desired by chemical manufacturers, the supply chain, and environmental and health activists alike such as streamlined reporting processes, improved compliance with reporting requirements, improved transparency in evaluation processes, reduced burdens on small enterprises, and support for the supply chain as they transition to safer chemistries.

Whether TSCA reform will be achieved in the near future, however, remains an open question. As this book went to press, a pair of TSCA reform bills were moving through both chambers of Congress. The "Frank L. Lautenberg Chemical Safety for the 21st Century Act" passed out of a Senate committee and was recommended for a full vote in April 2015, and the "TSCA Modernization Act of 2015" was passed nearly unanimously by the House in June 2015. Both bills include meaningful reforms praised by stakeholders on all sides, including revising TSCA's "unreasonable risk" requirement, prioritizing chemicals for evaluation, increasing EPA's ability to require safety testing for chemicals, and requiring greater documentation for confidential business information (CBI) claims. However, some public health activists and industry watchdogs are concerned that the reforms are too limited and potentially dangerous if they preempt states' abilities to enact their own regulations. Despite the difficult prospects of achieving significant reform in today's contentious political climate and the fears that watered-down reforms might prevent more meaningful change down the road, many remain optimistic. In the words of a longtime environmental health activist, "the chemical management system is so outdated and obsolete that everyone recognizes it has to change."

Environmental regulation is too often critiqued for harming economic growth and business innovation, and the chemical industry consistently claims that regulation dampens innovation.[19] Environmental sociologists, however, have shown that the claim that pollution is economically necessary is a "diversion of attention" from pursuing environmental sustainability and that the jobs-versus-environment trade-off is a false dichotomy.[20] Recent evidence suggests that tightening environmental regulations creates incentives for innovation in the chemical manufacturing sector, increasing the number of new chemicals introduced to the market and allowing safer chemicals to overcome barriers to market entry.[21] For example, stricter rules on phthalates in the European Union and the United States in the late 1990s were followed by "exponential growth in the number of patented inventions for phthalate alternatives."[22] Additionally, labeling requirements can help nudge product manufacturers away from unnecessary chemicals and educate consumers about the chemicals in what they buy. California's recent SB 1019, signed into law in September 2014, requires upholstered furniture to contain a label identifying whether it contains added flame retardants.[23]

Beyond TSCA reform, the time is right for other changes at the EPA. First, the agency should demand greater data disclosure from chemical manufacturers and users. Conflict of interest disclosure is standard in scientific publications, but no such disclosures are required for data submitted to the EPA for review. Because "conflict of interest inevitably shapes judgment," data submitted to government agencies for review should include a clear disclosure of who paid for the study and what restrictions came with the funding.[24] Several changes could also improve how the agency deals with CBI: charging a nominal fee for each separate CBI claim, a reasonable requirement because CBI claims involve additional effort and protocols; disallowing CBI claims for basic information, such as chemical structure and health-related studies; limiting the longevity of CBI claims; and establishing penalties for improper CBI requests. These changes would reduce the number and duration of CBI claims, making more data publicly accessible. The EPA should also work to gain access to data submitted to European authorities under REACH. Although many companies submitting data under REACH also operate in or import chemicals into the United States, the EPA and REACH officials have determined that they cannot share relevant data with each other under current law.[25] This blocks the EPA from access to all the "best available" science.

In the absence of data, the EPA sometimes relies on conservative assumptions and models in line with the exposure-proxy risk formula, and officials have suggested they may try to do the same with existing chemicals if "industry partners" do not provide needed data.[26] This reliance on conservative assumptions is appropriate and should be extended to existing chemicals. As EPA representatives repeatedly told me, "exposure controls can, will, and do fail." Once exposure occurs, especially for persistent and bioaccumulative chemicals, it is difficult or impossible to reverse. Additionally, risk assessment should move in the direction of cumulative exposure assessment, taking into account the fact that multiple chemicals share mechanisms of action and can potentially contribute to the same health effect. This is admittedly challenging, but cumulative risk assessment is a goal the agency has acknowledged.[27] The EPA should continue to develop strategies for assessing and regulating existing chemicals in the face of incomplete data. For example, as part of their risk assessment plans for 2013, the agency will use data from four flame retardants to complete assessments on eight others with insufficient data.[28] The agency can evaluate the challenges and successes of these evaluations after they are completed to develop similar programs for other classes of chemicals.

Additionally, improved communication between EPA scientists and risk managers could strengthen risk management decisions by ensuring that the science produced in the agency is policy relevant.[29] Environmental chemists, toxicologists, exposure scientists, and epidemiologists should play a key role in shaping the context of risk assessment and management.[30] According to EPA representatives, communication has apparently improved greatly in the past decade between the agency's Office of Research and Development (ORD) and the Program offices that have regulatory capacity, but gaps still exist. In interviews, ORD scientists shared stories of being told a year ahead of time that a Program office would like a carcinogenicity study (even though such studies typically take two years of laboratory work and another year to analyze the results), or of having their results reanalyzed by hired contractors in search of a specific finding. In a time of tightening agency budgets, data developed at the EPA should be efficiently targeted at filling data gaps to protect public health and the environment.[31]

Many scientists in academia and independent research institutes also consider how their research could fill regulatory data gaps when choosing topics and research questions. Expansion of this type of thinking would allow researchers to more effectively contribute to environmental health improvements. For example, instead of focusing on high-dose research, toxicity studies can be done with chemical doses that are empirically grounded in ongoing regulatory discussions or with doses that are environmentally relevant. Alternately, choosing to study orders of magnitude around an established regulatory reference dose could evaluate whether it truly represents a safe margin of error or whether it is inadequately protective.

The story of flame retardant research, regulation, and activism also demonstrates that scientists can contribute to policy making in other ways, by collaborating with advocates or by making their results accessible to a broad audience, not just their scientific peers. A public health scientist described a collaboration with environmental activists in which they jointly acquired a portable piece of laboratory equipment as "a win-win situation, where we got to use the device for scientific purposes, but we also developed sort of the methodology that the activists can use." Many flame retardant researchers have testified in public hearings or have agreed to be interviewed by journalists for mainstream news stories. Others have provided behind-the-scenes council and scientific review, part of what sociologist Scott Frickel and his collaborators call the "shadow mobilization" of scientist-advocates.[32] These activities are not regularly rewarded in academia; they do not automatically receive credit in one's quest for tenure, and they can even damage a scientist's reputation in some professional circles. Such norms may be changing, albeit slowly, and there are models of recently tenured and senior researchers who actively conduct policy-relevant and collaborative work. These collaborations can often strengthen the scientific process, for example, by providing access to a population of willing volunteers.[33] These collaborations also are of great benefit to activists, who use these scientific connections to develop advocacy arguments.

Additionally, through these collaborations, scientists can share their knowledge of the scientific process with nonscientists. As researchers themselves know all too well, scientific findings can be later proven to be wrong or findings can contradict each other. What at one time is assumed to be the state of the science can be shown later to be incorrect, and the consequences or outcomes of science-based thinking can be unintended and unpredictable. Acknowledgment of this is apparent in scientists' frequent admonitions that more research is needed, and a key role of the scientific community should be to educate the rest of us about the limitations and strengths of the scientific process.

Changes are also needed in the chemical industry generally and the flame retardant industry specifically. This case study has shown that aggressive product defense and lobbying activities harm the reputation of the industry's research and increase skepticism toward their science-based arguments. Industry's use of the science often involves inaccurate SST and can be transparently asymmetrical. For example, industry representatives frequently praise animal toxicology research when it shows no link between exposure and effect but tend to employ sweeping dismissals and hypercritical evaluations when such research finds evidence of toxicity. This asymmetry is visible to nonindustry stakeholders, and it inspires skepticism about how industry deals with science generally. While the authority of individual scientists has declined generally since World War II as a result of their engagement in social movements and political debates, the reputation of industry science is even less favorable, tarnished by cases of outright fraud and documented financial biases and conflicts of interest.[34] Although these past events and patterns cannot be undone, if the chemical industry expects stakeholders to respect industry-funded research in the future, it would do well to start respecting—and not attacking—quality research that is conducted elsewhere.

Product defense patterns are not universal across companies, of course. As I was told during my visit at FlameCorp by a senior scientist, "The [FlameCorp] attitude is, as you've probably heard . . . a hundred times, is to engage, not defend . . . You have to show data where you have the right data to address the issues, and if your data is sound, that should be enough . . . But again, being more proactive, to me, is the right way to do things. You know, we're not the bad guys." Some stakeholders may be skeptical of this industry engagement, but the trends toward corporate engagement are evident and often productive, and many environmental activists want better strategies to engage with willing industry partners.

Flame retardant manufacturers could improve engagement and contribute to the scientific evaluation of their chemicals by doing more exposure testing. The chemical industry favors the exposure-centric risk formula when it comes to evaluating chemical risk, but it develops (or at least publicizes) surprisingly little exposure data. Only one-fifth of the chemicals inventoried in U.S. commerce have any exposure data, and the computerized models that estimate exposure are imperfect at best.[35] An EPA scientist suggested that flame retardant companies should test for their newly developed flame retardants in environments near manufacturing plants and in biological samples from their manufacturing workers. These data would show whether the chemicals are persisting and bioaccumulating, and thus would either support or disprove industry claims that newly developed chemicals have a preferable environmental profile.

Additionally, the data produced by the industry should be shared publicly. Many studies conducted by industry are never published, and the results are shared with the EPA only in summary form. Most data that are submitted to the EPA are claimed as CBI and thus are not publicly available. Industry representatives regularly argue that academic and government data should be publicly available for reanalysis, and they frequently use the Freedom of Information Act (FOIA) to get raw data from federally funded studies for their own reanalysis. But industry data are not subject to FOIA, and I am aware of no instances in which a company has voluntarily shared its own raw data with an academic scientist for reanalysis. Environmental policy experts Wendy Wagner and David Michaels call this "bimodal" oversight: "Public research is subject to increased scrutiny, while private research remains largely insulated from outside review and meaningful agency oversight."[36] Correcting this double standard would allow for independent verification of industry findings, potentially increasing the legitimacy of industry-funded research.

The advocacy world would be well served to acknowledge the nuances of other institutions, including the nuances of science. For example, one activist said that the names of chemicals are alienating to nonscientists—tris(1,3-dichloro-2-propyl)-phosphate is quite a mouthful—and suggested calling them "toxic chemicals" without using scientific names. This communication strategy may be appealing to some segments of the public, but it will be alienating to scientists and many regulators, and it will give industry ammunition in asserting that the activists are uninformed. Additionally, technical details in the flame retardant story are important. Though activists regularly refer to "Tris" flame retardants, there are actually several different Tris flame retardants, and in the field of chemistry Tris can also refer to "three of something" or to entirely different chemical products. Though verbal slipperiness when referring to multiple Tris flame retardants may serve the advocacy goals of broadening regulation and public awareness of flame retardants, imprecision in activist language allows other stakeholders to accuse the activists of scientific impropriety.

Additionally, technical details in the flame retardant story are important. For example, the application- or product-specific nature of flame retardants has been the cause of confusion and conflict between flame retardant or product manufacturers aware of the specificity of individual flame retardants and environmental activists who hold up the huge number of existing flame retardants as a reason to restrict individual chemicals. While many activists recognize this, some imprecision remains. This creates tensions especially in the spaces where environmental activists and technical industry representatives come together, such as the EPA's Design for the Environment program or in standards and code development processes. Thus far, activist participation in these processes has been limited, likely due to a combination of limited technical expertise, lack of knowledge about which standards are most relevant and how to participate, few incentives to participate in terms of funding or public awareness, and limited time and resources. Though activists have achieved notable victories with California's Technical Bulletin 117 and the international candle standard for electronics, there are hundreds of codes and standards relevant to flame retardants. It is unrealistic to suggest that activists involve themselves in every code and standard process, but a seat at the table for nonindustry representatives might encourage these governing bodies to incorporate more health or environmental concerns alongside of the performance characteristics that currently dominate the processes.

Finally, the blue-green coalition between environmental health activists and firefighters shows significant promise. The first of its kind between the fire safety and environmental sectors, this alliance has already played a key role in state regulations and is poised to play a similarly important role if federal TSCA reform moves forward. Other alliances with the health care, transportation, and aviation industries may also increase the relevance and reach of environmental health campaigns, especially if they can extend beyond the single-chemical focus of flame retardant activism.

The long-term risks from ubiquitous chemical exposure are largely unknown, but toxicology and epidemiology research increasingly shows connections with various biomarkers, conditions, and diseases. Even in the absence of overwhelming environmental causation, the growing focus on the issue demands more responsive decision-making tools. Understanding these issues requires close attention to how stakeholders develop risk definitions and use uncertain science for desired policy goals. As the story of flame retardants demonstrates, making policy based on science will inevitably include political, economic, and social considerations because science will always be interpreted through political, economic, and social filters. Controversy is unlikely to subside as long as chemicals represent a multi-billion-dollar industry on the one hand, and an uncertain threat to public and environmental health on the other. Recognizing that these controversies involve debates over the future of our environment and health, not just the state of the science, will allow institutions to clearly outline their goals, expectations, and assumptions, improving communication between stakeholders and leading to strengthened public health and environmental regulation.

PLAYING THE FIELD: METHODOLOGICAL REFLECTIONS

In 2012, I attended the 13th International Workshop on Brominated and Other Flame Retardants held in Winnipeg, Canada, with over a hundred scientists and environmental policy makers, plus a handful of environmental activists, industry representatives, and journalists.[1] By this time, I had been studying the flame retardant world for over three years, and I had met and interviewed many of the people in attendance. I was familiar both with the general areas of research and with the more subtle references to behind-the-scenes relationships or legends about research practices.

At a cocktail hour the first evening, the crowd examined posters on flame retardant research and networked outside a large auditorium at the host university. I got a beer and said hello to a researcher I had interviewed a while back and seen several times since then. He gave me a hug and asked for my thoughts on the conference: what was it like being here "as an outsider? I don't want to say that you're unbiased, because everyone has their bias. But an outsider?" I responded by talking about how nice it was to have so many perspectives come together. He agreed, and expressed his enthusiasm for a panel to come the next day on policy implications. Another scientist standing next to us chimed in, saying that it wouldn't be that interesting without an industry voice: could they invite an industry representative in attendance to make it more interesting?

Throughout the conference, I occupied a narrow and unusual role as a researcher. I was simultaneously a social scientist among dozens of toxicologists and chemists; an outsider to the field with significant knowledge of that field; and a researcher with connections not just to the scientists

and regulators in attendance, but to the chemical industry being criticized and the environmental advocates driving that critique. These multiple fields—and my multiple roles—came together at events like the 2012
BFR workshop in ways that highlight both the difficulties and benefits of
my multi-sited, multi-method research process.

In this methodological afterward, I reflect on my role and experiences
as a researcher. I offer a detailed description of my methodological
choices, data sources, and researcher anxieties. The purpose of this afterward is thus twofold: to offer specific information on what I did for those
wanting to evaluate my methods or complete their own similar research,
and to reflect on the implications of my methodology for others interested
in environmental risk, science policy interfaces, studying elites, and qualitative research generally.

STUDYING THE FIELD

The flame retardant story I have examined in this book represents a case
study of institutional responses to, and stakeholder controversies over,
environmental risks. In the social sciences, a case study is "the detailed
examination of an aspect of a historical episode to develop or test historical explanations that may be generalizable to other events."[2] To investigate
this case study, I conducted in-depth interviews and short-term participant and nonparticipant observation at five sites, supplemented by a
detailed literature review, archival research, and content analysis of published documents and testimony. I began my preliminary research in
2009 and engaged in serious data collection through 2012, though I continued collecting data and conducting interviews through the summer of
2014. I received institutional review board (IRB) approval for the project
at Brown University (1006000211) and at Whitman College (13-14/02).

Instead of choosing a single research site to investigate controversies
over these chemicals, I examined the full field of flame retardant controversies, following work on Bourdieuan conceptions of scientific, interdisciplinary, and trans-scientific fields as social arenas of struggles for
resources and authorities.[3] Thus, instead of choosing one research site to
investigate controversies over these chemicals, I chose to examine the full

field of flame retardant controversies. This represents an example of *relational* ethnography, participant observation of fields, boundaries, or processes rather than preidentified places, groups, or organizations.[4]

In my research, I followed flame retardant chemicals from their development and production in chemical industry laboratories, to the EPA where they are regulated and approved, to academic research laboratories where they are studied by scientists of various disciplines, and finally into the public sphere of contention, where environmental activists and industry-funded lobbyists fight over different visions of how chemicals should be regulated. Sociologist Rebecca Gasior Altman calls this "a lifecycle approach to studying chemical politics," which follows the chemicals from point to point rather than assuming chemical politics can be captured in any single location or analytical field.[5]

In addition to the breadth of the field in this case, an added difficulty was that the events I studied were playing out in real time. My study period saw new regulations, legislative hearings, scientific findings, increased activism, and massive media attention. Significant new chapters in the flame retardant story were emerging even as I finalized edits on this book. Rather than focus on any one individual outcome, I have followed events as they continued to unfold to explain how the field of stakeholders contributed and responded to a series of connected scientific and policy developments. There were trade-offs that came with this approach: it is hard to continually analyze a stream of unfolding regulatory challenges, public relations campaigns, and legal battles, and it precludes the option of studying scientific or regulatory closure or consensus formation.[6] I tried to weave ongoing stories into this book as I wrote, but undoubtedly the flame retardant story continued to evolve after these pages went to print.

To capture the field of flame retardant controversies, I selected five sites based on a combination of methodological, theoretical, and practical considerations, a necessarily common practice in this type of multi-sited research.[7] I aimed to conduct participant observation and in-depth interviews with a broad range of stakeholders in order to map the field of actors and how they interact, and to compare and contrast how they selectively use scientific knowledge to arrive at and justify their characterizations of risk and hazard. My sites covered industry, academia, the nonprofit sector, and government. I wanted to observe risk assessment and scientific interpretation in process and to be able to move beyond relying entirely on single interviews with individual respondents. To this end, I chose sites with expected access to individuals engaged in developing, interpreting, and using scientific data related to flame retardants for the purposes of assessing chemical risk.

Multi-sited ethnographic research is especially well-suited for translocal and interdisciplinary topics, and allows the researcher to have different roles across different sites.[8] This was certainly the case in my research. Across my research sites, I was an observer, intern, and research fellow. I was never paid or reimbursed for my work in any way and hence had no conflicts of interest, though I benefited greatly from my research sites in various ways. In addition to the invaluable opportunities to learn from informants and observe institutional practices, and the personal value of meeting new colleagues and friends, one site provided me with temporary housing in the form of house-sitting jobs; at others I enjoyed potluck meals or working lunches; and three sites provided me with a desk and access to a work computer.

Participant Observation

I conducted participant observation at five different sites: a flame retardant manufacturing company's research and development (R&D) laboratory, the Design for the Environment (DfE) program at the EPA, a research program within the EPA's Office of Research and Development (ORD), an academic environmental chemistry laboratory, and an environmental health nonprofit organization. Though my role at each site was slightly different, in each setting I took daily field notes and gathered internal and publicly available documents to supplement my observations.

Flame Retardant Company R&D Laboratory

I conducted participant observation in October 2011 at the R&D laboratory of one of the three main flame retardant chemical companies in the United States, which I identify by the pseudonym "FlameCorp." I met a

FlameCorp advocacy representative in 2010, stayed in contact with him as my research progressed, and asked for the opportunity to visit FlameCorp for lengthier observations. He was open to the idea, and in coordination with FlameCorp managers and the R&D laboratory's manager, we developed a plan for my visit. In addition to my IRB approval, I signed a confidentiality agreement with the company to protect its identity and all confidential business information. Per this agreement, I maintain ownership over my notes and transcripts, and I send copies of my papers to my FlameCorp contact before publication; company representatives can comment on my work but cannot require changes or prevent publication in any way.

During my time at FlameCorp, I attended internal meetings and observed laboratory practice, including pouring foams, incorporating flame retardants into plastics, foam burn tests of foams with and without flame retardants, burn tests of fabrics with experimental flame retardants, and analyses of chemicals. I also conducted in-depth interviews with sixteen FlameCorp representatives, including scientists, government relations officials, and laboratory managers.

EPA's Design for the Environment Program

I was a research fellow at DfE, located at the EPA headquarters in Washington, D.C., from October to December 2011. I met a DfE representative at a flame retardant scientific conference, scheduled a lunch meeting, and asked to come to the DfE office for several months as an intern. I offered to do whatever work would be useful in exchange for the opportunity to observe their alternatives assessments on flame retardants and conduct interviews. After e-mail exchanges and a conference call with DfE leaders, my visit was approved. I received permission to identify DfE as the place where I conducted my research.

I worked at the DfE office full time, attending meetings, helping to draft and revise documents (including ongoing alternatives assessments), compiling PowerPoint presentations and fact sheets, and listening in on conference calls. I also presented my research to approximately thirty EPA employees from the Office of Pollution Prevention and Toxics. During this research period, I conducted interviews with seventeen EPA scientists and regulators, and with additional industry, regulatory, and fire safety stakeholders in the Washington, D.C., area.

EPA's Office of Research and Development

I was a visiting researcher at the EPA's Environmental Public Health Division within ORD, located in Chapel Hill, North Carolina, from January to April 2012. Unlike my work at DfE where I was more of a participant in the program, at the Environmental Public Health Division I was hosted by a researcher who did not work on flame retardants but was well connected to researchers at the EPA and the National Institute of Environmental Health Sciences (NIEHS). I received permission to identify the Environmental Public Health Division as the location of my research.

At this site, I focused on learning how advances in exposure and toxicological sciences contribute to risk assessment. I conducted twenty-three interviews with EPA and NIEHS researchers who had worked on flame retardant toxicology or exposure research, or who were working on how to improve risk assessment with emerging forms of science like high-throughput screening. I attended numerous meetings and research conferences hosted by the EPA and NIEHS. I also contributed to a collaborative statistical project, analyzing government biomonitoring data to look for associations between multiple-chemical exposure and self-reported health. Finally, my location in North Carolina allowed me to conduct additional interviews with furniture and supply chain representatives located in the Southeast.

As a visiting researcher at both ORD and DfE, I believe that many other EPA representatives were more willing to meet with me because I could write to them from an epa.gov e-mail address. The agency is regularly criticized for a lack of transparency, and formal interviews or conversations with journalists typically require extensive preapproval. In contrast, when I wrote to people from my epa.gov e-mail address and described my role at DfE or ORD as a visiting researcher, they may have been more likely to meet with me.

Academic Environmental Chemistry Laboratory

I shadowed researchers in an environmental chemistry laboratory over a two-month period. This laboratory is headed by a tenured professor who has published extensively on flame retardant exposure and toxicity. I had met this researcher at earlier flame retardant conferences, and when I requested permission to observe laboratory practices, this scientist was open to the idea. The laboratory includes approximately ten graduate students, several postdocs, and several undergraduate research assistants conducting research on flame retardant chemicals, exposure, and health effects.

I shadowed six researchers and spent a full "lab day" with each of them, ranging from three to eight hours. Most of these days were in the university's laboratory facilities, though one day was in the field collecting samples from participants' homes. Most of the time I simply observed and asked questions, though on one observation a researcher trained me to conduct simple timed research tasks under direct supervision. I also attended relevant workshops and academic lectures with these researchers.

Environmental Health Nonprofit

I worked as an intern at an environmental health nonprofit from June to July 2009. This organization is a state-based organization with a national and international presence in the environmental health field. It focuses on a variety of environmental health issues, conduct and contribute to exposure assessments, and have supported flame retardant regulation in their state in the past. I contacted the director of the organization in early 2009 and offered to come work for them as an unpaid intern in exchange for the opportunity to observe and participate in their work and interview some of the staff.

While working at the organization, I attended internal meetings, answered phones, helped organize public events and educational activities, assisted in environmental sampling, and wrote and edited documents and research summaries. I conducted in-depth interviews with employees of this organization as well as with representatives from other nonprofits, firefighters, and state-level environmental and public health regulators in the area.

In addition to working directly with this organization, I also connected with the broader network of flame retardant activists. For the duration of my research period I received e-mail lists from the national coordination effort on flame retardants. I also attended strategic planning meetings, listened to conference calls, and received hundreds of mailing list e-mails. At one activist conference, I volunteered to be the official note taker, which allowed me to take detailed notes without being overly intrusive. Some of these sources of data inform my analysis but are not referenced directly because I committed to not discussing strategy or ongoing planning activities.

Interviews

Multi-sited ethnography is often more interview-based than traditional, single-site ethnographies.[9] My research aligned with this pattern. I conducted 116 in-depth interviews between June 2009 and July 2014. Twelve additional interviews with flame retardant activists and firefighters were conducted by graduate students in the Contested Illnesses Research Group at Brown University. I conducted roughly two-thirds of my interviews between October 2011 and April 2012. In addition to interviewees working at each of my research sites, I sought out additional interviews with scientists, state and federal regulators and legislators, industry representatives, fire safety experts, and activists from many environmental and health social movement organizations. I identified potential interviewees through publications on flame retardant topics, references in the media, their organizations' involvement in flame retardant campaigns, and recommendations from other interviewees.

In total, I interviewed approximately twenty individuals from each of five sectors, corresponding with my research sites: environmental, health, or consumer activists (n=18), industry representatives (n=23), federal regulators (n=20), academic researchers (n=17), and government researchers (n=23). Additionally I interviewed firefighters and other fire safety specialists (n=10), state-level legislators (n=5), state-level regulators (n=7), and other researchers in nonprofit or consulting positions (n=5). Of course, these are fluid boundaries. For example, some of the activists have doctorates in biological sciences, and some of the regulators have also worked in industry.

I used semistructured interviews to allow for qualitative investigation, with questions organized around seven themes: respondents' professional trajectories, their work on flame retardants, the relationship between activism and science, industry and production, exposure research, risk and hazard assessment, and environmental regulation. I respected any requests to go off the record; when I was uncertain, I followed up with respondents about specific quotes.

I conducted interviews in person whenever possible, with twenty-three interviews conducted over the phone. I recorded the interviews on a digital recorder, though eighteen interviewees declined to be recorded (three industry representatives and fifteen government representatives). I took detailed notes during each interview and typed these up soon after each interview, always with names and identifying details removed. I saved the list of interviewees on a password-protected document on a password-protected computer.[10] A research assistant transcribed approximately twenty-five interviews, and I transcribed the rest. For those that I did not transcribe myself, I listened to the full interview while reading the transcript, making corrections as needed.

I do not identify individuals by name unless referring to a publicly available document or public event, such as individuals who gave public presentations at scientific conferences. I identify DfE and the Environmental Public Health Division as the EPA offices in which I worked, but do not identify any other research site by name. As I noted, FlameCorp is one of the three main flame retardant manufacturing companies in the United States. When I mention one of these three companies by name based on publicly available documents (e.g., press releases or media quotations), I may or may not be talking about FlameCorp.

Confidentiality is particularly difficult in closed communities or among small groups of stakeholders familiar with each other's work. Researchers distinguish between *external confidentiality*, the ability of outsiders to identify participants, and *internal confidentiality*, "the ability for research subjects involved in the study to identify each other in the final publication of the research."[11] Although IRBs and researchers are generally most concerned with external confidentiality, internal confidentiality is certainly a concern among flame retardant stakeholders. I was frequently reminded of this throughout my research. While signing the informed consent document at the start of our interview, one scientist asked how I would protect his confidentiality: "This is a small enough world that someone who knows could instantly tell, even if you don't mention the name." Medical sociologist Karen Kaiser notes that, "Although meticulous data cleaning can remove personal identifiers such as names, the contextual identifiers in individuals' life stories will remain."[12] In the case of flame retardants, there is the added complication that not all stakeholders respect or like each other, and some might be eager to identify damaging information about other stakeholders. To protect the internal confidentiality of my research participants, I use general descriptors (e.g., activist, industry representative). I also do not describe individuals in ways that could identify them to colleagues within research sites, even when describing their specific positions or experiences could deepen or strengthen my analysis. Anytime I use a specific job description, multiple people self-identified with that title in interviews. I also avoid identifying individuals by gender when that might decrease internal confidentiality to their colleagues.

Individual respondents in all sites were incredibly open with me and in some cases shared information that could be professionally compromising. Multiple interviewees commented on the issue of confidentiality during interviews. One scientist prefaced a controversial statement by saying, "Since this conversation is confidential, I'll be a little more candid than I might be otherwise." Similarly, an industry representative said that confidentiality tempered his concern about talking with me: "Frankly, I'm still a little hesitant to talk . . . Having this agreement that we signed, that [no] statements are going to be attributed to me or [my company] certainly helps." My assertion that their confidentiality would be protected allowed many respondents to share things they otherwise would not have, and I hope all of my research participants trust that I have fulfilled this goal.

Additionally, I respected all requests to go off the record or turn off my recorder during the interview. Over a dozen respondents from industry or the EPA requested this at some point in their interview. Sometimes these requests related to anecdotal accounts of how decisions had been made or how chemicals had been evaluated. In other instances, the off the record information was much more personal; several interviewees, for example, shared personal experiences with cancer. On occasion during my observations, I learned information—about specific chemicals, companies, or people—that I could not repeat or even put in my field notes. For example, if I ever accidentally overheard any confidential business information during my time at EPA, I could not use that information. Some information about campaign strategies and targets developed during activist meetings was confidential as well.

Beyond interviews and observations, I gathered data from several other sources. I set up Google Alerts for "flame retardant" and "PBDE" to cover emerging stories on the topic, listened to dozens of conference calls and webinars, and subscribed to several relevant e-mail mailing lists. I monitored leading scientific journals such as *Environmental Health Perspectives* and *Environmental Science & Technology* for new publications on flame retardants and environmental health. I also conducted several sweeping PubMed searches for articles, and regularly sought out new articles through the reference lists of papers I read. I attended scientific, regulatory, and activist conferences on flame retardants, environmental policy, and environmental health. I recorded web pages from all types of stakeholder websites, gathered public documents, and read publications from across the field of stakeholders. Following the flame retardant story for so many years also allowed me to keep up to date on the constantly evolving corpus of scientific publications.

Analysis and Report-Back

After transcribing the interviews, I imported interview summaries, transcripts, and field notes into NVivo 10.0, a software program for analyzing qualitative data.[13] I developed a hierarchical list of codes based on interview questions and general topics, and added to and modified this list in the early stages of coding, for a final total of 104 codes. Throughout the book, any quoted statements come directly from transcripts or my field notes. I always wrote my field notes soon after the events, so the quotations accurately reflect what I observed and heard, though it is possible that field note quotations are occasionally not verbatim.

Throughout my research process, I have been committed to sharing my findings with my field sites and interview participants. This is anchored in my commitment to *reflexive research ethics*, an ethical approach

I developed with colleagues at Brown University, Northeastern University, and the University of California at Berkeley.[14] Reflexive research ethics involves self-conscious, interactive, and iterative reflection on research practices and relationships with research participants, and highlights the relational nature of environmental health and justice research. In this project, reflexive research meant that I shared findings with field sites during and after my research visits. In addition to freely talking about my work one on one during my field research, I gave formal presentations on my research at FlameCorp and the EPA, wrote a detailed observation memo for the environmental health organization, and shared drafts of papers with key informants at all my research sites. As part of additional confidentiality agreements, FlameCorp and another industry organization requested that I share prepublication drafts of papers with them for review, but not revision or control. I also e-mailed copies of all publications to each and every interview participant (so long as I could find a working e-mail address). To protect their confidentiality, I sent these e-mails to interviewee clusters by field site with all addresses in blind carbon copy mode.

Also tying into reflexive research ethics are questions of access and ethics. From the start, I struggled with how to think about individuals as spokespeople for their organizations, agencies, or companies. I interviewed individuals, but throughout this book I often use people's words to make sense of institutions. This was particularly difficult for me regarding interviews with industry scientists. On the one hand, the researchers and experts that make up the flame retardant industry emphasized that their chemicals provide a social good, and they saw their work as grounded in the best-available science. On the other hand, the flame retardant industry as a corporate institution knowingly manufactures dangerous chemicals, is insufficiently diligent in research on chemical hazard and exposure, and prioritizes profits over caution regarding controversial chemicals. How could I as an external researcher reconcile these competing perspectives?

Social theorist Luc Boltanski calls this a *hermeneutic contradiction*: institutions are at once more than and reducible to individuals.[15] Individuals within institutions can also have an incomplete picture of the work of their employer. For example, FlameCorp scientists sometimes embraced the exposure-centric risk formula advocated by the chemical industry's marketing and advocacy bodies, and other times expressed opinions or shared stories that were out of alignment. In interviews, FlameCorp scientists objected to the idea that they were complicit in the continued use of toxic chemicals, and they emphasized their identities as environmentalists, responsible citizens, and parents. I do not aim to discredit these individuals or to suggest that they are greedy mouthpieces for immoral corporate practices. Instead, my multi-site, multi-method approach allows for the triangulation of individuals' perspectives with observations and content analyses to develop a more accurate and nuanced understanding of how different sets of stakeholders define risk, translate science to buttress this definition, and operationalize this definition and understanding of science with innumerable minute and consequential decisions about chemical risk and safety.

This understanding was only possible through my multi-sited methodology, but this approach has its own set of limitations. First, I have significant breadth of data, but uneven depth. I observed only one environmental health organization, one academic laboratory, and one company, though I conducted many interviews with individuals outside of these specific organizations. I have dozens of interviews with some stakeholder types, such as regulatory scientists, but only a handful of interviews with others, such as supply chain representatives. A different approach would have been to shrink the field but achieve greater depth— for example, to seek out ethnographic visits at multiple nongovernmental organization offices, or to be sure to interview many representatives of the organizations that develop fire codes and standards.

Another potential limitation of my methodology is that for my sites of participant observation, I initially sought and obtained consent from those in charge: laboratory managers, office leaders, or organization directors. Only after I had obtained consent to enter into a field site did I request the consent of each individual interview participant. As sociologist Nancy Plankey-Videla reflects regarding her own organizational ethnography, this challenges notions of informed consent.[16] Though I gave each individual a consent form and clearly stated that participation in general and responding to each specific question was completely voluntary, respondents may have felt compelled to speak with me. In one instance, a researcher in the academic chemistry laboratory made comments about being compelled to speak with me, comments that I believe were made in jest but that nonetheless reveal the power imbalances at play. As my field notes from that day state, "At one point he joked about Professor X making him meet with me, so everything he said would be biased. But I think he was joking . . ." These issues of consent are a significant concern in this type of institutionally based ethnography.

STUDYING UP AND STUDYING THE FIELD

Despite these limitations, my multi-sited research let me observe, understand, and discuss the full range of stakeholders involved in flame retardant controversies. Especially important, I would argue, were my interactions with industry. Compared to the frequency with which researchers study social movements, social scientists rarely "study up" by gaining in-depth research access to elite institutions such as chemical companies.[17] Part of this is likely due to industry's hesitancy to let outsiders observe and potentially critique their operations. (Interestingly, I can say without a doubt that my analysis is *less* critical of the flame retardant industry than it would have been had I not had the opportunity to interview FlameCorp scientists and other industry representatives, because learning about their practices and perspectives early in my research process convinced me that an agnotology perspective could not completely explain what I was seeing: the story was more complicated than just the intentional manufacturing of doubt.) But I also suspect that social scientists do not ask for this access as often as they might, anticipating a negative response from industry, or approaching anything that could be perceived as industry collaboration with a skeptical eye. Indeed, to gain access I had to agree to share prepublication drafts of all of my papers, something I did not have to do for my other research sites. This may be something other researchers are not willing to do. In my research, I found that industry representatives were frequently willing to speak with me, especially if I had "proper" introductions. FlameCorp especially was accommodating and welcoming in their efforts to engage my research. I encourage social scientists to seek greater access to industry, to better understand the hows and whys of industry decision making, not just the decisions that eventually become visible in the public sphere.

It is possible that my visit at FlameCorp was tightly controlled to portray a particularly positive image of the company. The laboratory members were told who I was and that I was coming, and a couple of them joked about hiding information from me during my visit. During one interview, a jovial scientist poked his head into an in-process interview. He looked at the man I was meeting with and said, "Remember what I told you to tell her," to which my interviewee quipped, "Oh yeah, I know nothing." I laughed and said jokingly, "We can stop the interview right now. We're done." All three of us can be heard laughing on the recording, and then we continued with my questions. I would argue that the humor surrounding my participants' jokes about withholding information demonstrates the high degree of openness that I was privy to during my visit.

Another challenge in my research was the general lack of trust and respect between some of my field sites. Industry representatives were often distrustful of environmental activists. One industry spokesperson said he had been willing to speak with me because I didn't sound like a "rabid environmentalist." Comments about environmentalists as "the green people" were common at FlameCorp. To an even greater degree, most environmental and health advocates were skeptical of the chemical manufacturers, following years of intense lobbying and inaccurate SST on the part of chemical companies. In interviews, activists described the flame retardant industry as "dirty bastards," "perfect villains," "drug dealers," "wolves in sheep's clothing," and "professional naysayers." At the end of one meeting, I spoke at length with an environmental activist about my research. During our discussion she questioned my attention to the industry's position, saying, "You talk about them as though they are rational human beings." She disagreed that the companies and their representatives could be investigated in this way.

LISTENING TO THE FIELD

It was often difficult engaging with these stakeholders who had so little respect for each other, especially when I do have my own biases toward environmental health protection, and when I believed that my interviewees all around were genuinely nice, thoughtful, and well-meaning people. In spite of these difficulties, I would argue that this research process was particularly valuable because of the controversy that characterizes the flame retardant story.

My multi-sited qualitative research lets me humanize the industry by talking to the individuals actually making day-to-day decisions on how to develop new chemicals or defend existing chemicals. Researchers at FlameCorp were adamant that they did not knowingly develop or continue to produce products if they knew them to be harmful.[18] As one scientist told me, "I have family, I have children. I don't want to put myself at risk by sitting, you know, in a house with harmful chemicals. I'm a human being too. I want to live." Another industry representative said, "I certainly would not feel comfortable, though, being in a position where I felt like I was defending something that was inappropriate. You know, I've got kids. I consider myself an environmentalist." These quotes highlight the importance of carefully listening to individual stakeholders when trying to understand industry decision making, and of understanding that individuals are at best imperfect spokespeople for the institutions they are called on to represent.

Additionally, I encountered a variety of perspectives on the severity of the issue of chemical exposure. Some respondents were not worried about their exposure to most chemicals. An industry scientist told me that the flame retardants his company made were undoubtedly safe, an EPA chemist told me that certain industrial chemicals were so safe he would eat them, and a regulatory scientist said that flame retardants were far from the most pressing environmental issue of the day. But others expressed opposite views. An EPA official told me that chemical exposure in general was a more pressing problem than climate change. Advocates regularly described flame retardants as poison. An environmental activist told me that after being biomonitored, he became so concerned about his chemical exposure that he holds his breath when city buses drive by rather than inhale the exhaust.

Making sense of these dramatically different positions required me to evaluate my own biases and do what I could to examine the state of the science on chemicals such as flame retardants. As a social scientist studying a highly technical scientific controversy, I was regularly confronted by the question of whether I knew "enough" about the topic to investigate it from a sociological perspective. I have always enjoyed the natural sciences, and as a graduate student took additional coursework in toxicology and environmental health to better understand the science around flame retardants. But in many ways I remained a novice in the fields I studied, reliant on key informants, summary documents, and close readings of texts to understand material.

In the appendix to her book on the molecular and genetic turns of environmental health research, sociologist Sara Shostak writes about occupying a similar position, asking "How Much Science Does a Social Scientist Need to Know?"[19] She found that in some cases her lack of technical knowledge was an advantage because it forced people to describe ideas, practices, and assumptions carefully. I had some of the same experiences: scientists asked me whether I had "sufficient background" for my research, or wondered aloud about my knowledge of organic chemistry. But like Shostak, I often found my relative lack of knowledge to be a benefit, not a hindrance, to understanding people's work and beliefs. Even when my knowledge was not questioned by the scientists and government officials with whom I spoke, I regularly confronted myself with this question. In the end, I know that I learned enough about the science of flame retardants that I was easily able to maintain a satisfactory level of discourse, understanding, and evaluation. Still, it is instructive to realize how we as researchers can doubt our own capabilities.

My methodological reflections thus point to several conclusions. First, multi-sited studies of environmental controversies need to include not only multiple field sites but multiple types of sites. In particular, industry and government sites contribute greatly to a well-rounded picture of environmental controversies. Second, residency or full-time observation at field sites, even if short in duration, is integral to obtaining a more concrete understanding of different players. Third, the negotiation of trust is a key element of research on controversial topics, requiring honest engagement with all stakeholders and carefully maintained guarantees of confidentiality. Finally, trust is an ongoing practice that can be continuously deepened. Though I expect each set of stakeholders can find something in this book that makes them uncomfortable, I hope that they all feel I have truthfully represented their words and experiences. I am exceedingly grateful to the institutions that welcomed me to observe and to the individuals who shared their time with me.

NOTES

ACKNOWLEDGMENTS

1. Alissa Cordner, "Defining and Defending Risk: Conceptual Risk Formulas in Environmental Controversies," *Journal of Environmental Studies and Sciences* 5, no. 3 (2015): 241–250; Alissa Cordner, "Strategic Science Translation and Environmental Controversies," *Science, Technology, & Human Values* 40, no. 6 (2015): 915–938; published online, DOI: 10.1177/0162243915584164.

1. UNCERTAIN SCIENCE AND THE FIGHT FOR ENVIRONMENTAL HEALTH

1. Give Toxics the Boot, "Homepage," 2014, http://givetoxicstheboot.org/; Lynne Peeples, "Firefighters Sound Alarm on Toxic Chemicals," *Huffington Post*, March 27, 2014, http://www.huffingtonpost.com/2014/03/27/firefighters-toxic-chemicals-regulation-flame-retardants_n_5034976.html.
2. Arlene Blum and Bruce N. Ames, "Flame-Retardant Additives as Possible Cancer Hazards," *Science* 195, no. 4273 (1977): 17–23, http://www.jstor.org/stable/1743512; Arlene Blum et al., "Children Absorb Tris-BP Flame Retardant from Sleepwear: Urine Contains the Mutagenic Metabolite, 2,3-Dibromopropanol," *Science* 201, no. 4360 (1978): 1020–1023.
3. Philip H. Abelson, "The Tris Controversy," *Science* 197, no. 4299 (1977): 113; Spring Mills Inc. v. CPSC, "Springs Mills, Inc. v. Consumer Products Safety Commission," 434 F. Supp. 416, 1977.
4. Marian Deborah Gold, Arlene Blum, and Bruce N. Ames, "Another Flame Retardant, Tris-(1,3-Dichloro-2-Propyl)-Phosphate, and Its Expected Metabolites Are Mutagens," *Science* 200, no. 19 (1978): 785–787; U.S. EPA, "Tris(2,3-Dibromopropyl) Phosphate Significant New Use Rule," *Federal Register* 52, no. 2703 (1987).
5. On Blum's early career, see Arlene Blum, *Breaking Trail: A Climbing Life* (New York: Scribner, 2005). On her return to flame retardants, see Arlene Blum, "Halogenated

Flame Retardants: A Global Concern," presentation at Flame Retardant Dilemma, February 10, 2012, Berkeley, Calif.; http://www.greensciencepolicy.org/sites/default /files/frd120210-ABlum.pdf.

6. Robin Dodson et al., "After the PBDE Phase-Out: A Broad Suite of Flame Retardants in Repeat House Dust Samples from California," *Environmental Science & Technology* 46, no. 24 (2012): 13056–13066.

7. Sheldon Krimsky, *Hormonal Chaos: The Scientific and Social Origins of the Environmental Endocrine Hypothesis* (Baltimore: Johns Hopkins University Press, 2000).

8. Ulrich Beck, *World Risk Society* (Malden, Mass.: Polity, 1999); Ulrich Beck, *Risk Society: Towards a New Modernity, Theory, Culture and Society* (Newbury Park, Calif.: Sage, 1992); Anthony Giddens, "Risk and Responsibility," *The Modern Law Review* 62, no. 1 (1999): 1–10; Ulrich Beck, Anthony Giddens, and Scott Lash, *Reflexive Modernization: Politics, Tradition and Aesthetics in the Modern Social Order* (Stanford, Calif.: Stanford University Press, 1994).

9. Margarita V. Alario and William R. Freudenburg, "Environmental Risks and Environmental Justice, Or How Titanic Risks Are Not So Titanic after All," *Sociological Inquiry* 80, no. 3 (July 12, 2010): 500–512; Paul Mohai, David Pellow, and J. Timmons Roberts, "Environmental Justice," *Annual Review of Environment and Resources* 34 (2009): 405–430.

10. U.S. EPA, "What Is the TSCA Chemical Substance Inventory?" U.S. EPA, retrieved December 10, 2010, http://www.epa.gov/oppt/newchems/pubs/inventory.htm.

11. Peter Egeghy et al., "The Exposure Data Landscape for Manufactured Chemicals," *Science of the Total Environment* 414, no. 1 (2012): 159–166; Richard Judson et al., "The Toxicity Data Landscape for Environmental Chemicals," *Environmental Health Perspectives* 117, no. 5 (2009): 685–695.

12. Nancy I. Maxwell, *Understanding Environmental Health* (Boston: Jones and Bartlett, 2009).

13. Krimsky, *Hormonal Chaos.*

14. Åke Bergman et al., "The Impact of Endocrine Disruption: A Consensus Statement on the State of the Science," *Environmental Health Perspectives* 121, no. 4 (2013): A104–A106.

15. Michael R. Edelstein, *Contaminated Communities: The Social and Psychological Impacts of Residential Toxic Exposure* (Boulder, Colo.: Westview, 1988).

16. Linda S. Birnbaum et al., "Consortium-Based Science: The NIEHS's Multipronged, Collaborative Approach to Assessing the Health Effects of Bisphenol A," *Environmental Health Perspectives* 120, no. 12 (2012): 1640–1644.

17. Arden Pope and Douglas Dockery, "Health Effects of Fine Particulate Air Pollution: Lines That Connect," *Journal of the Air and Waste Management Association* 56 (2006): 742.

18. David Carpenter and Pamela Miller, "Environmental Contamination of the Yupik People of St. Lawrence Island, Alaska," *Journal of Indigenous Research* 1, no. 1 (2011): 1–3.

19. LaSalle Leffall, Margaret Kripke, and Suzanne Reuben, *President's Cancer Panel: Reducing Environmental Cancer Risk* (Bethesda, Md.: U.S. Department of Health and Human Services, 2010).

20. Sara Shostak, *Exposed Science: Genes, the Environment, and the Politics of Population Health* (Berkeley: University of California Press, 2013); Sarah A. Vogel, *Is It Safe? BPA and the Struggle to Define the Safety of Chemicals* (Berkeley: University of California Press, 2013).

21. Ted Smith, David A. Sonnenfeld, and David Pellow, *Challenging the Chip: Labor Rights and Environmental Justice in the Global Electronics Industry* (Philadelphia: Temple University Press, 2006).

22. Joel A. Tickner, "From Reaction to Prevention," presented at the American Bar Association 39th National Spring Conference on the Environment—Chemicals Regulation. June 11, 2010, Baltimore.

23. Commonweal, "Wingspread Statement on the Precautionary Principle," accessed March 9, 2013. http://www.commonweal.org/programs/wingspread-statement.html.; see also the 1982 World Charter for Nature, and the 1992 Rio Declaration.

24. Ruth Alcock and Jerry Busby, "Risk Migration and Scientific Advance: The Case of Flame-Retardant Compounds," *Risk Analysis* 26, no. 2 (2006): 369–381.

25. Arthur Mol, *The Refinement of Production: Ecological Modernization Theory and the Chemical Industry* (Utrecht, the Netherlands: Van Arkel, 1995); Arthur Mol, David A. Sonnenfeld, and Gert Spaargaren, *The Ecological Modernisation Reader: Environmental Reform in Theory and Practice* (London: Routledge, 2009).

26. Chemtura, "Chemtura Corporation: Homepage," retrieved February 20, 2013, www.chemtura.com.

27. In 2013, companies released over 4.1 billion pounds of chemicals included on the Toxics Release Inventory. U.S. EPA, "Toxics Release Inventory Release Reports," retrieved April 28, 2015, http://iaspub.epa.gov/triexplorer/tri_release.chemical.

28. J. Timmons Roberts, "Emerging Global Environmental Standards: Prospects and Perils," in *Globalization and the Evolving World Society*, ed. Proshanta Nandi and Shahid Shahidullah (Boston: Leiden, 1998), 144–163.

29. Gary E. R. Hook, "Responsible Care and Credibility," *Environmental Health Perspectives* 104, no. 11 (1996): 1138–1139; Caroline E. Scruggs et al., "The Role of Chemical Policy in Improving Supply Chain Knowledge and Product Safety," *Journal of Environmental Studies and Sciences* 4 (2015): 132–141.

30. Kenneth Gould, David Pellow, and Allan Schnaiberg, "Interrogating the Treadmill of Production: Everything You Wanted to Know about the Treadmill but Were Afraid to Ask," *Organization & Environment* 17, no. 3 (2004): 296–316; Allan Schnaiberg and Kenneth Alan Gould, *Environment and Society: The Enduring Conflict* (New York: St. Martin's, 1994).

31. Eric O. Wright, "Interrogating the Treadmill of Production: Some Questions I Still Want to Know About and Am Not Afraid to Ask," *Organization & Environment* 17, no. 3 (2004): 317–322; Kenneth Gould, Allan Schnaiberg, and Adam Weinberg, *Local*

Environmental Struggles: Citizen Activism in the Treadmill of Production (New York: Cambridge University Press, 1996).

32. U.S. EPA, "Risk Assessment Portal: Basic Information," *U.S. EPA*, accessed April 28, 2015, http://epa.gov/riskassessment/basicinformation.htm#risk.

33. Scott Frickel and Kelly Moore, *The New Political Sociology of Science: Institutions, Networks, and Power* (Madison: University of Wisconsin Press, 2006).

34. Gil Eyal, "For a Sociology of Expertise: The Social Origins of the Autism Epidemic," *American Journal of Sociology* 118, no. 1 (2013): 1–45; Edward Woodhouse, "Nanoscience, Green Chemistry, and the Privileged Position of Science," in *The New Political Sociology of Science*, ed. Frickel and Moore, 148–181.

35. Scott Frickel, "Just Science? Organizing Scientist Activism in the U.S. Environmental Justice Movement," *Science as Culture* 13, no. 4 (2004): 449–469; Jurgen Habermas, *Toward a Rational Society: Student Protest, Science, and Politics* (Boston: Beacon, 1970); David Michaels and Celeste Monforton, "Manufacturing Uncertainty: Contested Science and the Protection of the Public's Health and Environment," *American Journal of Public Health* 95, Suppl. 1 (2005): S39–48; Abby Kinchy, *Seeds, Science, and Struggle: The Global Politics of Transgenic Crops* (Cambridge, Mass.: MIT Press, 2012).

36. Rachel Morello-Frosch et al., "Embodied Health Movements: Responses to a 'Scientized' World," in *The New Political Sociology of Science: Institutions, Networks, and Power*, ed. Scott Frickel and Kelly Moore (Madison: University of Wisconsin Press, 2006), 244–271; Lois Gibbs, "Citizen Activism for Environmental Health: The Growth of a Powerful New Grassroots Health Movement," *Annals of the American Academy of Political and Social Science* 584 (2002): 97–109.

37. Kelly Moore, *Disrupting Science: Social Movements, American Scientists, and the Politics of the Military, 1945–1975* (Princeton, N.J.: Princeton University Press, 2008).

38. David Hess, *Alternative Pathways in Science and Industry: Activism, Innovation, and the Environment in an Era of Globalization* (Cambridge, Mass.: MIT Press, 2007).

39. Steven Epstein, *Impure Science: AIDS, Activism, and the Politics of Knowledge, Medicine and Society* (Berkeley: University of California Press, 1996); Phil Brown, *Toxic Exposures: Contested Illnesses and the Environmental Health Movement* (New York: Columbia University Press, 2007); Sabrina McCormick, "Democratizing Science Movements: A New Framework for Mobilization and Contestation," *Social Studies of Science* 37, no. 4 (2007): 609–623; Rachel Morello-Frosch et al., "Toxic Ignorance and Right-to-Know in Biomonitoring Results Communication: A Survey of Scientists and Study Participants," *Environmental Health* 8, no. 1 (2009): 6; Eyal, "For a Sociology of Expertise"; Emma Lavoie et al., "Chemical Alternatives Assessment: Enabling Substitution to Safer Chemicals," *Environmental Science & Technology* 44, no. 24 (2010): 9244–9249.

40. Andrew Szasz, *Shopping Our Way to Safety: How We Changed from Protecting the Environment to Protecting Ourselves* (Minneapolis: University of Minnesota Press,

2007); Frickel, "Just Science?"; Morello-Frosch et al., "Toxic Ignorance"; Gibbs, "Citizen Activism."

41. Harry M. Collins and Robert Evans, *Rethinking Expertise* (Chicago: University of Chicago Press, 2007).

42. National Research Council, *Science and Decisions: Advancing Risk Assessment* (Washington, D.C.: Committee on Improving Risk Analysis Approaches Used by the U.S. EPA. National Academies Press, 2009), 4, http://www.nap.edu/catalog.php?record_id =12209.

43. Soraya Boudia and Nathalie Jas, "Introduction: The Greatness and Misery of Science in a Toxic World," in *Powerless Science? Science and Politics in a Toxic World*, ed. Soraya Boudia and Nathalie Jas (New York: Berghahn, 2014), 24.

44. Daniel L. Kleinman and Sainath Suryanarayanan, "Dying Bees and the Social Production of Ignorance," *Science, Technology, & Human Values* 38, no. 4 (May 3, 2012): 219.

45. Scott Frickel et al., "Undone Science: Charting Social Movement and Civil Society Challenges to Research Agenda Setting," *Science, Technology, & Human Values* 35, no. 4 (2010): 444–476; David Hess, "The Potentials and Limitations of Civil Society Research: Getting Undone Science Done," *Sociological Inquiry* 79, no. 3 (2009): 306–327.

46. For notable exceptions, see Gabrielle Hecht, *Being Nuclear: Africans and the Global Uranium Trade* (Cambridge, Mass.: MIT Press, 2012); Adriana Petryna, *Life Exposed: Biological Citizens after Chernobyl* (Princeton, N.J.: Princeton University Press, 2002); Vogel, *Is It Safe?*

47. Scott Frickel and Michelle Edwards, "Untangling Ignorance in Environmental Risk Assessment," in *Powerless Science?* ed. Boudia and Jas, 215–233.

48. Daniel Sarewitz, "How Science Makes Environmental Controversies Worse," *Environmental Science & Policy* 7 (2004): 398.

2. HOT TOPICS: FLAME RETARDANTS IN THE PUBLIC SPHERE

1. Ami Zota et al., "Elevated House Dust and Serum Concentrations of PBDEs in California: Unintended Consequences of Furniture Flammability Standards?," *Environmental Science & Technology* 42, no. 21 (2008): 8158–8164; Linda S. Birnbaum and Daniele F. Staskal, "Brominated Flame Retardants: Cause for Concern?," *Environmental Health Perspectives* 112, no. 1 (2004): 9–17; Ronald A. Hites, "Polybrominated Diphenyl Ethers in the Environment and in People: A Meta-Analysis of Concentrations," *Environmental Science & Technology* 38, no. 4 (2004): 945–956.

2. Pat Thomas, "Flame Retardants Are the New Lead," *The Ecologist*, June 13, 2014, http://www.theecologist.org/blogs_and_comments/commentators/2431814/flame_retardants_are_the_new_lead_our_children_must_be_protected.html.

3. National Research Council, *Science and Decisions*, 4.

4. Sandra Steingraber, *Living Downstream: An Ecologist Looks at Cancer and the Environment* (Reading, Mass.: Addison-Wesley, 1997), 72–73.

5. Noah M. Sachs, "Rescuing the Strong Precautionary Principle from Its Critics," *University of Illinois Law Review*, no. 4 (2011): 1285–1338.

6. Commonweal, "Wingspread Statement on the Precautionary Principle."

7. Vytenis Babrauskas et al., "Flame Retardants in Building Insulation: A Case for Re-Evaluating Building Codes," *Building Research and Information* 40, no. 6 (2012): 738–755.

8. U.S. Department of Energy, "Average Material Consumption for a Light Vehicle," Table 4.16, in *Transportation Energy Data Book*, 31st ed. (Washington, D.C.: DOE, 2012).

9. Birnbaum and Staskal, "Brominated Flame Retardants," 9.

10. U.S. EPA, "An Alternatives Assessment for the Flame Retardant Decabromodiphenyl Ether (DecaBDE)," U.S. EPA, Design for the Environment, accessed April 28, 2014, http://www2.epa.gov/sites/production/files/2014-05/documents/decabde_final.pdf.

11. Sergei Levchik, "Modes of Flame Retardant Action: Can Halogen-Free FRs Replace Brominated FRs?" presented at the 12th Workshop on Brominated and Other Flame Retardants, June 6–7, 2011, Boston.

12. The element phos*phorous* must not be confused with organo-phos*phate* chemicals. Organo-phosphates are a class of chemical compounds that contain phosphorous in certain chemical arrangements. They are toxicants that were developed as nerve agents, pesticides, and herbicides, and they act by permanently inhibiting a neurotransmitter. Some flame retardants such as TDCPP are organo-phosphates. On the categorical division of flame retardants, see U.S. EPA, "An Alternatives Assessment for DecaBDE," 3–2.

13. Birnbaum and Staskal, "Brominated Flame Retardants," 9.

14. Aluminum hydroxide is also known as aluminum trihydrate. For information on chemical usage, see: Reporterlink, "World Flame Retardants Market," 2013; Joseph Serbaroli, *Plastics 101: A Primer on Flame Retardants for Thermoplastics*, Ampacet Technical Services Report (Tarrytown, N.Y.: Ampacet, 2005), retrieved April 28, 2015, www.ampacet.com/usersimage/File/tutorials/FlameRetardants.pdf.

15. U.S. EPA, "An Alternatives Assessment for DecaBDE," 3–26.

16. POPs are compounds containing carbon (i.e., organic chemicals) that do not readily biodegrade and that persist and bioaccumulate in humans and animals. POPs are governed by the international Stockholm Convention on Persistent Organic Pollutants, enacted in 2001, which aims to identify and phase out global use of POPs. The Stockholm Convention currently has 151 signatories. The United States is neither a signatory nor a party to the Convention. Stockholm Convention, "Stockholm Convention" (Chatelaine, Switzerland: Secretariat of the Stockholm Convention, 2012); European Chemicals Agency, *Guidance on Information Requirements and Chemical Safety Assessment* (Helsinki: European Chemicals Agency, 2008), retrieved April 28, 2015, http://echa.europa.eu/documents/10162/13632/information_requirements_r11

_en.pdf; U.S. EPA, "Persistent Bioacccumulative and Toxic (PBT) Chemical Program," retrieved April 28, 2015, http://www.epa.gov/pbt.

17. National Cancer Institute, "Dictionary of Cancer Terms," 2013, http://www.cancer.gov/dictionary.

18. I use the name Chemtura to refer to both Chemtura and Great Lakes Solutions. Chemtura is the parent company, and most respondents referred to the company as Chemtura, not Great Lakes Solutions.

19. Within the flame retardant industry, different companies have different reputations in regulatory, scientific, and activist circles, based on product development efforts, profitability, advocacy efforts, and perceptions of the progressivity of their research and advocacy work. Although I learned much about how FlameCorp is generally perceived by other spheres and could confirm these perceptions with observational and interview data, I do not provide information that might distinguish FlameCorp from its competitors and thus identify the company.

20. Melody Voith, "Stronger Demand in Fourth Quarter," *Chemical and Engineering News* 88, no. 5 (2010): 9.

21. Global Information Inc., "US Flame Retardant Demand to Reach 938 Million Pounds in 2016, Growing 4.6% Annually," press release, October 1, 2012, http://www.ireachcontent.com/news-releases/us-flame-retardant-demand-to-reach-938-million-pounds-in-2016-growing-46-annually-172172821.html; RnR Market Research, "Flame Retardant Industry to Touch $10.340 Million by 2019," press release 2, 2014, http://beforeitsnews.com/science-and-technology/2014/08/flame-retardant-industry-to-touch-10340-million-by-2019-2712888.html.

22. Reporterlink, "World Flame Retardants Market."

23. Mehran Alaee and Richard J. Wenning, "The Significance of Brominated Flame Retardants in the Environment: Current Understanding, Issues and Challenges," *Chemosphere* 46, no. 5 (2002): 579–582.

24. Michel le Bras, "Preface," in *Fire Retardancy of Polymers: New Applications of Mineral Fillers*, ed. Michel le Bras, Charles Wilkie, and Serge Bourbigot (Cambridge: Royal Society of Chemistry, 2005), v–vii.

25. Brian Hector MacGillivray, Ruth Alcock, and Jerry Busby, "Is Risk-Based Regulation Feasible? The Case of Polybrominated Diphenyl Ethers (PBDEs)," *Risk Analysis* 31, no. 2 (2011): 266–281; Vogel, *Is It Safe?* 17–18.

26. Oregon Department of Environmental Quality, "Fact Sheet," 2003; Wisconsin Department of Natural Resources, "Asbestos," 2007.

27. United States Census, "Historical Census of Housing Table: House Heating Fuel," *Census of Housing*, 2011; U.S. Office on Smoking and Health, *Smoking and Health: A Report of the Surgeon General* (Washington, D.C., 1979), http://profiles.nlm.nih.gov/NN/B/C/M/D/.

28. B. Y. Welke, "The Cowboy Suit Tragedy: Spreading Risk, Owning Hazard in the Modern American Consumer Economy," *Journal of American History* 101, no. 1 (2014):

97–121; Peter Kerr, "Demand Increases for Fire-Safe Clothing," *New York Times*, June 11, 1983, http://www.nytimes.com/1983/06/11/style/demand-increases-for-fire-safe-clothing.html.

29. National Safety Council, *Injury Facts* (Itasca, Ill.: NSC, 2011).

30. Michael J. Healey, "The Stuff Upholstery's Made Of" (Loudon, Tenn.: Polyurethane Foam Association, 2015), accessed April 30, 2015, http://www.pfa.org/affiliates/rr.html.

31. Prior to 1974, fire mortality was calculated through an analysis of death certificates, potentially under-counting fire deaths. The passage of the Federal Fire Prevention and Control Act of 1974 authorized the creation of the National Fire Incidence Reporting System, which allowed for the systematic collection of data about fire rates, causes, deaths and injuries, and damages. The National Fire Protection Association started reporting detailed fire statistics in 1977 using this much improved data source. For these statistics and more information, see Ben Evarts, *Trends and Patterns of U.S. Fire Losses in 2010* (Quincy, Mass.: National Fire Protection Association, 2011), http://www.nfpa.org/itemDetail.asp?categoryID=2454&itemID=55589&URL=Research/Statistical reports/Overall fire statistics/; National Fire Protection Association, "Home Fires," 2013, http://www.nfpa.org/research/reports-and-statistics/fires-by-property-type/residential/home-fires.

32. U.S. Consumer Products Safety Commission, *Children's Sleepwear Regulations: 16 CFR 1615 & 1616* (U.S., 2001), http://www.cpsc.gov//PageFiles/103092/regsumsleepwear.pdf; Flammable Fabrics Act of 1953, 15 U.S.C. §§ 1191–1204 (1953); Pub. L. 83–88; 67 Stat. 11 (June 30, 1953); http://www.cpsc.gov/PageFiles/98982/FLAMMABLE%20FABRICS%20ACT.txt.

33. CPSC *Children's Sleepwear Regulations*; California Department of Consumer Affairs, *Technical Bulletin 117–2012: Requirements, Test Procedure, and Apparatus for Testing the Smolder Resistance of Upholstered Furniture* (Sacramento: California Bureau of Electronic and Appliance Repair, Home Furnishings and Thermal Insulation, 2012), http://www.bearhfti.ca.gov/about_us/tb117_finaldraft.pdf; U.S. Department of Transportation, *Federal Motor Vehicle Safety Standard 302: Flammability of Interior Materials* (Washington, D.C.: U.S. Department of Transportation, 1972), http://www.nhtsa.gov/cars/rules/import/fmvss/index.html.

34. Heather M. Stapleton et al., "Novel and High Volume Use Flame Retardants in US Couches Reflective of the 2005 PentaBDE Phase Out," *Environmental Science & Technology* 46, no. 24 (2012): 13432–13439.

35. Michael R. Reich, "Environmental Politics and Science: The Case of PBB Contamination in Michigan," *American Journal of Public Health* 73, no. 3 (1983): 302–313.

36. Joyce Egginton, *The Poisoning of Michigan* (New York: Norton, 2009); Reich, "Environmental Politics and Science."

37. Luther Carter, "Michigan's PBB Incident: Chemical Mix-Up Leads to Disaster," *Science* 192, no. April (1976): 240–243.

38. Egginton, *The Poisoning of Michigan*, 15.

39. European Union, "Restriction of the Use of Certain Hazardous Substances (RoHS). Directive 2011/65/EU," 2011.

40. Michigan Department of Community Health, "PBBs in Michigan: Frequently Asked Questions 2011 Update," 2011, accessed April 30, 2015, http://www.michigan.gov /documents/mdch_PBB_FAQ_92051_7.pdf; Rollins School of Public Health, "PBB Registry Research Findings" (Atlanta: Rollins School of Public Health, Emory University, 2013), accessed April 30, 2015, http://pbbregistry.emory.edu/Research/Research%20 Findings.html.

41. Rollins School of Public Health, "PBB Registry Research Findings"; National Toxicology Program, *Report on Carcinogens*, 12th ed. (Research Triangle Park, N.C.: U.S. Department of Health and Human Services, 2011), 347–348.

42. Chanley M. Small et al., "Reproductive Outcomes among Women Exposed to a Brominated Flame Retardant in Utero," *Archives of Environmental and Occupational Health* 66, no. 4 (2011): 201–208.

43. Abelson, "Tris Controversy," 113.

44. Blum and Ames, "Flame-Retardant Additives."

45. Blum et al., "Children Absorb Tris-BP."

46. Abelson, "Tris Controversy," 113; Blum and Ames, "Flame-Retardant Additives"; U.S. Consumer Products Safety Commission, "News from CPSC: CSPS Bans TRIS-Treated Children's Garments," Release 77–030 (Washington, D.C.: Office of Information and Public Affairs, 1977), http://www.cpsc.gov/cpscpub/prerel/prhtml77/77030.html; David O'Brien, *What Process Is Due? Courts and Science-Policy Disputes* (New York: Russel Sage Foundation, 1987).

47. O'Brien, *What Process*; Spring Mills, "Springs Mills, Inc."

48. U.S. EPA, "Tris(2,3-Dibromopropyl) Phosphate," 2703.

49. Mark Hosenball, "Karl Marx and the Pajama Game," *Mother Jones*, November 1979, http://www.motherjones.com/politics/1979/11/karl-marx-and-pajama-game.

50. Michael deCourcy Hinds, "Reagan Signs Law on Pajama Makers," *New York Times*, January 1, 1983; An Act to Provide for the Payment of Losses Incurred as a Result of the Ban on the Use of the Chemical Tris in Apparel, Fabric, Yarn, or Fiber, and for Other Purposes, Pub. L. No. 97-395, 96 Stat. 2001 (1982). https://www.govtrack.us /congress/bills/97/s823/text.1982.

51. Jayvee Brand, Inc., et al., v. United States of America. No 82–1167. United States Court of Appeals, District of Columbia Circuit Court (1983).

52. Gold et al., "Another Flame Retardant."

53. Kellyn Betts, "Discontinued Pajama Flame Retardant Detected in Baby Products and House Dust," *Environmental Science & Technology* 43, no. 19 (2009): 7159.

54. U.S. Consumer Products Safety Commission, *Children's Sleepwear Regulations: 16 CFR 1615 & 1616.*

55. Healther Stapleton et al., "Detection of Organophosphate Flame Retardants in Furniture Foam and U.S. House Dust," *Environmental Science & Technology* 43, no. 19 (2009): 7490–7495.

56. National Fire Protection Association, "History," 2014.

57. Annys Shin, "Fighting for Safety," *Washington Post*, January 26, 2008, http://www
.washingtonpost.com/wp-dyn/content/article/2008/01/25/AR2008012503170.html.

58. The NASFM website in 2011 posted a statement on its funding sources: "NASFM en-
thusiastically seeks partnerships with business to tackle the many complex challenges
posed by fire . . . NASFM and its members preserve their independence and objectivity." National Association of State Fire Marshals, "Policy Regarding Corporate Support of the National Association of State Fire Marshals," 2011, http://www.firemarshals
.org/organization/finance.html. For information on Sparber's work with NASFM, see
Shin, "Fighting for Safety."

59. Patricia Callahan and Sam Roe, "Big Tobacco Wins Fire Marshals as Allies in Flame
Retardant Push," *Chicago Tribune*, May 8, 2012, http://www.chicagotribune.com/ct
-met-flames-tobacco-20120508-story.html.

60. Callahan and Roe, "Big Tobacco Wins."

61. R.J. Reynolds, "Strategic Plan 1996," 1996, 3, http://legacy.library.ucsf.edu/tid
/bus6odoo (italics in original).

62. R. J. Reynolds, "Strategic Plan 1996."

63. National Fire Protection Association, "States That Have Passed Fire-Safe Cigarette
Laws," last modified August 26, 2011, http://www.nfpa.org/safety-information
/for-consumers/causes/smoking/coalition-for-fire-safe-cigarettes/states-that-have
-passed-fire-safe-cigarette-laws.

64. Callahan and Roe, "Big Tobacco Wins."

65. Center for Responsive Politics, "Sparber & Assoc." (Washington, D.C.: Center for
Responsive Politics, 2012). From 1998–2000, Sparber and Associates was also paid
$170,000 in lobbying fees from the Methyl Bromide Working Group, an industry group
representing Albemarle, Ameribrom (now part of ICL-IP), Great Lakes (now Chemtura),
and TriCal Inc. (a soil fumigation company) in their quest to keep the ozone-depleting
pesticide methyl bromide on the market: CorpWatch, "Methyl Bromide Working
Group," last modified March 31, 1997, http://corpwatch.org/article.php?id=906.

66. Wilson Center, "Wilson Center Experts: Peter O'Rourke," accessed April 30, 2015,
http://www.wilsoncenter.org/staff/peter-orourke.

67. National Association of State Fire Marshals, "News," *NASFM News*, February (2008),
http://www.firemarshals.org/pdf/Feb-08-News.pdf.

68. Center for Responsive Politics, "Sparber & Assoc."

69. National Association of State Fire Marshals, "National Furniture Flammability Standard," June 6, 2014, http://www.firemarshals.org/pdf/NASFM_upholstered_furniture
_flammability_resolution_2013–4.pdf.

70. Callahan and Roe, "Big Tobacco Wins."

71. U.S. EPA, *Polybrominated Diphenyl Ethers (PBDEs) Action Plan* (Washington,
D.C.: U.S. EPA, 2009), http://www.epa.gov/oppt/existingchemicals/pubs/actionplans
/pbdes_ap_2009_1230_final.pdf.

72. Alaee and Wenning, "Significance of Brominated Flame Retardants," 579–580.
73. Vincent J. de Carlo, "Studies on Brominated Chemicals in the Environment," *Annals of the New York Academy of Sciences* 320 (1979): 678–681.
74. U.S. Congress, "Toxic Substances Control Act," *15 U.S.C 2601–2692*, 1976.
75. Kerstin Nylund et al., "Analysis of Some Polyhalogenated Organic Pollutants in Sediment and Sewage Sludge," *Chemosphere* 24, no. 12 (1992): 1721–1730.
76. U.S. EPA, *Polyhalogenated Dibenzo-P-Dioxins/Dibenzofurans; Testing and Reporting Requirements*, 1987.
77. I was unable to interview these Swedish scientists, so I cannot independently confirm this story. One U.S.-based scientist told a different version of this story: the Swedish researchers first identified PBDEs in wildlife samples and then looked for them in banked Swedish breast milk samples.
78. Daiva Meironyté, Koidu Norén, and Åke Bergman, "Analysis of Polybrominated Diphenyl Ethers in Swedish Human Milk. A Time-Related Trend Study, 1972–1997," *Journal of Toxicology and Environmental Health* 58, no. 6 (1999): 329–341.
79. Koidu Norén and Daiva Meironyté, "Certain Organochlorine and Organobromine Contaminates in Swedish Human Milk in Perspective of Past 20–30 Years," *Chemosphere* 40 (2000): 1111–1123.
80. Meironyté et al., "Analysis of Polybrominated Diphenyl Ethers." Per Eriksson, Eva Jakobsson, and Anders Fredriksson, "Brominated Flame Retardants: A Novel Class of Developmental Neurotoxicants in Our Environment?" *Environmental Health Perspectives* 109, no. 9 (2001): 903–908.
81. Author's analysis, April 2015. Based on similar graphs published in: Phil Brown and Alissa Cordner, "Lessons Learned from Flame Retardant Use and Regulation Could Enhance Future Control of Potentially Hazardous Chemicals," *Health Affairs* 30, no. 5 (2010): 906–914.
82. Ruthann A. Rudel et al., "Phthalates, Alkylphenols, Pesticides, Polybrominated Diphenyl Ethers, and Other Endocrine-Disrupting Compounds in Indoor Air and Dust," *Environmental Science & Technology* 37, no. 20 (2003): 4543–4553.
83. Andreas Sjödin et al., "Serum Concentrations of Polybrominated Diphenyl Ethers (PBDEs) and Polybrominated Biphenyl (PBB) in the United States Population: 2003–2004," *Environmental Science & Technology* 42, no. 4 (2008): 1377–1384; Zota et al., "Elevated House Dust."
84. Thomas A. McDonald, "Polybrominated Diphenylether Levels among United States Residents: Daily Intake and Risk of Harm to the Developing Brain and Reproductive Organs," *Integrated Environ Assess Management* 1, no. 4 (2005): 343–354.
85. U.S. EPA, *PBDEs Action Plan*.
86. Ruth Alcock and Jerry Busby, "Risk Migration and Scientific Advance," 369–381.
87. U.S. EPA, *Polybrominated Diphenyl Ethers (PBDEs) Project Plan* (Washington, D.C.: U.S. EPA, 2006), http://www.epa.gov/oppt/pbde/pubs/pbdestatus1208.pdf.

88. U.S. EPA, *Polybrominated Diphenyl Ethers (PBDEs)*; U.S. EPA, "Existing Chemicals Action Plans" (Washington, D.C.: U.S. EPA, 2012).

89. U.S. Congress, "Toxic Substances Control Act."

90. U.S. EPA, *Essential Principles for Reform of Chemicals Management Legislation* (Washington, D.C.: U.S. EPA, 2009).

91. U.S. EPA, "Voluntary Children's Chemical Evaluation Program (VCCEP)," last updated August 31, 2010, http://www.epa.gov/oppt/vccep/.

92. U.S. EPA, "VCCEP Status: Pentabromodiphenyl Ether" (Environmental Protection Agency, 2010), http://www.epa.gov/oppt/vccep/pubs/chem22.html; U.S. EPA, "VCCEP Status: Octabromodiphenyl Ether" (2010), http://www.epa.gov/oppt/vccep/pubs/chem23.html.

93. Myrto Petreas et al., "High Body Burdens of 2,2',4,4'-Tetrabromodiphenyl Ether (BDE-47) in California Women," *Environmental Health Perspectives* 111, no. 9 (March 10, 2003): 1175–1179; Jianwen She et al., "PBDEs in the San Francisco Bay Area: Measurements in Harbor Seal Blubber and Human Breast Adipose Tissue," *Chemosphere* 46, no. 5 (February 2002): 697–707.

94. State of California, *AB 302: Polybrominated Diphenyl Ether*, 2003, http://www.leginfo.ca.gov/pub/03-04/bill/asm/ab_0301-0350/ab_302_bill_20030811_chaptered.html.

95. National Caucus of Environmental Legislators, "Status: PBDE Legislation," 2007, www.ncel.net/articles/PBDE.Legislation.Laws.Website.doc.

96. Marshall Moore, "Testimony Regarding Oversight of EPA Authorities and Actions to Control Exposures to Toxic Chemicals," U.S. Senate Committee Hearing on Environment and Public Works. July 24, 2014.

97. Design for the Environment, *Furniture Flame Retardancy Partnership: Environmental Profiles of Chemical Flame-Retardant Alternatives for Low-Density Polyurethane Foam* (Washington, D.C., 2005), http://www.epa.gov/dfe/pubs/flameret/altrep-v1/altrepv1-f1c.pdf; Design for the Environment, *Flame Retardants Used in Flexible Polyurethane Foam: An Alternatives Assessment Update* (Washington, D.C., 2015), http://www.epa.gov/saferchoice/flame-retardants-used-flexible-polyurethane-foam.

98. This timeline shares much in common with a case of the pharmaceutical phenylpropanolamine discussed by David Michaels. In both cases, the industry objected to scientific assessments of their products and delayed regulatory action until after a replacement product had been developed. David Michaels, "Manufactured Uncertainty: Contested Science and the Protection of the Public's Health and Environment," in *Agnotology: The Making and Unmaking of Ignorance*, ed. Robert Proctor and Londa Schiebinger (Stanford, Calif.: Stanford University Press, 2008), 97.

99. U.S. EPA, *PBDEs Project Plan*.

100. Congeners are varieties of the same chemical. There are 209 possible PBDE congeners that differ by the number of bromine atoms attached to carbon rings. U.S. EPA, *PBDEs Action Plan*.

101. Charles Auer, "Letter from Charles Auer, U.S. EPA, to Susan Lewis, American Chemistry Council. August 25th" (Washington, D.C.: U.S. EPA, 2005), http://www.epa.gov/oppt/vccep/pubs/chem21.html.

102. James Gulliford, "Letter from James Gulliford, U.S. EPA, to Nancy Sandrof, American Chemistry Council" (Washington, D.C.: U.S. EPA, 2008), http://www.epa.gov/oppt/vccep/pubs/decaltr.pdf.

103. Chris Bryant, "Letter from Chris Bryant, American Chemistry Council, to James Gulliford, U.S. EPA, July 17, 2008," 2008, http://www.epa.gov/oppt/vccep/pubs/090508.pdf; Charles Auer, "Letter from Charles M., U.S. EPA, to Chris Bryant, American Chemistry Council, September 5, 2008," 2008, http://www.epa.gov/oppt/vccep/pubs/090508.pdf.

104. U.S. EPA, "DecaBDE Phase-out Initiative" (Washington, D.C.: U.S. EPA, 2009).

105. The restriction of DecaBDE in 2008 through the European Union's Restriction of Hazardous Substances in Electronics may also have played a role. European Union, "Restriction of the Use of Certain Hazardous Substances (RoHS)." Directive 2011/65/EU, 2011.

106. U.S. Office of the Inspector General, *EPA's Voluntary Chemical Evaluation Program Did Not Achieve Children's Health Protection Goals*, Report No. 11-P-0379 (Washington, D.C.: U.S. EPA, 2011).

107. U.S. EPA, "Existing Chemicals Action Plans."

108. The "chemicals of concern" list was eventually withdrawn from review by the Office of Information and Regulatory Affairs after waiting for over 1,200 days, despite a requirement that rules be reviewed within 90 days. Richard Denison, "Stymied at Every Turn: EPA Withdraws Two Draft TSCA Proposals in the Face of Endless Delay at OMB," *EDF Health,* September 6, 2013, http://blogs.edf.org/health/2013/09/06/stymied-at-every-turn-epa-withdraws-two-draft-tsca-proposals-in-the-face-of-endless-delay-at-omb/; U.S. Office of Information and Regulatory Affairs, "Reginfo.gov: EPA/OCSPP TSCA Chemicals of Concern List" (Washington, D.C.: Office of Information and Regulatory Affairs, 2013). On the DecaBDE Alternatives Assessment, see U.S. EPA, "An Alternatives Assessment for DecaBDE."

109. U.S. EPA, "Certain Polybrominated Diphenylethers: Significant New Use Rule and Test Rule, 40 CFR Parts 721, 795, and 799," *Federal Register* 77, no. 63 (2012): 19862–19898.

110. U.S. Office of Information and Regulatory Affairs, "View Rule: RIN 2070-AJ08," *Reginfo.gov,* 2015, http://reginfo.gov/public/do/eAgendaViewRule?pubId=201210&RIN=2070-AJ08; Michael Hawthorne, "Toxic Flame Retardant May Get a Reprieve," *Chicago Tribune,* December 20, 2012, http://articles.chicagotribune.com/2012-12-20/news/ct-met-flames-chemical-rules-20121220_1_flame-retardants-pbdes-ban-toxic-chemicals.

111. Boeing Company, "Boeing/OIRA Discussion of Proposed EPA Regulations Regarding DecaBDE: February 16, 2011," presentation to U.S. Office of Information

and Regulatory Affairs, 2011, http://www.whitehouse.gov/sites/default/files/omb/assets /oira_2070/2070_02162011–1.pdf; Lawrence Cullen, "Letter from Lawrence Cullen, Counsel for iGPS, to Maria Doa, Environmental Protection Agency." July 31, 2012.

112. Sabine Kemmlein, Dorte Herzke, and Robin J. Law, "Brominated Flame Retardants in the European Chemicals Policy of REACH—Regulation and Determination in Materials," *Journal of Chromatography A* 1216, no. 3 (2009): 320–333.

113. Alissa Sasso, "ECHA Raises Its Sights: Several Recent Additions to the REACH Candidate List Set Precedents," Environmental Defense Fund, January 29, 2013, http://blogs.edf.org/nanotechnology/2013/01/29/echa-raises-its-sights-several-recent-additions-to-the-reach-candidate-list-set-precedents.

114. Kellyn Betts, "Glut of Data on 'New' Flame Retardant Documents Its Presence All Over the World," *Environmental Science & Technology* 43, no. 2 (2009): 236–237; Hites, "PBDEs in the Environment."

115. Centers for Disease Control and Prevention, *Fourth National Report on Human Exposure to Environmental Chemicals* (Atlanta: CDC, 2009).

116. McDonald, "PBDEs among United States Residents," 343; Douglas Fischer et al., "Children Show Highest Levels of Polybrominated Diphenyl Ethers in a California Family of Four: A Case Study," *Environmental Health Perspectives* 114, no. 10 (2006): 1581–1584; Lucio G. Costa and Gennaro Giordano, "Developmental Neurotoxicity of Polybrominated Diphenyl Ether (PBDE) Flame Retardants," *Neurotoxicology* 28, no. 6 (2007): 1047–1067.

117. Kellyn Betts, "Could Flame Retardants Deter Electronics Recycling?," *Environmental Science & Technology* 35, no. 3 (2001): 58A; Andreas Sjodin and Hakan Carlsson, "Flame Retardants in Indoor Air at an Electronics Recycling Plant and at Other Work Environments," *Environmental Science & Technology* 35, no. 3 (2001): 448–454; Susan D. Shaw et al., "Persistent Organic Pollutants Including Polychlorinated and Polybrominated Dibenzo-P-Dioxins and Dibenzofurans in Firefighters from Northern California," *Chemosphere* 91, no. 10 (2013): 1386–1394; Joseph G. Allen et al., "Exposure to Flame Retardant Chemicals on Commercial Airplanes," *Environmental Health* 12 (2013): 17.

118. Aimin Chen et al., "Prenatal Polybrominated Diphenyl Ether Exposures and Neurodevelopment in U.S. Children through 5 Years of Age: The HOME Study," *Environmental Health Perspectives* 122, no. 8 (2014): 856–862; Ami Zota, Gary Adamkiewicz, and Rachel Morello-Frosch, "Are PBDEs an Environmental Equity Concern? Exposure Disparities by Socioeconomic Status," *Environmental Science & Technology* 44, no. 15 (2010): 5691–5692.

119. Megan K. Horton et al., "Predictors of Serum Concentrations of Polybrominated Flame Retardants among Healthy Pregnant Women in an Urban Environment: A Cross-Sectional Study," *Environmental Health* 12 (2013): 23.

120. Ruthann A. Rudel and Laura J. Perovich, "Endocrine Disrupting Chemicals in Indoor and Outdoor Air," *Atmospheric Environment* 43, no. 1 (2009): 170–181.

121. Horton et al., "Predictors of Serum Concentrations," 23

122. Kellyn Betts, "Unwelcome Guest: PBDEs in Indoor Dust," *Environmental Health Perspectives* 116, no. 5 (May 2008): A202–A208; Birnbaum and Staskal, "Brominated Flame Retardants," 12–16; U.S. EPA, *Toxicological Review of Decabromodiphenyl Ether (BDE-299) in Support of Summary Information on the Integrated Risk Information System. EPA 635-R-07-008F* (Washington, D.C.: U.S. EPA, Integrated Risk Information System, 2008), www.epa.gov/ncea/iris/toxreviews/0035tr.pdf.

123. John Biesemeier et al., "An Oral Developmental Neurotoxicity Study of Decabromodiphenyl Ether (DecaBDE) in Rats," *Birth Defects Research* 92 (2011): 17–35.

124. This is not the case for all health effects. For example, PBDE exposure has been associated with hypothyroidism in animals but hyperthyroidism in humans. Jonathan Chevrier et al., "Polybrominated Diphenylether (PBDE) Flame Retardants and Thyroid Hormone during Pregnancy," *Environmental Health Perspectives* 118, no. 10 (2010): 1444–1449.

125. Henrik Viberg, Anders Fredriksson, and Per Eriksson, "Neonatal Exposure to Polybrominated Diphenyl Ether (PBDE 153) Disrupts Spontaneous Behavior, Impairs Learning and Memory, and Decreases Hippocampal Cholinergic Receptors in Adult Mice," *Toxicology and Applied Pharmacology* 192, no. 2 (2003): 95–106; Henrik Viberg et al., "Neurobehavioral Derangements in Adult Mice Receiving Decabrominated Diphenyl Ether (PBDE 209) during a Defined Period of Neonatal Brain Development," *Toxicological Sciences* 76, no. 1 (2003): 112–120; Deborah C. Rice et al., "Developmental Delays and Locomotor Activity in the C57BL6/J Mouse Following Neonatal Exposure to the Fully-Brominated PBDE, Decabromodiphenyl Ether," *Neurotoxicology and Teratology* 29, no. 4 (2007): 511–520.

126. Chen et al., "Prenatal PBDE Exposures"; Julie B. Herbstman et al., "Prenatal Exposure to PBDEs and Neurodevelopment," *Environmental Health Perspectives* 118, no. 5 (2010): 712–719; Brenda Eskenazi et al., "In Utero and Childhood Polybrominated Diphenyl Ether (PBDE) Exposures and Neurodevelopment in the CHAMACOS Study," *Environmental Health Perspectives* 121, no. 2 (2013): 257–262.

127. N. Abdelouahab, Y. Ainmelk, and L. Takser, "Polybrominated Diphenyl Ethers and Sperm Quality," *Reproductive Toxicology* 31, no. 4 (May 2011): 546–550; Kim G. Harley et al., "PBDE Concentrations in Women's Serum and Fecundability," *Environmental Health Perspectives* 118, no. 5 (2010): 699–704; Chevrier et al., "PBDE Flame Retardants."

128. U.S. EPA, *Flame Retardants in Printed Circuit Boards* (Washington, D.C., 2014), http://www2.epa.gov/sites/production/files/2015-01/documents/pcb_updated_draft_report.pdf.

129. U.S. EPA, "Hexabromocyclododecane (HBCD) Action Plan," 2010, http://www.epa.gov/oppt/existingchemicals/pubs/actionplans/RIN2070-AZ10_HBCD action plan_Final_2010-08-09.pdf; Stockholm Convention, "Reports and Decisions," 2014, http://chm.pops.int/ConventionoftheParties(COP)/Decisions/tabid/208/Default.aspx.

130. Design for the Environment, *Flame Retardant Alternatives for Hexabromocyclododecane (HBCD)* (Washington, D.C., 2014), http://www.epa.gov/dfe/pubs/projects/hbcd/hbcd-full-report-508.pdf.

131. Veena Singla, "HBCD Alternatives Assessment: Narrow Focus Misses Large Problems," Green Science Policy Institute, October 10, 2013, http://greensciencepolicy.org/hbcd-alternatives-assessment-narrow-focus-misses-large-problems/.

132. Dodson et al., "After the PBDE Phase-Out," 13056.

133. Dodson et al., "After the PBDE Phase-Out"; Stapleton et al., "Novel and High Volume Use"; Stapleton et al., "Detection of Organophosphate Flame Retardants."

134. Laura Dishaw et al., "Is the PentaBDE Replacement, Tris(1,3-Dichloro-2-Propyl)phosphate (TDCPP), a Developmental Neurotoxicant? Studies in PC12 Cells," *Toxicology and Applied Pharmacology* 256, no. 3 (2011): 281–289.

135. John D. Meeker and Heather M. Stapleton, "House Dust Concentrations of Organophosphate Flame Retardants in Relation to Hormone Levels and Semen Quality Parameters," *Environmental Health Perspectives* 118, no. 3 (2010): 318–323.

136. Heather Patisaul et al., "Accumulation and Endocrine Disrupting Effects of the Flame Retardant Mixture Firemaster 550 in Rats: An Exploratory Assessment," *Journal of Biochemical Molecular Toxicology* 27, no. 2 (2012): 124–136.

137. Jonathan S. Bearr, Heather M. Stapleton, and Carys L. Mitchelmore, "Accumulation and DNA Damage in Fathead Minnows (Pimephales Promelas) Exposed to 2 Brominated Flame-Retardant Mixtures, Firemaster 550 and Firemaster BZ-54," *Environmental Toxicology and Chemistry* 29, no. 3 (2010): 722–729.

138. State of California, "Chemicals Known to the State to Cause Cancer or Reproductive Toxicity" (Sacramento: California EPA, 2012); New York State, "A6195–2011: Prohibits the Sale of Child Products Containing Tris," 2011; Washington State Department of Ecology, "Rule Development—Children's Safe Products Reporting Rule," 2012, http://www.ecy.wa.gov/programs/swfa/rules/ruleChildSafe.html; Israeli Chemicals Limited, "ICL to Increase Production of Polymeric Flame Retardant at West Virginia Facility," *PR Newswire*, November 13, 2012, http://www.prnewswire.com/news-releases/icl-to-increase-production-of-polymeric-flame-retardant-at-west-virginia-facility-179095971.html.

139. U.S. EPA, "TSCA Work Plan Chemicals" (Washington, D.C.: U.S. EPA, 2014), http://www.epa.gov/oppt/existingchemicals/pubs/workplans.html.

140. Inorganic Phosphorous and Nitrogen Flame Retardants Association, "About Us," accessed October 24, 2012. http://www.pinfa.org/about-us.html.

141. Donald Demko, "So You Want a Flame-Retarded Formulation," *Plastics Engineering* 59, no. 1 (2003).

142. California Legislature, "SB-1019: Upholstered Furniture: Flame Retardant Chemicals," 2014, http://leginfo.legislature.ca.gov/faces/billNavClient.xhtml?bill_id=201320140SB1019.

143. Consumer Products Safety Commission, *Upholstered Furniture Full Scale Chair Tests—Open Flame Ignition Results and Analysis* (Bethesda, Md.: CPSC Division

of Combustion and Fire Sciences, 2012), http://www.cpsc.gov/PageFiles/93436
/openflame.pdf.

144. Susan Shaw et al., "Halogenated Flame Retardants: Do the Fire Safety Benefits Justify
the Risks?," *Reviews on Environmental Health* 25, no. 4 (2010): 261–305; Anna Stec,
"Influence of Fire Retardants on Fire Toxicity," presented at the American Chemical
Society Meeting, March 27, 2012, San Diego.

145. National Commission on Fire Prevention and Control, *America Burning* (Washington, D.C., 1973).

3. DEFENDING RISK AND DEFINING SAFETY

Parts of this chapter previously appeared in Alissa Cordner, "Defining and Defending Risk."

1. Lavoie et al., "Chemical Alternatives Assessment." DfE also has a labeling program
for safer consumer products, recently rebranded as the Safer Choice program. U.S.
EPA, "Safer Choice," accessed April 30, 2015, http://www2.epa.gov/saferchoice.

2. U.S. EPA, "Design for the Environment Program Alternatives Assessment Criteria for
Hazard Evaluation" (Washington, D.C.: EPA, Office of Pollution Prevention and
Toxics, 2011), http://www.epa.gov/dfe/alternatives_assessment_criteria_for_hazard
_eval.pdf.

3. National Research Council, *Risk Assessment in the Federal Government: Managing the
Process* (Washington, D.C., 1983).

4. Scott Frickel, Richard Campanella, and M. Bess Vincent, "Mapping Knowledge In-
vestments in the Aftermath of Hurricane Katrina: A New Approach for Assessing
Regulatory Agency Responses to Environmental Disaster," *Environmental Science &
Policy* 12, no. 2 (2009): 119–133.

5. Judith Bradbury, "The Policy Implications of Differing Concepts of Risk," *Science,
Technology, & Human Values* 14, no. 4 (1989): 380–399; William Freudenburg and
Susan Pastor, "Public Responses to Technological Risks: Toward a Sociological Per-
spective," *Sociological Quarterly* 33, no. 3 (1992): 389–512.

6. On disciplinary perspectives on risk, see Catherine Althaus, "A Disciplinary Perspec-
tive on the Epistemological Status of Risk," *Risk Analysis* 25, no. 3 (2005): 567–588;
Peter Taylor-Gooby and Jens Zinn, "Current Directions in Risk Research: New
Developments in Psychology and Sociology," *Risk Analysis* 26, no. 2 (2006): 397–411.
On risk perception, see Paul Slovic, "Perception of Risk," *Science* 236, no. 4799
(1987): 280–285; Peter Sandman, "Hazard versus Outrage in the Public Perception of
Risk," in *Effective Risk Communication: The Role and Responsibility of Government
and Nongovernment Organizations*, ed. V. T. Covello, David McCallum, and Maria
Pavlova (New York: Plenum, 1989), 45–49. On organizational risk studies, see Lee
Clarke, "Explaining Choices among Technological Risks," *Social Problems* 35, no. 1
(1988): 22–35.

7. Terje Aven, *Misconceptions of Risk* (Chichester, U.K.: Wiley, 2010).

8. Kelly Joyce, "Is Tuna Safe? A Sociological Analysis of Federal Fish Advisories," in *Mercury Pollution: A Transdisciplinary Treatment*, ed. Sharon L. Zuber and Michael C. Newman (Boca Raton, Fla.: CRC, 2011), 72–73.

9. Ortwin Renn, *Risk Governance: Coping with Uncertainty in a Complex World* (London: Earthscan, 2008).

10. Ulrich Beck, "Climate for Change, or How to Create a Green Modernity?" *Theory, Culture, and Society* 27 (2010): 261.

11. Terje Aven, "Foundational Issues in Risk Assessment and Risk Management," *Risk Analysis* 32, no. 10 (2012): 1647–1656.

12. Aven, *Misconceptions of Risk*.

13. Althaus, "Disciplinary Perspective," 567.

14. Sheldon Krimsky and Dominic Golding, *Social Theories of Risk* (Westport, Conn.: Praeger, 1992); Jens Zinn, "Introduction: The Contribution of Sociology to the Discourse on Risk and Uncertainty," in *Social Theories of Risk and Uncertainty: An Introduction*, ed. Jens Zinn (Malden, Mass.: Blackwell, 2008), 1–17.

15. William Freudenburg, "Perceived Risk, Real Risk: Social Science and the Art of Probabilistic Risk Assessment," *Science* 242, no. 4875 (1988): 44–49; Kathleen Tierney, "Toward a Critical Sociology of Risk," *Sociological Forum* 14, no. 2 (1999): 215–242.

16. Kinchy, *Seeds, Science, and Struggle*, 80.

17. Frickel and Moore, *New Political Sociology of Science*; Kelly Moore et al., "Science and Neoliberal Globalization: A Political Sociological Approach," *Theory and Society* 40, no. 5 (2011): 505–532.

18. National Research Council, *Science and Decisions*.

19. For example, susceptibility, vulnerability, and sensitivity are often defined based on each other without a stand-alone definition.

20. U.S. EPA, *Guidelines for Exposure Assessment* (Washington, D.C.: U.S. EPA, Risk Assessment Forum, 1992).

21. U.S. EPA, *Community Air Screening How-To Manual*, EPA-744-B-04-001 (Research Triangle Park, N.C.: U.S. EPA, Office of Air Quality Planning and Standards, 2004), http://www.epa.gov/oppt/cahp/pubs/howto.htm.

22. U.S. EPA, "Pesticides: Glossary" (Washington, D.C.: EPA, Office of Pesticide Programs, 2012).

23. U.S. EPA, "A Dictionary of Technical and Legal Terms Related to Drinking Water" (Washington, D.C.: EPA, n.d.).

24. New Jersey Department of Environmental Protection, *Strategies for Addressing Cumulative Impacts in Environmental Justice Communities* (Trenton: N.J. Department of Environmental Protection, 2009), http://www.nj.gov/dep/ej/docs/ejac_impacts_report200903.pdf.

25. Holger Hoffmann-Riem and Brian Wynne, "In Risk Assessment, One Has to Admit Ignorance," *Nature* 416, no. 6877 (2002): 123.

26. This is similar to Zinn's technical-objectivist risk definition, or Aven's risk definition relating consequences, severity, and uncertainty about an activity. Zinn, "Introduction," 1–17; Aven, *Misconceptions of Risk*, 227.

27. Charles Perrow, *Normal Accidents: Living with High-Risk Technologies* (New York: Basic Books, 1984).

28. Ira Richards, *Principles and Practice of Toxicology in Public Health* (Boston: Jones and Bartlett, 2008).

29. U.S. EPA, "New Chemicals: Assessing Risk" (Washington, D.C.: EPA, 2010), http://www.epa.gov/oppt/newchems/pubs/assess.htm, emphasis in original.

30. U.S. EPA, "IRIS Glossary" (Washington, D.C.: EPA, 2011), http://ofmpub.epa.gov/sor_internet/registry/termreg/searchandretrieve/glossariesandkeywordlists/search.do?details=&glossaryName=IRIS%20Glossary.

31. Langdon Winner, *The Whale and the Reactor* (Chicago: University of Chicago Press, 1986).

32. For this reason, some risk assessment methodologies, like the IRIS assessments, also include an estimation of their "confidence" in their assessment, based on the quantity and quality of the data included.

33. Rose Hoban, "Q&A with NIEHS Head Linda Birnbaum," *North Carolina Health News*, April 2, 2012, http://www.northcarolinahealthnews.org/2012/04/02/q-a-with-niehs-head-linda-birnbaum-on-bpa/.

34. Edward J. Calabrese and Linda A. Baldwin, "Toxicology Rethinks Its Central Belief," *Nature* 421, no. 6924 (2003): 691–2; Rylee Phuong Do et al., "Non-Monotonic Dose Effects of in Utero Exposure to di(2-Ethylhexyl) Phthalate (DEHP) on Testicular and Serum Testosterone and Anogenital Distance in Male Mouse Fetuses," *Reproductive Toxicology* 34, no. 4 (December 2012): 614–621.

35. Edward J. Calabrese and Linda A. Baldwin, "Defining Hormesis," *Human and Experimental Toxicology* 21, no. 2 (2002): 91–97.

36. Laura N. Vandenberg et al., "Hormones and Endocrine-Disrupting Chemicals: Low-Dose Effects and Nonmonotonic Dose Responses," *Endocrine Reviews* 33, no. 3 (2012): 378–455.

37. Krimsky, *Hormonal Chaos*.

38. U.S. EPA, "Endocrine Disruptor Screening Program for the 21st Century: EDSP21 Work Plan" (Washington, D.C.: U.S. EPA, Office of Chemical Safety and Pollution Prevention, 2011), http://www.epa.gov/endo/pubs/edsp21_work_plan_summary_overview_final.pdf.

39. Åke Bergman et al., *The State-of-the-Science of Endocrine Disrupting Chemicals* (Geneva: World Health Organization, 2012), http://www.who.int/ceh/publications/endocrine/en/.

40. Susan Bell, *DES Daughters: Embodied Knowledge and the Transformation of Women's Health Politics* (Philadelphia: Temple University Press, 2009).

41. Charles Schmidt, "Uncertain Inheritance: Transgenerational Effects of Environmental Exposures," *Environmental Health Perspectives* 121, no. 10 (2013): 298–303.

42. U.S. EPA, "Office of Children's Health Protection" (Washington, D.C.: EPA, 2013), http://www2.epa.gov/children.

43. On evidence-based toxicology, see: Sebastian Hoffman and Thomas Hartung, "Toward an Evidence-Based Toxicology," *Human and Experimental Toxicology* 25 (2006): 497–513.

44. Chevrier et al., "PBDE Flame Retardants"; Costa and Giordano, "Developmental Neurotoxicity"; Herbstman et al., "Prenatal Exposure to PBDEs."

45. Shostak, *Exposed Science.*

46. Gerda Egger et al., "Epigenetics in Human Disease and Prospects for Epigenetic Therapy," *Nature* 429, no. 6990 (2004): 457–463.

47. Christopher Williams, "Polymeric Flame Retardants: Possible Less Hazardous Alternatives to Decabromodiphenyl Ether?" presented at the 13th Workshop on Brominated and Other Flame Retardants, Winnipeg, Canada, June 4–5, 2012.

48. Maxwell, *Understanding Environmental Health*, 19.

49. On exposure pathways and PBDEs, see Joseph Allen et al., "Personal Exposure to Polybrominated Diphenyl Ethers (PBDEs) in Residential Indoor Air," *Environmental Science & Technology* 41, no. 13 (2007): 4574–4579.

50. American Chemistry Council, "ACC Responds to Stapleton Study on the Presence of Flame Retardants in Children's Products" (Washington, D.C.: American Chemistry Council, 2011), http://www.americanchemistry.com/Media/PressReleasesTranscripts /ACC-news-releases/ACC-Responds-to-Stapleton-Study-on-the-Presence-of-Flame -Retardants-in-Childrens-Products.html.

51. Wendy Epseland and Mitchell Stevens, "Commensuration as a Social Process," *American Journal of Sociology* 24 (1998): 313–343.

52. VECAP, "The Voluntary Emissions Control Action Programme," 2014, http://www .vecap.info/.

53. U.S. EPA, "Design for the Environment Program Alternatives Assessment Criteria."

54. U.S. EPA, "Design for the Environment Program."

55. E. Gottmann et al., "Data Quality in Predictive Toxicology: Reproducibility of Rodent Carcinogenicity Experiments," *Environmental Health Perspectives* 109, no. 5 (2001): 509–514.

56. Paul Anastas and John Warner, *Green Chemistry: Theory and Practice* (Oxford: Oxford University Press, 1998).

57. U.S. EPA, "Alternatives Assessment for the DecaBDE," 4-29–4-33.

58. U.S. EPA, "Alternatives Assessment for the DecaBDE," 4–268.

59. U.S. EPA, "Summary of the Toxic Substances Control Act: 15 U.S.C. §2601 et Seq.," 2014, http://www2.epa.gov/laws-regulations/summary-toxic-substances-control-act.

60. U.S. EPA, "TSCA Workplan Chemicals: Methods Document" (Washington, D.C.: EPA, 2012), http://www.epa.gov/oppt/existingchemicals/pubs/wpmethods.pdf.

61. U.S. EPA, "TSCA Workplan Chemicals," 6.

62. William Ruckelshaus, "Risk in a Free Society," *Risk Analysis* 4, no. 3 (1984): 157–162.

63. Brown, *Toxic Exposures.*

64. Naomi Oreskes and Erik M. Conway make similar arguments about the tobacco and fossil fuel industries. *Merchants of Doubt: How a Handful of Scientists Obscured the Truth on Issues from Tobacco Smoke to Global Warming* (New York: Bloomsbury, 2010).

65. American Sustainable Business Council, "Poll of Small Business Owners on Toxic Chemicals" (Washington, D.C.: American Sustainable Business Council, 2012); Safer Chemicals Healthy Families, "Presentation of Findings from a Survey of 825 Voters in 75 Swing Congressional Districts," Washington, D.C.: Mellman Group (2010), http://www.saferchemicals.org/PDF/resources/schf-poll-final.pdf.

66. Rebecca Gasior Altman et al., "Pollution Comes Home and Gets Personal: Women's Experience of Household Chemical Exposure," *Journal of Health and Social Behavior* 49, no. 4 (2008): 417–435.

67. Environmental Working Group, *Body Burden: The Pollution in Newborns* (Washington D.C., 2005), http://www.ewg.org/research/body-burden-pollution-newborns.

68. Frickel et al., "Mapping Knowledge Investments," 132.

69. Vogel, *Is It Safe?*

70. Tony Iallonardo, "'Safe Chemicals Act of 2011' Introduced Today Legislation Would Protect American Families from Toxic Chemicals," Safer Chemicals Healthy Families, April 14, 2011.

71. For example, it is a common misconception that chemicals included in alternatives assessment reports are preferred alternatives. In fact, DfE explicitly states that this is not the case: "The alternatives included in this assessment are viable and functional but not necessarily preferable." U.S. EPA, "Alternatives Assessment for DecaBDE," 1–2.

72. David Magnus, "Risk Management versus the Precautionary Principle: Agnotology as a Strategy in the Debate over Genetically Engineered Organisms," in *Agnotology: The Making and Unmaking of Ignorance*, ed. Robert Proctor and Londa Schiebinger (Stanford, Calif.: Stanford University Press, 2008), 252.

73. David Kriebel and Joel Tickner, "Reenergizing Public Health through Precaution," *American Journal of Public Health* 91, no. 9 (2001): 1351–1355.

74. Magnus, "Risk Management," 262.

4. STRATEGIC SCIENCE TRANSLATION

Parts of this chapter appeared previously in Cordner, "Strategic Science Translation."

1. BFR Workshop, "13th Workshop on Brominated and Other Flame Retardants," June 4–6, 2012, Winnipeg, Canada. https://web.archive.org/web/20140517004744/http://bfr2012.org/.

2. Part of this study was published as Sarah C. Marteinson et al., "The Flame Retardant B-1,2-Dibromo-4-(1, 2-Dibromoethyl)cyclohexane: Fate, Fertility, and Reproductive Success in the American Kestrels (*Falco sparverius*)," *Environmental Science & Technology* 46, no. 15 (2012): 8440–8447.

3. Abby J. Kinchy and Daniel Lee Kleinman, "Organizing Credibility: Discursive and Organizational Orthodoxy on the Borders of Ecology and Politics," *Social Studies of Science* 33, no. 6 (2003): 380.

4. Pierre Bourdieu, *Science of Science and Reflexivity* (Chicago: University of Chicago Press, 2004); Robert King Merton and Norman William Storer, *The Sociology of Science: Theoretical and Empirical Investigations* (Chicago: University of Chicago Press, 1973); Michael Polanyi, "The Republic of Science: Its Political and Economic Theory," *Minerva* 38 (2000): 1–32; Mathieu Albert and Daniel Lee Kleinman, "Bringing Pierre Bourdieu to Science and Technology Studies," *Minerva* 49 (2011): 263–273.

5. Thomas Gieryn, *Cultural Boundaries of Science* (Chicago: University of Chicago Press, 1999); Thomas Gieryn, "Boundary-Work and the Demarcation of Science from Non-Science: Strains and Interests in Professional Ideologies of Scientists," *American Sociological Review* 48, no. 6 (1983): 781–795; Sheila S. Jasanoff, "Contested Boundaries in Policy-Relevant Science," *Social Studies of Science* 17, no. 2 (May 1987): 195–230, http://www.jstor.org/stable/284949.

6. Kelly Moore, "Organizing Integrity: American Science and the Creation of Public Interest Organizations, 1955–1975," *American Journal of Sociology* 101, no. 6 (May 1996): 1592–1627.

7. Althaus, "A Disciplinary Perspective," 569.

8. Andy Stirling and David Gee, "Science, Precaution, and Practice," *Public Health Reports* 117, no. 6 (2002): 521–533.

9. Matthias Gross, "The Unknown in Process: Dynamic Connections of Ignorance, Non-Knowledge and Related Concepts," *Current Sociology* 55 (2007): 751.

10. There are socially valued areas of ignorance as well; for example, some archaeologists argue that artifacts should not be studied if they were not legally and ethically obtained, and institutional review boards are predicated on the idea that some types of research are unacceptable. Robert Proctor and Londa Schiebinger, *Agnotology: The Making and Unmaking of Ignorance* (Stanford, Calif.: Stanford University Press, 2008).

11. National Research Council, *Science and Decisions*, 4.

12. Frickel and Edwards, "Untangling Ignorance," 229.

13. Alissa Cordner and Phil Brown, "Moments of Uncertainty: Ethical Considerations and Emerging Contaminants," *Sociological Forum* 28, no. 3 (2013): 469–494.

14. Linsey McGoey, "Strategic Unknowns: Towards a Sociology of Ignorance," *Economy and Society* 41, no. 1 (2012): 1–16.

15. Frickel et al., "Undone Science"; Hess, "The Potentials and Limitations."

16. Joanna Kempler, Jon F. Merz, and Charles L. Bosk, "Forbidden Knowledge: Public Controversy and the Production of Nonknowledge," *Sociological Forum* 26, no. 3 (2011): 475–500.

17. Proctor and Schiebinger, *Agnotology*, 8.

18. Gerald E. Markowitz and David Rosner, *Deceit and Denial: The Deadly Politics of Industrial Pollution* (Berkeley: University of California Press, 2002); David Michaels,

Doubt Is Their Product: How Industry's Assault on Science Threatens Your Health (New York: Oxford University Press, 2008); Oreskes and Conway, *Merchants of Doubt*.

19. Sheldon Krimsky, "The Funding Effect in Science and Its Implications for the Judiciary," *Journal of Law and Policy* 8, no. 1 (2005): 43–68; Richard Smith, "Medical Journals Are an Extension of the Marketing Arm of Pharmaceutical Companies," *PLoS Medicine* 2, no. 5 (2005): e138. This pattern exists in BPA research, though no comparable study of flame retardant funding exists: Frederick vom Saal and Claude Hughes, "An Extensive New Literature Concerning Low-Dose Effects of Bisphenol A Shows the Need for a New Risk Assessment," *Environmental Health Perspectives* 113, no. 8 (2005): 926–933.

20. Nicholas Shapiro, "Un-Knowing Exposure: Toxic Emergency Housing, Strategic Inclusivity and Governance in the US Gulf South," in *Knowledge, Technology, and Law*, ed. Emilie Cloatre and Martyn Pickersgill (Hoboken, N.J.: Taylor and Francis, 2014), 192.

21. Pierre Bourdieu, "The Peculiar History of Scientific Reason," *Sociological Forum* 6, no. 1 (1991): 3.

22. Albert and Kleinman, "Bringing Pierre Bourdieu," 266.

23. Abby J. Kinchy, "Anti-Genetic Engineering Activism and Scientized Politics in the Case of 'Contaminated' Mexican Maize," *Agriculture and Human Values* 27, no. 4 (2010): 505–517; Michaels and Monforton, "Manufacturing Uncertainty"; Morello-Frosch et al., "Embodied Health Movements."

24. Frickel et al., "Undone Science," 464.

25. Moore, *Disrupting Science*; Scott Frickel and Neil Gross, "A General Theory of Scientific/Intellectual Movements," *American Sociological Review* 70, no. 2 (2005): 204–232; Shostak, *Exposed Science*.

26. Frickel et al., "Undone Science"; Hess, "The Potentials and Limitations."

27. George Devereux, *From Anxiety to Method in the Behavioral Sciences, 1966* (The Hague: Mouton, 1966).

28. Bourdieu, "Peculiar History," 3.

29. Vogel, *Is It Safe?*, 85.

30. U.S. EPA, "Good Laboratory Practices Standards" (Washington, D.C.: U.S. EPA, 2013).

31. Craig S. Barrow and James W. Conrad Jr., "Assessing the Reliability and Credibility of Industry Science and Scientists," *Environmental Health Perspectives* 114, no. 2 (2006): 153–155; Jennifer Sass, "Credibility of Scientists: Conflict of Interest and Bias," *Environmental Health Perspectives* 114, no. 3 (2006): A147–148.

32. U.S. EPA, "ToxCast: Screening Chemicals to Predict Toxicity Faster and Better" (Washington, D.C.: U.S. EPA, 2013).

33. In science and technology studies, "black boxes" refer to scientific objects that are presented as "completed projects, not messy constellations," offering a shorthand for systems involving complicated inputs and outputs. Sergio Sismondo, *An Introduction*

to Science and Technology Studies, 2nd ed. (Malden, Mass.: Blackwell, 2010); Bruno Latour, *Science in Action: How to Follow Scientists and Engineers through Society* (Cambridge, Mass.: Harvard University Press, 1987).

34. Shostak, *Exposed Science*, 19.
35. U.S. EPA, "Endocrine Disruptor Screening Program."
36. Vogel, *Is It Safe?*
37. Sarewitz, "How Science Makes Environmental Controversies Worse."
38. Author's analysis in Web of Science, March 23, 2015.
39. Chemtura's flame retardants are part of the Great Lakes Solutions division of the company.
40. Heather M. Stapleton et al., "Alternate and New Brominated Flame Retardants Detected in U.S. House Dust," *Environmental Science & Technology* 42, no. 18 (2008): 6910–6916.
41. Recent publications document significant levels of Firemaster 550 components and other flame retardants in couches and household dust: Stapleton et al., "Novel and High Volume Use"; Dodson et al., "After the PBDE Phase-Out."
42. This study has since been published: Patisaul et al., "Accumulation and Endocrine Disrupting Effects."
43. Albert and Kleinman, "Bringing Pierre Bourdieu."
44. Aaron L. Panofsky, "Field Analysis and Interdisciplinary Science: Scientific Capital Exchange in Behavior Genetics," *Minerva* 49, no. 3 (2011): 295–316.
45. Chantal Gagnon, "Political Translation," in *Handbook of Translation Studies*, vol. 12, ed. Yves Gambier and Luc van Doorslaer (Amsterdam: John Benjamins, 2010), 252–256.
46. To be clear, I am not following actor-network theory's use of translation as the ability of actors to construct shared meanings and the ability of ideas or objects to move and change within networks. Michel Callon, "Some Elements of a Sociology of Translation: Domestication of the Scallops and the Fishermen of St. Brieuc Bay," in *Power, Action, and Belief: A New Sociology of Knowledge?*, ed. J. Law (London: Routledge, 1986), 196–223; Sismondo, *An Introduction to Science and Technology Studies.*
47. SRP, "Community Engagement and Research Translation," last modified February 18, 2011, http://www.niehs.nih.gov/research/supported/dert/programs/srp/outreach/index.cfm.
48. This definition does not fully align with postmodern translation studies, which asserts that "contrary to the prevalent requirement that they [translators] do otherwise, they will always be visible as they leave marks of the decisions they have made." Ben van Wyke, "Ethics and Translation," in *Handbook of Translation Studies*, vol. 1, ed. Yves Gambier and Luc van Doorslaer (Amsterdam: John Benjamins, 2010), 113.
49. Translation is discussed by Harry M. Collins and Robert Evans as a component of interactional expertise needed to move activity between different fields. "The Third Wave of Science Studies: Studies of Expertise and Experience," *Social Studies of Science* 32, no. 2 (2002): 235–296.
50. Albert and Kleinman, "Bringing Pierre Bourdieu," 271.

51. Panofsky, "Field Analysis and Interdisciplinary Science," 298.
52. U.S. EPA, "DecaBDE Phase-Out Initiative" (Washington D.C.: U.S. EPA, 2009).
53. Rebecca Williams, "The Environment Report: Is Fire Safety Putting Us at Risk?" (Ann Arbor: University of Michigan, 2010).
54. For studies finding effects from DecaBDE exposure, see Viberg et al., "Neurobehavioral Derangements in Adult Mice"; Rice et al., "Developmental Delays and Locomotor Activity." For industry critiques of these studies, see Marcia Hardy, Marek Banasik, and Todd Stedeford, "Toxicology and Human Health Assessment of Decabromodiphenyl Ether," *Critical Reviews in Toxicology* 38, Suppl. 3 (2009): 1–44.
55. Biesemeier et al., "An Oral Developmental Neurotoxicity Study."
56. Biesemeier et al., "An Oral Developmental Neurotoxicity Study," 30.
57. Bromine Science and Environment Forum, "Who We Are," 2013. http://www.bsef .com/who-we-are.
58. Wendy Wagner and David Michaels, "Equal Treatment for Regulatory Science: Extending the Controls Governing the Quality of Public Research to Private Research," *American Journal of Law and Medicine* 30 (2004): 125.
59. Biesemeier et al., "An Oral Developmental Neurotoxicity Study," 34.
60. U.S. EPA, "Comments on the Design for the Environment (DfE) Program Alternatives Assessment for the Flame Retardant Decabromodiphenyl Ether," Vol. 1 and 2 (Washington, D.C.: U.S. EPA, 2012).
61. U.S. EPA, "Comments on the DfE Program Alternatives Assessment for DecaBDE," 1:36, 2:245–320.
62. U.S. EPA, "Comments on the DfE Program Alternatives Assessment for DecaBDE," 1:43, 2:246, 2:260.
63. U.S. EPA, "Alternatives Assessment for DecaBDE," vii.
64. U.S. EPA, "Comments on the DfE Program Alternatives Assessment for DecaBDE," 1:63.
65. U.S. EPA, "Design for the Environment Program Alternatives Assessment Criteria," 4–285.
66. For example, the Biesemeier study is not included on the Green Science Policy Institute's list of almost three-hundred references for flame retardant toxicity. Green Science Policy Institute, "Bibliography," accessed April 30, 2015, http://greensciencepolicy .org/bibliography/#health.
67. Author's analysis using citations in Google Scholar, March 26, 2015.
68. Latour, *Science in Action*.
69. Spencer Weart, "Global Warming: How Skepticism Became Denial," *Bulletin of the Atomic Scientists* 67, no. 1 (2011): 41–50.
70. Steve Wing, "The Limits of Epidemiology," *Journal of Radiological Protection* 1, no. 2 (1994): 74–86.
71. Krimsky, "Funding Effect."
72. Michael Carolan, "The Bright- and Blind-Spots of Science: Why Objective Knowledge Is Not Enough to Resolve Environmental Controversies," *Critical Sociology* 34 (2008): 725.

73. Sheldon Krimsky, "The Weight of Scientific Evidence in Policy and Law," *American Journal of Public Health* 95, Suppl. 1 (2005): S129–136.
74. U.S. EPA, "Design for the Environment Program Alternatives Assessment Criteria."
75. U.S. EPA, "Design for the Environment Program Alternatives Assessment Criteria," 23; Clean Production Action, "GreenScreen v. 1.2," 2013.
76. Arnold Schecter et al., "Polybrominated Diphenyl Ethers Contamination of United States Food," *Environmental Science & Technology* 38, no. 20 (2004): 5306–5311; Marcia Hardy, "Comment on 'Polybrominated Diphenyl Ethers Contamination of United States Food,'" *Environmental Science & Technology* 39, no. 7 (2005): 2414.
77. Gagnon, "Political Translation."
78. Chaille Brindley, "Pallet Wars Round II: State and Federal Pallet Bans Point to Increased Government Scrutiny," *Pallet Enterprise*, June 1, 2010, http://www.palletenterprise.com/articledatabase/view.asp?articleID=3119.
79. Merton and Storer, *The Sociology of Science;* Gordon Gauchat, "Politicization of Science in the Public Sphere: A Study of Public Trust in the United States, 1974–2010," *American Sociological Review* 72, no. 2 (2012): 167–187.
80. Julie Goodman et al., "Fecundability and Serum PBDE Concentrations in Women," *Environmental Health Perspectives* 118, no. 8 (2010): A330; Kim Harley et al., "PBDE Concentrations in Women: Harley et al. Respond," *Environmental Health Perspectives* 118, no. 8 (2010): A330–A331.
81. Joe DiGangi et al., "San Antonio Statement on Brominated and Chlorinated Flame Retardants," *Environmental Health Perspectives* 118, no. 12 (2010): A516–A518.
82. U.S. EPA, "Alternatives Assessment for DecaBDE."
83. U.S. EPA, "Comments on the DfE Program Alternatives Assessment for DecaBDE."
84. Sarewitz, "How Science Makes Environmental Controversies Worse," 385.
85. Krimsky, "Weight of Scientific Evidence."
86. Oreskes and Conway, *Merchants of Doubt*; Epseland and Stevens, "Commensuration"; Vogel, *Is It Safe?*

5. NEGOTIATING SCIENCE, POLITICIZING SCIENCE

1. Dr. Rice is now retired.
2. State of Maine, *An Act to Reduce Contamination of Breast Milk and the Environment from Release of Brominated Chemicals in Consumer Products*, 2004, http://www.mainelegislature.org/legis/bills/bills_121st/billtexts/LD179001-1.asp.
3. Rice et al., "Developmental Delays and Locomotor Activity"; Deborah C. Rice et al., "Behavioral Changes in Aging but Not Young Mice after Neonatal Exposure to the Polybrominated Flame Retardant decaBDE," *Environmental Health Perspectives* 117, no. 12 (2009): 1903–1911; Maine Department of Environmental Protection and Maine Center for Disease Control and Prevention, *Brominated Flame Retardants: Third Annual Report to the Maine Legislature* (Augusta, Maine, 2007).

4. Kevin Miller, "DEP Urges Legislative Ban on Fire Retardant," *Bangor Daily News*, February 16, 2007, http://archive.bangordailynews.com/2007/02/16/dep-urges -legislative-ban-on-fire-retardant/.

5. Miller, "DEP Urges Legislative Ban."

6. U.S. EPA, "Draft Toxicological Reviews of Polybrominated Diphenyl Ethers (PBDEs): In Support of the Summary Information in the Integrated Risk Information System (IRIS)," *Federal Register* 71, no. 246 (2006), http://www.epa.gov/fedrgstr/EPA- RESEARCH/2006/December/Day-22/r21969.htm.

7. U.S. House of Representatives, "Science under Siege: Scientific Integrity at the Environmental Protection Agency" (Washington, D.C.: U.S. Government Printing Office, 2008).

8. Sharon Kneiss, "Letter from Sharon Kneiss, American Chemistry Council, to George Gray, U.S. EPA Assistant Administrator for Research and Development. May 3, 2007," 2007, 1, http://www.ewg.org/research/epa-axes-panel-chair-request-chemical-industry -lobbyists/review-panel-timeline. A second major concern outlined in the ACC's letter was that the reference dose was based on "a study that fails to meet the Agency's standards for accurate and reliable data." The ACC argued that the "so called 'Viberg study'" did not conform to the EPA's guidelines for neurotoxicity tests or Good Laboratory Practices [p. 6]. It is important to note that each of the five external reviewers, including Dr. Rice, also raised significant concerns about the Viberg data, but concluded that it was the best available evidence for most evaluations in the IRIS report. U.S. EPA, "External Peer Review: Toxicological Review for Polybrominated Diphenyl Ethers (PBDEs) Human Health Assessment (March 2007)," 2007, http:// www.ewg.org/research/epa-axes-panel-chair-request-chemical-industry-lobbyists /review-panel-timeline.

9. Kneiss, "Letter from Sharon Kneiss," May 3, 2007, 6.

10. Kneiss, "Letter from Sharon Kneiss," May 3, 2007, emphasis in original.

11. George Gray, "Letter from George Gray, U.S. EPA, to Nancy Sandrof, American Chemistry Council," 2007, http://www.ewg.org/research/epa-axes-panel-chair-request -chemical-industry-lobbyists/review-panel-timeline.

12. U.S. EPA, *External Peer Review: Toxicological Review for Polybrominated Diphenyl Ethers (PBDEs) Human Health Assessment* (August 2007); U.S. House of Representatives, "Science under Siege."

13. U.S. EPA, *External Peer Review: Toxicological Review for Polybrominated Diphenyl Ethers (PBDEs) Human Health Assessment* (November 2007), http://static.ewg.org /files/NovemberPBDEcomments.pdf?_ga=1.29053630.365782455.1407193628.

14. U.S. EPA, "Letter from George Gray, U.S. EPA, to Nancy Sandrof, American Chemistry Council," 2008, http://www.ewg.org/research/epa-axes-panel-chair-request-chemical -industry-lobbyists/review-panel-timeline.

15. John Baldacci, "Letter from John Baldacci, State of Maine, to Stephen Johnson, EPA," 2008, http://www.ewg.org/research/epa-axes-panel-chair-request-chemical-industry -lobbyists/review-panel-timeline.

16. Environmental Working Group, "EPA Axes Panel Chair at Request of Chemical Industry Lobbyists: Review Panel Timeline," 2008, http://www.ewg.org/research/epa-axes-panel-chair-request-chemical-industry-lobbyists/review-panel-timeline.

17. U.S. House of Representatives, "Science under Siege."

18. U.S. House of Representatives, "Science under Siege," 104, 212, 245.

19. Union of Concerned Scientists, *Voices of Scientists at the EPA: Human Health and the Environment Depend on Independent Science* (Cambridge, Mass., 2008), 1, http://www.ucsusa.org/assets/documents/scientific_integrity/epa-survey-brochure.pdf. These feelings remained potent at EPA even as I conducted my fieldwork in 2010–2012. Respondents talked about the "dark days of the Bush administration" and emphasized the power that political appointees and administrative directives had on the agency's work and effectiveness.

20. Environmental Working Group, "EPA Axes Panel Chair at Request of Chemical Industry Lobbyists: 17 Conflicted Reviewers," 2008, http://www.ewg.org/research/epa-axes-panel-chair-request-chemical-industry-lobbyists/17-conflicted-reviewers.

21. Environmental Working Group, "EPA Axes Panel Chair."

22. U.S. House of Representatives, "Science under Siege."

23. John D. Dingell and Bart Stupak, "Letter from John D. Dingell and Burt Stupak, Subcommittee on Oversight and Investigations, to Stephen Johnson, EPA," 2008, http://www.ewg.org/sites/default/files/110-ltr_031308_EPA_BPA.pdf.

24. U.S. House of Representatives, "Science under Siege."

25. Ronnie Greene, "Ouster of Scientist from EPA Panel Shows Industry Clout," *Center for Public Integrity*, February 13, 2013, http://www.publicintegrity.org/2013/02/13/12199/ouster-scientist-epa-panel-shows-industry-clout.

26. Similarly, Klein and Kleinman argue that the social construction of technology involves negotiation between technology designers and users: Hans K. Klein and Daniel L. Kleinman, "The Social Construction of Technology: Structural Considerations," *Science, Technology, & Human Values* 27, no. 1 (2002): 28–52.

27. The EPA has twelve offices and ten regions around the country. The Office of Research and Development, located in North Carolina's Research Triangle Park, is the research arm of the EPA and has no regulatory components. The other federal offices, located in Washington, D.C., are regulatory or programmatic in nature. The regulatory offices, also called "program offices," fulfill specific statutory responsibilities. The Office of Chemical Safety and Pollution Prevention works to reduce risks from pesticides and industrial chemicals, and it includes three suboffices: the Office of Pesticide Programs, the Office of Science Coordination and Policy, and the Office of Pollution Prevention and Toxics, which regulates new and existing chemicals under TSCA and also includes environmental stewardship programs like the DfE programs (see http://epa.gov/aboutepa/ocspp.html for more information).

28. TSCA covers industrial chemicals, those that are used for manufacturing, production, or research related to goods and services. Many chemical substances are not covered by TSCA, including pesticides, food, food additives, cosmetics, tobacco,

chemicals for military purposes, pharmaceuticals, naturally occurring substances, mixtures of chemicals, by-products or intermediate chemicals from the manufacturing process, or export-only chemicals. U.S. EPA, "TSCA Chemical Substance Inventory" (Washington, D.C.: U.S. EPA, 2011).

29. U.S. Congress, "Toxic Substances Control Act."

30. U.S. EPA, "Train Sees New Toxic Substances Law as 'Preventive Medicine,'" press release, October 21, 1976, http://www2.epa.gov/aboutepa/train-sees-new-toxic-subst ances-law-preventive-medicine.

31. U.S. EPA, *Essential Principles for Reform*.

32. U.S. Government Accountability Office, *Chemical Regulation: Options Exist to Improve EPA's Ability to Assess Health Risks and Manage Its Chemical Review Program* (Washington, D.C.: Government Accountability Office, 2005), http://www.gao.gov /products/GAO-05-458.

33. I downloaded the TSCA inventory, available at http://www.epa.gov/oppt/ex istingchemicals/pubs/tscainventory. The inventory contains the chemical name, TSCA identification number, CAS number (a unique chemical identifier assigned by the Chemical Abstracts Service of the American Chemical Society), qualification for certain regulatory exemptions, and information about any regulatory actions, with chemicals separated by CBI status. See also U.S. EPA, "Summary of Accomplishments: New Chemicals Program Activities through September 30, 2012" (Washington, D.C.: U.S. EPA, 2012).

34. The use of halogenated chlorofluoroalkanes in aerosol spray containers was banned in 1978, in advance of the Montreal Protocol. Some uses of the most potent form of dioxin, TCDD (2,3,7,8-tetrachlorodibenzo-p-dioxin), were banned in 1980, and TCDD was further regulated through the 1985 Resource Conservation and Recovery Act. Polychlorinated biphenyls (PCBs) were banned as part of the original TSCA statute. The EPA banned most uses of asbestos in 1989, but most of that regulation was vacated by a 5th Circuit Court of Appeals ruling in 1991. Hexavalent chromium was banned in 1990, though only in certain water treatment applications. U.S. Government Accountability Office, *Chemical Regulation*.

35. Sarah A. Vogel and Jody A. Roberts, "Why the Toxic Substances Control Act Needs an Overhaul, and How to Strengthen Oversight of Chemicals in the Interim," *Health Affairs* 30, no. 5 (2011): 898–905.

36. U.S. EPA, "Summary of Accomplishments."

37. Author's analysis, see note 33 in this chapter.

38. Rebecca Jones, "Evaluation of Health Hazard Endpoints," presented at the Chemical Assessment and Management Workshop, September 2010, Beijing, People's Republic of China.

39. Author's analysis, see note 33 in this chapter.

40. The European Union's chemical regulation system REACH requires significantly more data on chemical risk; however, the EPA and the European Agency have determined that they cannot share relevant data with each other, and the chemical industry

has resisted data sharing efforts, meaning that the EPA does not have access to all the data submitted to European authorities. Maria Hegstad, "EPA Presses Industry to Provide Flame Retardants' Data for TSCA Review," *Inside EPA*, Feburary 2, 2013.

41. U.S. EPA, *Polymer Exemption Guidance Manual* (Washington, D.C., 1997), http://www.epa.gov/oppt/newchems/pubs/polyguid.pdf.

42. Richard Denison and Marianne Engleman, "Letter from Richard Denison and Marianne Engleman to Cass Sunstein and OIRA" (Washington, D.C.: Environmental Defense Fund, 2012), http://blogs.edf.org/nanotechnology/files/2012/02/TSCA-Letter-to-OMB-re-CBI-02-29-12-Final.pdf.

43. Wagner and Michaels, "Equal Treatment for Regulatory Science."

44. Author's analysis, see note 33 in this chapter.

45. U.S. EPA, "EPA Releases Formerly Confidential Chemical Information" (Washington, D.C.: U.S. EPA, 2011).

46. U.S. EPA, *Essential Principles for Reform*.

47. American Chemistry Council, "10 Principles for Modernizing TSCA" (Washington, D.C.: American Chemistry Council, 2012), http://www.americanchemistry.com/Policy/Chemical-Safety/TSCA/10-Principles-for-Modernizing-TSCA.pdf.

48. U.S. Senate, "Oversight of EPA Authorities and Actions to Control Exposures to Toxic Chemicals, Committee on Environment and Public Works. July 24, 2012," 2012; U.S. EPA, "Existing Chemicals Action Plans"; U.S. EPA, "TSCA Section 5(b)(4) Concern List" (Washington, D.C.: U.S. EPA, 2012).

49. U.S. EPA, "Analog Identification Method" (Washington, D.C.: U.S. EPA, 2012).

50. U.S. EPA, "New Chemicals: Chemicals Categories Report" (Washington, D.C.: U.S. EPA, 2012).

51. Anna Coutlakis, "Introduction to the TSCA New Chemicals Program: Scope of the Program and Authorities," presentation at the Chemical Assessment and Management Workshop, Beijing, People's Republic of China, September 2010.

52. U.S. EPA, "Summary of Accomplishments"; Coutlakis, "Introduction to the TSCA."

53. U.S. EPA, "Summary of Accomplishments."

54. Lavoie et al., "Chemical Alternatives Assessment."

55. U.S. EPA, "Design for the Environment Program Alternatives Assessment Criteria," 4.

56. Cheryl Hogue, "Assessing Alternatives to Toxic Chemicals," *Chemical and Engineering News* 91, no. 50 (2013): 19–20, http://cen.acs.org/articles/91/i50/Assessing-Alternatives-Toxic-Chemicals.html; National Research Council, *A Framework to Guide Selection of Chemical Alternatives* (Washington, D.C., 2015), http://www.nap.edu/catalog/18872/a-framework-to-guide-selection-of-chemical-alternatives.

57. U.S. EPA, "Alternatives Assessment for DecaBDE," 4-1–4-7.

58. For example, a substance is a "Moderate" carcinogen if there is "limited or marginal evidence of carcinogencity in animals" but a "High" carcinogen if it is identified as a "suspected human carcinogen" by the Globally Harmonized System carcinogen classification system. U.S. EPA, "Design for the Environment Program Alternatives Assessment Criteria," 12.

59. U.S. EPA, "Alternatives Assessment for DecaBDE."

60. Lavoie et al., "Chemical Alternatives Assessment," 9244.

61. Lavoie et al., "Chemical Alternatives Assessment," 9245.

62. Maia Jack, "Principles of Alternatives Assessment," Grocery Manufacturers Association, September 7, 2012, http://www.gmaonline.org/file-manager/20120907_Green%20Chemistry%20AA%20Coalition%20Document_Principles%20of%20Altern atives%20Assessment.pdf.

63. On precautionary action, see Michael Carolan, "The Precautionary Principle and Traditional Risk Assessment: Rethinking How We Assess and Mitigate Environmental Threats," *Organization & Environment* 20, no. 1 (2007): 5–24; Magnus, "Risk Management versus the Precautionary Principle"; Carolyn Raffensperger and Joel Tickner, *Protecting Public Health and the Environment: Implementing the Precautionary Principle* (Washington, D.C.: Island Press, 1999). On green chemistry, see Anastas and Warner, *Green Chemistry.* On false positives, see David Kriebel et al., "The Precautionary Principle in Environmental Science," *Environmental Health Perspectives* 109, no. 9 (2001): 871–876.

64. Plastic pallets are typically lighter and more durable than wood pallets, offering energy savings, but because they are plastic, they are highly flammable. Consequently, they are manufactured with high levels of flame retardants. Maine passed a restriction banning DecaBDE in plastic pallets by 2011, with a requirement that the industry fund an alternatives assessment to identify a preferable alternative, but this assessment found that there were no cost- and performance-effective alternatives to DecaBDE. Tensions between plastic and wood pallet manufacturers typically run high in stakeholder processes that bring them together. Brindley, "Pallet Wars Round II."

65. Bill Hall, "Defending against Product De-Selection Attacks: Where Do We Stand?," presented at the Flexible Vinyl Products Division Conference, Society of the Plastics Industry, Burlington, Vt., July 12, 2011.

66. U.S. EPA, "Design for the Environment Program Alternatives Assessment Criteria," 10.

67. Nhan Nguyen, "Overview of Occupational Exposure and Environmental Release Assessment," presented at the Chemical Assessment and Management Workshop, September 2010, Beijing, People's Republic of China.

68. George Gray et al., "The Annapolis Accords on the Use of Toxicology in Risk Assessment and Decision-Making: An Annapolis Center Workshop Report," *Toxicological Mechanisms and Methods* 11, no. 3 (2001): 225–231; Vogel, *Is It Safe?*, 85.

69. Dror Etzion and Gerald F. Davis, "Revolving Doors? A Network Analysis of Corporate Officers and U.S. Government Officials," *Journal of Management Inquiry* 17, no. 3 (2008): 157–161.

70. Edward T. Walker and Christopher M. Rea, "The Political Mobilization of Firms and Industries," *Annual Review of Sociology* 40 (2014): 281–304; Daniel Faber, *Capitalizing on Environmental Injustice: The Polluter-Industrial Complex in the Age of Globalization* (Lanham, Md.: Rowman & Littlefield, 2008).

71. Center for Science in the Public Interest, "Harvard University," *Integrity in Science*, accessed April 30, 2015, https://www.cspinet.org/integrity/nonprofits/harvard_university .html.

72. George Gray, "George M. Gray Curriculum Vitae," last modified February 2012, https://dtsc.ca.gov/LawsRegsPolicies/upload/Gray_CV.pdf.

73. Croplife America, "Update on the 2012 CLA Science Forum," *Science Forum*, 2012, http://www.croplifeamerica.org/sites/default/files/2012 CLA Science Forum Update on Weight of Evidence.pdf; George M. Gray and Joshua T. Cohen, "Rethink Chemical Risk Assessments," *Nature* 489 (2012): 27–28.

74. George Gray, "Best Practices for Sustainable Alternative Chemical Substitution," presentation at the HESI Annual Meeting, Alexandria, Va., 2013, http://www.hesiglobal .org/files/Gray-Sustainable Alternative Chemical Solutions.pdf.

75. "Todd Stedeford," LinkedIn, accessed April 30, 2015, http://www.linkedin.com/pub /todd-stedeford/5/16/321.

76. On the neurodevelopmental toxicity in rats, see Biesemeier et al., "An Oral Developmental Neurotoxicity Study." For examples of critical letters to the editor, see Marek Banasik et al., "Comment on 'Brominated Flame Retardants in Children's Toys: Concentration, Composition, and Children's Exposure and Risk Assessment,'" *Environmental Science & Technology* 44, no. 3 (2010): 1152–1153; Marcia Hardy et al., "Comment on 'Alternate and New Brominated Flame Retardants Detected in U.S. House Dust,'" *Environmental Science & Technology* 42, no. 24 (2008): 9453–9454. On the listing of TDCPP on Proposition 65, see State of California, "Chemicals Known to the State." For an example of attacks on other scientists, see Marek Banasik et al., "Comment on 'Assessing Chemical Risk,'" *Science* 1136 (2011), http://www .sciencemag.org/content/331/6021/1136.1.citation/reply#content-block.

77. Michael Hawthorne, "Chemical Firm's Champion Now EPA Expert," *Chicago Tribune*, September 10, 2012, http://articles.chicagotribune.com/2012-09-10/news/ct-met -flame-retardants-epa-20120910_1_flame-retardants-chemicals-epa-administrator-lisa -jackson.

78. U.S. EPA, "TSCA Work Plan for Chemical Assessments: 2014 Update," 2015, http:// www.epa.gov/oppt/existingchemicals/pubs/TSCA_Work_Plan_Chemicals_2014 _Update-final.pdf.

79. Hawthorne, "Chemical Firm's Champion."

80. Michaels, *Doubt Is Their Product*, xi.

81. Michaels, *Doubt Is Their Product*, 46.

82. Gradient, "Science and Strategies for Health and the Environment," accessed April 30, 2015, http://www.gradientcorp.com/who-we-are-and-what-we-do.html.

83. Michaels, *Doubt Is Their Product*, 3.

84. The phrase "junk science" was popularized by Peter Huber's book *Galileo's Revenge: Junk Science in the Courtroom* (New York: Basic Books, 1991). It is used to disparage research critical of the product being defended. Steve Milloy, founder and editor of www.junkscience.org, current contributor on Fox News, and scholar at the libertarian-

leaning Competitive Enterprise Institute, continues the fight against anti-industry science with almost-daily posts arguing for the safety of air pollution and DDT, critiques of the overreach by government agencies, and none-too-subtle attacks on individual scientists for their research methods, findings, and personalities.

85. Julie E. Goodman, "Neurodevelopmental Effects of Decabromodiphenyl Ether (BDE-209) and Implications for the Reference Dose," *Regulatory Toxicology and Pharmacology* 54, no. 1 (2009): 91–104.

86. Biesemeier et al., "Oral Developmental Neurotoxicity Study."

87. U.S. EPA, "Comments on the DfE Alternatives Assessment for DecaBDE," Vol. 1 and 2; U.S. EPA, *Response to Comments on the Draft Alternatives Assessment for Decabromodiphenyl Ether (decaBDE)* (Washington, D.C., 2014), http://www.epa.gov/oppt/dfe/pubs/projects/decaBDE/140127-response-to-comments-on-decabde-report.pdf.

88. U.S. EPA, "Comments on the DfE Alternatives Assessment for DecaBDE."

89. U.S. EPA, "Comments on the DfE Alternatives Assessment for DecaBDE," 1:35–69, 2:245–328.

90. Frank Ackerman, *Poisoned for Pennies: The Economics of Toxics and Precaution* (Washington, D.C.: Island Press, 2008).

91. Moore, *Disrupting Science*; Daniel S. Greenberg, *Science, Money, and Politics: Political Triumph and Ethical Erosion* (Chicago: University of Chicago Press, 2001).

92. Peter Groenewegen and Lois Peters, "The Emergence and Change of Materials Science and Engineering in the United States," *Science, Technology, & Human Values* 27, no. 1 (2002): 112–133.

93. Daniel Kleinman and Steven Vallas, "Science, Capitalism, and the Rise of the 'Knowledge Worker': The Changing Structure of Knowledge Production in the United States," *Theory and Society* 30, no. 4 (2001): 451–492.

94. Krimsky, "Funding Effect."

95. American Chemical Society, "Instructions to Authors for *Environmental Science & Technology*" (American Chemical Society, 2013).

96. German Toxicology Society, "Guide for Authors," *Toxicology*, accessed April 30, 2015. http://www.elsevier.com/journals/toxicology/0300–483X/guide-for-authors#54000.

97. Polanyi, "Republic of Science."

98. The largest and best-known case involved DuPont and perfluorinated chemicals. In 2006, the company settled with the EPA for over $16 million; U.S. EPA, "EPA Settles PFOA Case Against DuPont for Largest Environmental Administrative Penalty in Agency History" (Washington, D.C.: U.S. EPA, 2005). In another TSCA Section 8(e) settlement, 3M agreed to remove dozens of perfluorinated chemicals from the market; U.S. EPA, "EPA Settles Case Involving 3M Voluntary Disclosures of Toxic Substances Violations" (Washington, D.C.: U.S. EPA, 2006). And through a one-time compliance program called the Compliance Audit Program initiated in 1991, the EPA received 10,000 studies and over $22 million in penalties from over a hundred companies: Elizabeth C. Brown et al., *TSCA Deskbook* (Washington, D.C.: Environmental Law Institute, 1999).

99. Jason Plautz, "EPA to Withdraw 2 High-Profile Regulations," *Greenwire*, 2013, http://www.eenews.net/stories/1059986842.

100. Lisa Heinzerling, "Who Will Run the EPA?" *Yale Journal of Regulation* 30 (2013): 42.

101. Chris Hamby and Jim Morris, "EPA Proposal Has Been under Review for 638 Days," Center for Public Integrity, February 9, 2012, http://www.iwatchnews.org/2012/02/09/8109/chemicals-Concern-List-Stuck-Omb; Rena Steinzor and Michael Patoka, "The Bottleneck," *Environmental Forum*, January/February 2012, 36–40.

102. Freedom of Information Act, "Home Page" (Washington, D.C.: U.S. Department of Justice, 2013).

103. As Proctor and Schiebinger write in *Agnotology*, "Disclosures and even 'transparency' are double-edged swords, however, as shown by the tobacco industry's work to draft and organize passage of the Data Access Act of 1998 and the Data Quality Act of 2000 . . . the seemingly noble goal of transparency can be an instrument in the service of organized duplicity" (21). See also Annamaria Baba et al., "Legislating 'Sound Science': The Role of the Tobacco Institute," *American Journal of Public Health* 95, Suppl. 1 (2005): S20–S27.

104. Marcia Hardy et al., "Prenatal Developmental Toxicity of Decabromodiphenyl Ethane in the Rat and Rabbit," *Birth Defects Research Part B: Developmental and Reproductive Toxicology* 89, no. 2 (2010): 139–146.

105. Hardy et al., "Prenatal Developmental Toxicity," 146, emphasis added.

106. Marek Banasik et al., "Tetrabromobisphenol A and Model-Derived Risks for Reproductive Toxicity," *Toxicology* 260, no. 1–3 (2009): 150–152.

107. James Evans, "Industry Collaboration, Scientific Sharing and the Dissemination of Knowledge," *Social Studies of Science* 40, no. 5 (2010): 757–791.

108. Ruthann Rudel and Elizabeth Newton, "Letter to the Editor: Exposure Assessment for Decabromodiphenyl Ether (decaBDE) Is Likely to Underestimate General U.S. Population Exposure," *Journal of Children's Health* 2, no. 2 (2004): 171–173.

109. U.S. EPA, "Design for the Environment Program Alternatives Assessment Criteria."

110. Carolan, "Bright- and Blind-Spots of Science."

111. U.S. Office of the Inspector General, *EPA's Voluntary Chemical Evaluation Program Did Not Achieve Children's Health Protection Goals*, Report No. 11-P-0379.

112. Corporate Europe Observatory, "House of Mirrors: Burson-Marsteller Brussels Lobbying for the Bromine Industry," January 2005, http://archive.corporateeurope.org/lobbycracy/houseofmirrors.html.

113. Szasz, *Shopping Our Way to Safety*.

114. Schnaiberg and Gould, *Environment and Society*.

115. Dow Chemical Company, "ICL-IP Harnesses Dow's Polymeric Flame Retardant Technology in New Product Development," *Additives for Polymers*, March 2012, 2, http://ac.els-cdn.com/S0306374712700350/1-s2.0-S0306374712700350-main.pdf?_tid=7888e08c-1d93-11e4-9e4e-00000aabof01&acdnat=1407348174_05f557f4a8a07004dd277c68e57e6183.

6. SCIENCE FOR ADVOCACY

1. I focus on the work of organized groups and NGOs, not "the public" or "consumers," because I was told by many stakeholders that the pressure that mattered was organized by NGOs, and that individual consumers usually reflected an NGO-established agenda. As an EPA scientist said, "It's hard to see how consumers can respond to an issue without some assistance . . . Usually it gets into the news because the NGOs have written a report or something." Additionally, I use the descriptors "activist," "advocate," and "organizer" more or less interchangeably, though I also acknowledge the labels' different connotations. As an exception, several respondents specifically requested that I identify them as "advocates" and not another term.

2. Alissa Cordner and Phil Brown, "A Multisector Alliance Approach to Environmental Social Movements: Flame Retardants and Chemical Reform in the United States," *Environmental Sociology* 1, no. 1 (2015): 69–79.

3. The Green Science Policy Institute worked almost exclusively on flame retardants in the early years of is existence, but more recently it has expanded to other pollutants and regulatory issues. Green Science Policy Institute, "About" (Berkeley, Calif.: Green Science Policy Institute, 2015), accessed April 30, 2015, http://greensciencepolicy.org /about/.

4. DiGangi et al., "San Antonio Statement."

5. Triclosan is an antifungal, antimicrobial agent originally registered as a pesticide, now widely used in consumer products including antimicrobial soap and toothpaste.

6. Robert D. Benford and David A. Snow, "Framing Processes and Social Movements: An Overview and Assessment," *Annual Review of Sociology* 26 (2000): 611.

7. Fischer et al., "Children Show Highest Levels"; Craig Butt et al., "Metabolites of Organophosphate Flame Retardants and 2-Ethylhexyl Tetrabromobenzoate (EH-TBB) in Urine from Paired Mothers and Toddlers," *Environmental Science & Technology* 48, no. 17 (2014): 10432–10438.

8. Eskenazi et al., "In Utero and Childhood"; Herbstman et al., "Prenatal Exposure"; Chen et al., "Prenatal PBDE Exposures."

9. U.S. EPA, "Office of Children's Health Protection."

10. Sherry MacDonald, "Hospital Initiative Urges Development of Safer Interior Furnishings," Healthier Hospitals Initiative, July 11, 2012, http://healthierhospitals.org/media -center/press-releases/hospital-initiative-urges-development-safer-interior -furnishings.

11. Kaiser Permanente, "Kaiser Permanente Commits to Purchasing Furniture Free from Toxic Flame Retardant Chemicals," press release, June 3, 2015, http://share.kaiserper manente.org/article/kaiser-permanente-commits-to-purchasing-furniture-free-from -toxic-flame-retardant-chemicals/.

12. Alissa Cordner et al., "Firefighters and Flame Retardant Activism," *New Solutions* 24, no. 4 (2015): 511–534.

13. Brian Mayer, *Blue-Green Coalitions: Fighting for Safe Workplaces and Health Communities* (Ithaca, N.Y.: Cornell University Press, 2008).

14. Grace LeMasters et al., "Cancer Risk among Firefighters: A Review and Meta-Analysis of 32 Studies," *Journal of Occupational and Environmental Medicine* 48, no. 11 (2006): 1189–1202; A. L. Golden, S. B. Markowitz, and P. J. Landrigan, "The Risk of Cancer in Firefighters," *Occupational Medicine* 10, no. 4 (1995): 803–820; International Agency for Research on Cancer, "Painting, Firefighting, and Shiftwork," *IARC Monographs on the Evaluation of Carcinogenic Risks to Humans*, vol. 98, (Lyon, France: IARC, 2010).

15. International Association of Fire Fighters, "State Presumptive Disability Laws," accessed April 30, 2015, http://www.iaff.org/hs/phi/docs/PresumptiveDisabilityChart.pdf.

16. Stec, "Influence of Fire Retardants"; Shaw et al., "Halogenated Flame Retardants."

17. International Association of Fire Fighters, "About the IAFF," accessed April 30, 2015, http://www.iaff.org/about/default.asp.

18. For example, see International Association of Fire Fighters, "IAFF Calls on Canadian Government to Ban PBDEs," press release, November 2, 2007, http://www.iaff.org/canada/Updates/IAFF_PBDE_02112007.htm.

19. Scott Shane, "Tobacco Industry Tied to Firefighters: Donations Seen as Way to Weaken Support for Fire-Safe Cigarettes," *Baltimore Sun*, February 16, 1999, http://articles.baltimoresun.com/1999-02-16/news/9902160065_1_tobacco-fire-safe-cigarettes-cigarette-manufacturers; Shin, "Fighting for Safety"; Callahan and Roe, "Big Tobacco Wins."

20. John Martell, "Letter from John Martell, Matt Vinci, Dennis Sweeney, and Kelly Fox to Cal Dooley. Re: Telling the Truth About Chemical Flame Retardants," July 13, 2012, http://www.saferchemicals.org/PDF/fire_fighters_letter_to_chemical_industry.pdf.

21. Eddy Crews and Margaret Vaughn, "Memorandum from Eddy Crews, Associated Fire Fighters of Illinois, and Margaret Vaughn, Illinois Fire Fighters Association, to the Members of the Illinois Legislature," March 20, 2007, accessed April 30, 2015, http://www.mnceh.org/fire-fighter-organizations-across-us-support-banning-or-restricting-toxic-flame-retardant-deca-bde.

22. David Snow et al., "Frame Alignment Processes, Micromobilization, and Movement Participation," *American Sociological Review* 51, no. 4 (1986): 464–481.

23. I use the term *corporate advocates* or *activists*, rather than a term like corporate boosters, so that I can compare their strategies with those of other social movement organizations and better identify differences and commonalities. Just as environmental health activists pursue environmental health goals, corporate activists pursue corporate goals. This is a form of *corporate mobilization*, corporate advocacy strategies of "engagement in electoral politics, direct corporate lobbying, collective action through associations and coalitions, business campaigns in civil society, and political aspects of corporate responsibility." Walker and Rea, "Political Mobilization," 281.

24. I did not encounter any individuals or groups who independently (that is, without any financial support from or connection to the flame retardant industry) lobby or testify against flame retardant bans.

25. Aaron M. McCright and Riley E. Dunlap, "Challenging Global Warming as a Social Problem: An Analysis of the Conservative Movement's Counter-Claims," *Social Problems* 47, no. 4 (2000): 499–522; Walker and Rea, "Political Mobilization."

26. Edward T. Walker, "Privatizing Participation: Civic Change and the Organizational Dynamics of Grassroots Lobbying Firms," *American Sociological Review* 74, no. 1 (2009): 83–105.

27. Thomas P. Lyon and John W. Maxwell, "Astroturf: Interest Group Lobbying and Corporate Strategy," *Journal of Economics and Management Strategy* 13, no. 4 (2004): 563.

28. These figures include funding from flame retardant-specific organizations and manufacturers, based on my analysis from www.cal-access.ss.ca.gov. A detailed and widely-cited investigative piece in *Environmental Health News* in 2011 reported that the flame retardant industry spent $23.2 million dollars lobbying in California on flame retardants (Liza Gross, "Special Report: Flame Retardant Industry Spent $23 Million on Lobbying, Campaign Donations," *Environmental Health News*, November 16, 2011, http://www.environmentalhealthnews.org/ehs/news/2011/money-to -burn.). This piece incorrectly assumes that all lobbying money from the ACC and the California Manufacturers and Technology Association was spent on flame retardants, but reporting documents from these three firms reveal that all three also lobbied on other issues (cal-access.ss.ca.gov).

29. Gross, "Special Report"; Cal-Access, "Lobbying Activity."

30. Bromine Science and Environment Forum, "Who We Are."

31. Mervyn Susser, "Editorial: Goliath and Some Davids in the Tobacco Wars," *American Journal of Public Health* 87, no. 10 (1997): 1593–1594. Burson-Marsteller also runs the Alliance for Consumer Fire Safety in Europe (www.acfse.org). According to a 2005 exposé, "without Burson-Marsteller there would be no BSEF" (Corporate Europe Observatory, "House of Mirrors").

32. Cal-Access, "Lobbying Activity: Bromine Science and Environmental Forum" (Sacramento: California Secretary of State, 2013); National Institute on Money in State Politics, "Lobbyist Link: Bromine Science and Environmental Forum" (Helena, MT: National Institute on Money in State Politics, 2013); Bromine Science and Environment Forum, "Report of Lobbyist Employer, Form 635, 7/01/2007–9/30/2007," 2007, http://cal-access.ss.ca.gov/PDFGen/pdfgen.prg?filingid=1294651&amendid=1.

33. Corporate Europe Observatory, "House of Mirrors."

34. Gross, "Special Report: Flame Retardant Industry."

35. Citizens for Fire Safety, "Form 990. Return of Organization Exempt from Income Tax," 2010, doi:http://www.scribd.com/doc/48001474/fire-safety; Citizens for Fire Safety, "Citizens for Fire Safety Institute Webpage," 2012.

36. Grant D. Gillham was identified as the Executive Director from 2008–2010. In 2008, the list of officers also included David Sanders (Chemtura), Laura Ruiz (Albemarle), and Michael Spiegelstein (BSEF). In 2009–2010, the list of officers included Lloyd Moon (Chemtura), Paul Sawyer (Albemarle), Joel Tenney (ICL-IP), and Barbara Little (Albemarle). Citizens for Fire Safety, "Form 990. Return of Organization Exempt from Income Tax," 2010; Citizens for Fire Safety, "Form 990. Return of Organization Exempt from Income Tax," 2009, doi:http://www.scribd.com/doc/48001474/fire-safety; Citizens for Fire Safety, "Form 990. Return of Organization Exempt from Income Tax," 2008, doi:http://www.scribd.com/doc/48001474/fire-safety.

37. Citizens for Fire Safety, "Form 990," 2010.

38. Bryan Goodman, "North American Flame Retardant Alliance to Provide All Advocacy Services to Member Companies," press release, August 31, 2012, American Chemistry Council, http://www.americanchemistry.com/Media/PressReleasesTranscripts/ACC-news-releases/NAFRA-to-Provide-All-Advocacy-Services-to-Members.html.

39. Center for Responsive Politics, "American Chemistry Council," accessed March 26, 2015, https://www.opensecrets.org/orgs/summary.php?id=D000000365.

40. Goodman, "North American Flame Retardant Alliance."

41. Matthew Blais, *The Utility of CA TB 117: Does the Regulation Add Value?* (San Antonio: Southwest Research Institute, 2012). A later paper coauthored by Blais investigates the emissions from various combinations of flame retardant and non-flame retardant foams and fabrics. It appears to use the same materials and similar methods, but focuses on emissions released during burn tests, not the efficacy of TB 117. The authors disclose that the study was funded by the ACC and NAFRA. Matthew Blais and Karen Carpenter, "Flexible Polyurethane Foams: A Comparative Measurement of Toxic Vapors and Other Toxic Emissions in Controlled Combustion Environments of Foams with and without Fire Retardants," *Fire Technology* 51 (2013): 3–18.

42. Moore, Marshall. "Testimony Regarding Oversight of EPA Authorities and Actions to Control Exposures to Toxic Chemicals." U.S. Senate Committee Hearing on Environment and Public Works, July 24, 2014.

43. Gordon Nelson, "Testimony Regarding the Consumer Choice Fire Safety Act before the California State Senate Committee on Business, Professions, & Consumer Protection," April 25, 2011, http://24.104.59.141/channel/viewvideo/2353.

44. Sam Roe, "Doubts Cast on New Research Touted by Fire-Retardant Lobby," *Chicago Tribune*, December 30, 2012, http://articles.chicagotribune.com/2012-12-30/news/ct-met-flames-southwest-study-20121230_1_flame-retardants-vytenis-babrauskas-paper.

45. American Chemistry Council, "ACC Responds to Chicago Tribune Story on Flame Retardant Research" (Washington, D.C.: American Chemistry Council, 2012); Roe, "Doubts Cast."

46. American Chemistry Council, "ACC Responds to Chicago Tribune."

47. Gross, "Special Report: Flame Retardant Industry."

48. American Council on Science and Health, "American Council on Science and Health: Dispatch,"retrieved February 5, 2013, http://www.acsh.org/blog; American Chemistry Council, "American Chemistry Matters," 2013.

49. Safer Chemicals Healthy Families, "Safer Chemicals, Healthy Families: Who We Are," 2013.

50. Edward Walker, "Industry-Driven Activism," *Contexts* 9, no. 2 (2010): 45–49.

51. David Snow, "Social Movements as Challenges to Authority: Resistance to an Emerging Conceptual Hegemony," in *Authority in Contention*, ed. Daniel Myers and Daniel Cress (Boston: Elsevier, 2004), 3.

52. Edward Walker, Andrew W. Martin, and John D. McCarthy, "Confronting the State, the Corporation, and the Academy: The Influence of Institutional Targets on Social Movement Repertoires," *American Journal of Sociology* 114, no. 1 (2008): 35–76; Daniel Myers and Daniel Cress, *Authority in Contention* (Boston: Elsevier, 2004).

53. Hess, *Alternative Pathways*.

54. Phil Brown et al., "A Gulf of Difference: Disputes over Gulf War-Related Illnesses," *Journal of Health and Social Behavior* 42 (2001): 235–257; Phil Brown, Rachel Morello-Frosch, and Stephen Zavestoski, *Contested Illnesses: Citizens, Science, and Health Social Movements* (Berkeley: University of California Press, 2012). For example, some breast cancer activists have challenged the dominant epidemiological paradigm that breast cancer is primarily caused by lifestyle and genetic factors, arguing instead that environmental causation and chemical exposures play a significant role. Phil Brown et al., " 'A Lab of Our Own': Environmental Causation of Breast Cancer and Challenges to the Dominant Epidemiological Paradigm," *Science, Technology, & Human Values* 31, no. 5 (2006): 499–536.

55. Hess, *Alternative Pathways*.

56. Walker et al., "Confronting the State," 41.

57. Erika Schreder, "Hidden Hazards in the Nursery" (Seattle: Washington Toxics Coalition/Safer States, 2012); Heather M. Stapleton et al., "Identification of Flame Retardants in Polyurethane Foam Collected from Baby Products," *Environmental Science & Technology* 45, no. 12 (2011): 5323–5331.

58. Car seats are governed by a federal flammability standard, the U.S. Department of Transportation *Federal Motor Vehicle Safety Standard 302: Flammability of Interior Materials*.

59. Healthy Child Staff, "Graco Commits to Banning Toxic Flame Retardants from Children's Products," Healthy Child Healthy World, July 16, 2012, http://www.healthychild.org/blog/comments/press_release_graco_commits_to_banning_toxic_flame_retardants_from_children.

60. Many respondents who worked on flame retardants also worked on chemicals such as BPA and used this as an example of how change happens more quickly within the marketplace than at the FDA or EPA. An environmental health scientist said, "I honestly think that a lot of the changes we're going to see in the chemicals in our homes

are going to be driven by consumers, not by legislation . . . I mean let's look at BPA, for example . . . Companies are taking that out of their products, not because there's been any legislation at this point, but because it's gotten a lot of publicity and consumers are demanding it."

61. Scruggs et al., "Role of Chemical Policy," 132.

62. On the presence of flame retardants in baby products, see Stapleton et al., "Identification of Flame Retardants."

63. Michael Kirschner and Arlene Blum, "When Product Safety and the Environment Appear to Collide," *Conformity* (2009): 12–16, doi:http://conformity.com/artman /publish/feature_267.shtml; International Electrotechnical Commission, "TC 108 Voting Result," August 8, 2012, http://www.iec.ch/dyn/www/f?p=103:52:0::::FSP_ORG _ID,FSP_DOC_ID,FSP_DOC_PIECE_ID:1311,139509,269491; Green Science Policy Institute, "Another 'Candle Standard' Bites the Dust," April 6, 2015, http://greensci encepolicy.org/another-candle-standard-bites-the-dust/.

64. American Chemistry Council, "Key Facts: The Need for an Open Flame Test," accessed April 30, 2015, http://flameretardants.americanchemistry.com/FAQs/The -Need-for-an-Open-Flame-Test.html; California Department of Consumer Affairs, *Technical Bulletin 117-2013* (Sacramento, CA: California Bureau of Electronic and Appliance Repair, Home Furnishings and Thermal Insulation, 2013), http://www.bhfti.ca .gov/about/laws/propregs.shtml.

65. In contrast, legislation mandating fire-safe cigarettes, which was also passed state-by-state, was identical across all fifty states. When asked why this was the activist strategy, a fire safety organizer said that it "couldn't have passed" if the standards varied state to state, and that the standard was based on "a good, solid, scientific standard" developed by the National Institute of Safety and Technology.

66. Schreder, "Hidden Hazards," 11. This strategy extends beyond flame retardants to other chemicals including phthalates, parabens, and BPA. Amy Lubitow, "The Battle over Bisphenol-A: United States Chemical Policy and the New Networked Environmental Politics" (Ph.D., diss., Northeastern University, 2011).

67. Stephanie Lee, "Patchwork of Bills Cover Flame Retardants," *San Francisco Chronicle*, February 4, 2013, http://www.sfgate.com/health/article/Patchwork-of-bills-cover -flame-retardants-4353348.php.

68. David S. Meyer, "Protest and Political Opportunities," *Annual Review of Sociology* 30 (2004): 125–145.

69. Richard Denison, "States Act While Congress Fiddles," *EDF Health*, January 28, 2013, http://blogs.edf.org/health/2013/01/28/states-act-while-congress-fiddles/.

70. Safer Chemicals Healthy Families, "Who We Are."

71. Lubitow, "The Battle over Bisphenol-A."

72. Safer Chemicals Healthy Families, "National Stroller Brigade Today at U.S. Capital," accessed February 5, 2013, http://www.saferchemicals.org/2012/05/national-stroller -brigade-today-at-us-capitol.html.

73. Gieryn, *Cultural Boundaries*; Jasanoff, "Contested Boundaries."

74. McCormick, "Democratizing Science Movements," 610.

75. Brown et al., "'A Lab of Our Own,'" 502; Cordner and Brown, "Moments of Uncertainty," 478.

76. Alissa Cordner, Phil Brown, and Margaret Mulcahy, "Playing with Fire: The World of Flame Retardant Activism and Policy," in *Players and Arenas: The Interactive Dynamics of Protest*, ed. Jan Duyvendak and James M. Jasper (Amsterdam: Amsterdam University Press, 2014).

77. Bourdieu, "Peculiar History."

78. DiGangi et al., "San Antonio Statement."

79. Environmental Working Group, "Staff," accessed April 30, 2015, http://www.ewg.org/about-us/staff.

80. Stapleton et al., "Novel and High Volume Use."

81. Morello-Frosch et al., "Toxic Ignorance."

82. Szasz, *Shopping Our Way to Safety*.

83. Gyorgy Pataki, "Ecological Modernization as a Paradigm of Corporate Sustainability," *Sustainable Development* 17, no. 2 (2009): 85.

84. Shobita Parthasarathy, "Breaking the Expertise Barrier: Understanding Activist Strategies in Science and Technology Policy Domains," *Science and Public Policy* 37, no. 5 (2010): 355–367.

85. For example, the Women Firefighters Biomonitoring Collaborative Study is an ongoing study of women firefighters, led by the United Fire Service Women, the San Francisco Cancer Prevention Foundation, and researchers at the University of California at Berkeley and San Francisco, the Silent Spring Institute, and environmental health organizations.

86. Kleinman and Vallas, "Science, Capitalism."

87. Citizens for Fire Safety, "Form 990," 2008; "Form 990," 2009; "Form 990," 2010.

88. Environmental Working Group, "Form 990, Return of Organization Exempt from Income Tax," 2015, doi:http://www.scribd.com/doc/48001474/fire-safety; Environmental Working Group, "Staff."

89. Rachel Best, "Disease Politics and Medical Research Funding: Three Ways Advocacy Shapes Policy," *American Sociological Review* 77, no. 5 (2012): 780–803.

90. Alissa Cordner et al., "Reflexive Research Ethics for Environmental Health and Justice: Academics and Movement-Building," *Social Movement Studies* 11, no. 2 (2012): 161–176; Liam O'Fallon and Allen Dearry, "Community-Based Participatory Research as a Tool to Advance Environmental Health Sciences," *Environmental Health Perspectives* 110, Suppl. 2 (2002): 155–159.

CONCLUSION: THE PURSUIT OF CHEMICAL JUSTICE

1. Frickel and Moore, *New Political Sociology of Science*.

2. Kinchy, *Seeds, Science, and Struggle*.

3. Since the September 11 terrorist attacks, fire data are more closely tied to grant money from the federal government. According to fire scientists, this increases incentives to report data, but may also increase incentives to attribute fires to "unknown" causes.

4. Roe, "Doubts Cast."

5. California Department of Consumer Affairs, Technical Bulletin 117-2013.

6. Massachusetts Department of Fire Services, "Comprehensive Model Fire Code," 2014, http://www.mass.gov/eopss/agencies/dfs/dfs2/osfm/fire-prev/comprehensive-model -fire-code.html; Health Care Without Harm, "Health Care Sector Moves Away from Flame Retardants in Upholstered Furniture," press release, September 11, 2014, https:// noharm-uscanada.org/articles/press-release/us-canada/health-care-sector-moves -away-flame-retardants-upholstered.

7. Ackerman, *Poisoned for Pennies;* Epseland and Stevens, "Commensuration."

8. Richard Denison, "EDF Comments on National Academy of Sciences Workshop on 'Weight of Evidence' in Chemical Assessments" (Washington, D.C.: Environmental Defense Fund, 2013).

9. California Office of Environmental Health Hazard Assessment, "Evidence on the Carcinogenicity of Tris(1,3-Dichloro-2-Propyl) Phosphate," 2011, http://oehha.ca.gov /prop65/hazard_ident/pdf_zip/TDCPP070811.pdf.

10. Mohai et al., "Environmental Justice."

11. Andrew Lakoff and Erik Klinenberg, "Of Risk and Pork: Urban Security and the Politics of Objectivity," *Theory and Society* 39, no. 5 (2010): 503–525.

12. On climate denial, see Aaron M. McCright and Riley E. Dunlap, "The Politicization of Climate Change and the Polarization in the American Public's Views of Global Warming, 2001–2010," *Sociological Quarterly* 52, no. 2 (2011): 155–194; Weart, "Global Warming."

13. Carolan, "Precautionary Principle," 7–9.

14. Design for the Environment, *Flame Retardants Used in Flexible Polyurethane Foam.*

15. Anastas and Warner, *Green Chemistry;* U.S. EPA, "Green Chemistry" (Washington, D.C.: U.S. EPA, 2013).

16. Joel A. Tickner et al., "Advancing Safer Alternatives through Functional Substitution," *Environmental Science & Technology* 49, no. 2 (2015): 742–749.

17. Biomonitoring California, "Biomonitoring California," accessed April 30, 2015, http:// www.biomonitoring.ca.gov; State of Maine, *Toxic Chemicals in Children's Products,* http://www.mainelegislature.org/legis/statutes/38/title38ch16-Dseco.html.

18. Horton et al., "Predictors of Serum Concentrations," 1.

19. Pat Rizzuto, "Law Center Says Strict Chemical Controls Foster Innovation, Markets, Protect People." *Bloomberg BNA*, February 18, 2013.

20. William Freudenburg, "Privileged Access, Privileged Accounts: Toward a Socially Structured Theory of Resources and Discourses," *Social Forces* 84, no. 1 (2005): 89–114; Richard Feiock and Christopher Stream, "Environmental Protection versus Economic Development: A False Trade-Off?" *Public Administration Review* 61, no. 3 (2001): 313–321.

21. Baskut Tuncak, "How Stronger Laws Help Bring Safer Chemicals to Market" (Washington, D.C.: Center for International Environmental Law, 2013).

22. Tuncak, "How Stronger Laws Help," 1.

23. California Legislature, SB-1019.

24. Michaels, "Manufactured Uncertainty," 103.

25. Hegstad, "EPA Presses Industry."

26. James J. Jones, "TSCA Existing Chemicals Program," presented at the American Chemistry Council GlobalChem Conference, February 26, 2013, Washington, D.C.

27. National Research Council, *Science and Decisions*, 213–239.

28. U.S. EPA, "EPA Announces Chemicals for Risk Assessment in 2013, Focus on Widely Used Flame Retardants," 2013, http://yosemite.epa.gov/opa/admpress.nsf/bd4379a92 ceceeac8525735900400c27/c6be79994c3fd08785257b3b0054e2fa!OpenDocument.

29. National Research Council, *Science and Decisions*, 65–92.

30. National Research Council, *Toxicity Testing in the 21st Century: A Vision and a Strategy* (Washington, D.C.: Committee on the Toxicity Testing and Assessment of Environmental Agents/National Academies Press, 2007), http://www.nap.edu/catalog.php ?record_id=11970.

31. The EPA was directed to cut $472 million under the 2013 Sequester. Anthony Lacey, "Sequester Order Cuts EPA Less Than Anticipated but Key Programs Hit," *Inside EPA*, March 3, 2013, 2, http://insideepa.com/Inside-EPA-General/Public-Content-ACC /sequester-order-cuts-epa-less-than-anticipated-but-key-programs-hit/menu-id-1026 .html. The EPA's 2014 fiscal year budget was over $2 billion less than its 2010 budget: U.S. EPA, "EPA's Budget and Spending," 2014, http://www2.epa.gov/planandbudget/budget.

32. Scott Frickel, Rebekah Torcasso, and Annika Y. Anderson, "The Organization of Expert Activism: Shadow Mobilization in Two Social Movements," *Mobilization* 20, no. 3 (2015) 305–323.

33. Cordner et al., "Reflexive Research Ethics," 170.

34. On scientific practice, see Moore, *Disrupting Science*. On GLP practice, see Vogel, *Is It Safe?* On conflict of interest, see Krimsky, "The Funding Effect"; Wagner and Michaels, "Equal Treatment for Regulatory Science."

35. Egeghy et al., "The Exposure Data Landscape."

36. Wagner and Michaels, "Equal Treatment for Regulatory Science," 120.

APPENDIX. PLAYING THE FIELD: METHODOLOGICAL REFLECTIONS

1. BFR Workshop, 13th Workshop on Brominated and Other Flame Retardants, June 4–6, 2012, Winnipeg, Canada. https://web.archive.org/web/20140517004744/http://bfr2012 .org/.

2. Alexander George and Andrew Bennett, *Case Studies and Theory Development in the Social Sciences* (Cambridge, Mass.: MIT Press, 2005).

3. Bourdieu, "The Peculiar History"; Albert and Kleinman, "Bringing Pierre Bourdieu."

4. Matthew Desmond, "Relational Ethnography," *Theory and Society* 43, no. 5 (2014): 547–579.

5. Rebecca Gasior Altman, "Chemical Body Burden and Place-Based Struggles for Environmental Health and Justice" (Ph.D. diss., Sociology Department, Brown University, 2008), 17.

6. Uri Shwed and Peter Bearman, "The Temporal Structure of Scientific Consensus Formation," *American Sociological Review* 75, no. 6 (2010): 817–840; Wiebe E. Bijker, Thomas P. Hughes, and Trevor J Pinch, eds., *The Social Construction of Technological Systems: New Directions in the Sociology and History of Technology* (Cambridge, Mass.: MIT Press, 1987).

7. Ulf Hannerz, "Being There . . . and There . . . and There!," *Ethnography* 4, no. 2 (2003): 201–216.

8. George Marcus, "Ethnography In/of the World System: The Emergence of Multi-Sited Ethnography," *Annual Review of Anthropology* 24 (1995): 95–117; Hannerz, "Being There."

9. Hannerz, "Being There," 211.

10. If I was starting this project now, I would instead choose to encrypt the file, as I have learned that encryption is more secure than the password protection feature in Microsoft Office.

11. Martin Tolich, "Internal Confidentiality: When Confidentiality Assurances Fail Relational Informants," *Qualitative Sociology* 27, no. 1 (2004): 101.

12. Karen Kaiser, "Protecting Respondent Confidentiality in Qualitative Research," *Qualitative Health Research* 19, no. 11 (2009): 1634.

13. QSR International, NVivo, version 10, 2010.

14. Cordner et al., "Reflexive Research Ethics."

15. Luc Boltanski, *On Critique: A Sociology of Emancipation* (Cambridge: Polity, 2011).

16. Nancy Plankey-Videla, "Informed Consent as Process: Problematizing Informed Consent in Organizational Ethnographies," *Qualitative Sociology* 35, no. 1 (2012): 1–21.

17. Laura Nader, "Up the Anthropologist—Perspectives Gained from Studying Up," in *Reinventing Anthropology*, ed. Dell Hymes (New York: Pantheon, 1974), 284–311.

18. This is similar to discussions of tobacco industry research. As David Michaels writes, tobacco scientists "convince[d] themselves that the products they are defending are safe, and that the evidence of harm is inaccurate, or misleading, or trivial" ("Manufactured Uncertainty," 101).

19. Shostak, *Exposed Science*, 218.

REFERENCES

Abdelouahab, N., Y. Ainmelk, and L. Takser. "Polybrominated Diphenyl Ethers and Sperm Quality." *Reproductive Toxicology (Elmsford, N.Y.)* 31, no. 4 (May 2011): 546–550.

Abelson, Philip H. "The Tris Controversy." *Science* 197, no. 4299 (1977): 113.

Ackerman, Frank. *Poisoned for Pennies: The Economics of Toxics and Precaution.* Washington, D.C.: Island, 2008.

Alaee, Mehran, and Richard J. Wenning. "The Significance of Brominated Flame Retardants in the Environment: Current Understanding, Issues and Challenges." *Chemosphere* 46, no. 5 (February 2002): 579–582.

Alario, Margarita V., and William R. Freudenburg. "Environmental Risks and Environmental Justice, Or How Titanic Risks Are Not So Titanic after All." *Sociological Inquiry* 80, no. 3 (July 12, 2010): 500–512.

Albert, Mathieu, and Daniel Lee Kleinman. "Bringing Pierre Bourdieu to Science and Technology Studies." *Minerva* 49 (2011): 263–273.

Alcock, Ruth, and Jerry Busby. "Risk Migration and Scientific Advance: The Case of Flame-Retardant Compounds." *Risk Analysis* 26, no. 2 (2006): 369–381.

Allen, Joseph G., Heather M. Stapleton, Jose Vallarino, Eileen McNeely, Michael D. McClean, Stuart J. Harrad, Cassandra B. Rauert, and John D. Spengler. "Exposure to Flame Retardant Chemicals on Commercial Airplanes." *Environmental Health* 12, no. 1 (2013): 17.

Allen, Joseph, Michael McClean, Heather M. Stapleton, Jessica Nelson, and Thomas Webster. "Personal Exposure to Polybrominated Diphenyl Ethers (PBDEs) in Residential Indoor Air." *Environmental Science & Technology* 41, no. 13 (2007): 4574–4579.

Althaus, Catherine. "A Disciplinary Perspective on the Epistemological Status of Risk." *Risk Analysis* 25, no. 3 (2005): 567–588.

Altman, Rebecca Gasior. "Chemical Body Burden and Place-Based Struggles for Environmental Health and Justice." Ph.D. diss, Sociology Department, Brown University, 2008.

Altman, Rebecca Gasior, Rachel Morello-Frosch, Julia Green Brody, Ruthann Rudel, Phil Brown, and Mara Averick. "Pollution Comes Home and Gets Personal: Women's Experience of Household Chemical Exposure." *Journal of Health and Social Behavior* 49, no. 4 (2008): 417–435.

American Chemical Society. "Instructions to Authors for *Environmental Science & Technology.*" Modified May 2014. http://pubs.acs.org/paragonplus/submission /esthag/esthag_authguide.pdf.

American Chemistry Council. "10 Principles for Modernizing TSCA." American-chemistry.com. Accessed August 30, 2012. http://www.americanchemistry.com /Policy/Chemical-Safety/TSCA/10-Principles-for-Modernizing-TSCA.pdf.

——. "ACC Responds to Chicago Tribune Story on Flame Retardant Research." Americanchemistry.com, modified December 31, 2012. http://blog.americanchemistry .com/2012/12/acc-responds-to-chicago-tribune-story-on-flame-retardant -research.

——. "ACC Responds to Stapleton Study on the Presence of Flame Retardants in Children's Products." Americanchemistry.com, modified May 17, 2011. http://www .americanchemistry.com/Media/PressReleasesTranscripts/ACC-news-releases /ACC-Responds-to-Stapleton-Study-on-the-Presence-of-Flame-Retardants-in -Childrens-Products.html.

——. "American Chemistry Matters." Americanchemistry.com. Accessed April 30, 2015. http://blog.americanchemistry.com/.

——. "Key Facts: The Need for an Open Flame Test." Americanchemistry.com. Accessed April 30, 2015. http://flameretardants.americanchemistry.com/FAQs/The -Need-for-an-Open-Flame-Test.html.

American Council on Science and Health. "American Council on Science and Health: Dispatch." Accessed February 5, 2013. http://www.acsh.org/blog.

American Sustainable Business Council. "Poll of Small Business Owners on Toxic Chemicals." Washington, D.C.: American Sustainable Business Council, 2012.

An Act to Provide for the Payment of Losses Incurred as a Result of the Ban on the Use of the Chemical Tris in Apparel, Fabric, Yarn, or Fiber, and for Other Purposes, Pub. L. No. 97-395, 96 Stat. 2001 (1982). https://www.govtrack.us/congress/bills/97 /s823/text.

Anastas, Paul, and John Warner. *Green Chemistry: Theory and Practice.* Oxford: Oxford University Press, 1998.

Auer, Charles. "Letter from Charles Auer, U.S. EPA, to Susan Lewis, American Chemistry Council. August 25th." Washington, D.C.: U.S. Environmental Protection Agency, 2005. http://www.epa.gov/oppt/vccep/pubs/chem21.html.

——. "Letter from Charles Auer, U.S. EPA, to Chris Bryant, American Chemistry Council, September 5, 2008." 2008. http://www.epa.gov/oppt/vccep/pubs/090508.pdf.

Aven, Terje. "Foundational Issues in Risk Assessment and Risk Management." *Risk Analysis* 32, no. 10 (October 2012): 1647–1656.

——. *Misconceptions of Risk.* Chichester, U.K.: Wiley, 2010.

Baba, Annamaria, Daniel Cook, Thomas Mcgarity, and Lisa Bero. "Legislating 'Sound Science': The Role of the Tobacco Institute." *American Journal of Public Health* 95, Suppl. 1 (2005): S20–S27.

Babrauskas, Vytenis, Donald Lucas, David Eisenberg, Singla Veena, Michel Dedeo, and Arlene Blum. "Flame Retardants in Building Insulation: A Case for Re-Evaluating Building Codes." *Building Research and Information* 40, no. 6 (2012): 738–755.

Baldacci, John. "Letter from John Baldacci, State of Maine, to Stephen Johnson, EPA," 2008. http://www.ewg.org/research/epa-axes-panel-chair-request-chemical-industry -lobbyists/review-panel-timeline.

Banasik, Marek, Raymond Harbison, Arnold Hopp, and Todd Stedeford. "Comment on 'Assessing Chemical Risk.'" *Science* 1136 (2011): E-Letters. http://www.sciencemag .org/content/331/6021/1136.1.full/reply#sci_el_14544.

Banasik, Marek, Marcia Hardy, Raymond Harbison, Ching-Hung Hsu, and Todd Stedeford. "Tetrabromobisphenol A and Model-Derived Risks for Reproductive Toxicity." *Toxicology* 260, no. 1–3 (2009): 150–152.

Banasik, Marek, Marcia Hardy, Raymond Harbison, and Todd Stedeford. "Comment on 'Brominated Flame Retardants in Children's Toys: Concentration, Composition, and Children's Exposure and Risk Assessment.'" *Environmental Science & Technology* 44, no. 3 (2010): 1152–1153.

Barrow, Craig S., and James W. Conrad Jr. "Assessing the Reliability and Credibility of Industry Science and Scientists." *Environmental Health Perspectives* 114, no. 2 (2006): 153–155.

Bearr, Jonathan S., Heather M. Stapleton, and Carys L. Mitchelmore. "Accumulation and DNA Damage in Fathead Minnows (*Pimephales promelas*) Exposed to 2 Brominated Flame-Retardant Mixtures, Firemaster 550 and Firemaster BZ-54." *Environmental Toxicology and Chemistry* 29, no. 3 (2010): 722–729.

Beck, Ulrich. "Climate for Change, or How to Create a Green Modernity?" *Theory, Culture, and Society* 27 (2010): 254–266.

——. *Risk Society: Towards a New Modernity. Theory, Culture, and Society.* Newbury Park, Calif.: Sage, 1992.

——. *World Risk Society.* Malden, Mass.: Polity, 1999.

Beck, Ulrich, Anthony Giddens, and Scott Lash. *Reflexive Modernization: Politics, Tradition, and Aesthetics in the Modern Social Order.* Stanford, Calif.: Stanford University Press, 1994.

Bell, Susan. *DES Daughters : Embodied Knowledge and the Transformation of Women's Health Politics.* Philadelphia: Temple University Press, 2009.

Benford, Robert D., and David A. Snow. "Framing Processes and Social Movements: An Overview and Assessment." *Annual Review of Sociology* 26, no. 1 (2000): 611–639.

Bergman, Åke, Jerrold J. Heindel, Susan Jobling, Karen A. Kidd, and R. Thomas Zoeller. *The State-of-the-Science of Endocrine Disrupting Chemicals.* Geneva: World Health Organization, 2012.

Bergman, Åke, Jerrold J. Heindel, Tim Kasten, Karen A. Kidd, Susan Jobling, Maria Neira, R. Thomas Zoeller, et al. "The Impact of Endocrine Disruption: A Consensus Statement on the State of the Science." *Environmental Health Perspectives* 121, no. 4 (2013): A104–A106.

Best, Rachel. "Disease Politics and Medical Research Funding: Three Ways Advocacy Shapes Policy." *American Sociological Review* 77, no. 5 (2012): 780–803.

Betts, Kellyn. "Could Flame Retardants Deter Electronics Recycling?" *Environmental Science & Technology* 35, no. 3 (January 2001): 58A–59A.

——. "Discontinued Pajama Flame Retardant Detected in Baby Products and House Dust." *Environmental Science & Technology* 43, no. 19 (2009): 7159.

——. "Glut of Data on 'New' Flame Retardant Documents Its Presence All Over the World." *Environmental Science & Technology* 43, no. 2 (2009): 236–237.

——. "Unwelcome Guest: PBDEs in Indoor Dust." *Environmental Health Perspectives* 116, no. 5 (May 2008): A202–A208.

BFR Workshop. "13th Workshop on Brominated and Other Flame Retardants." June 4–6, 2012, Winnipeg, Canada. https://web.archive.org/web/20140517004744/http://bfr2012.org/. Accessed August 22, 2012. www.bfr2012.org.

Biesemeier, John, Melissa Beck, Hanna Silberberg, Nicole Myers, John Ariano, Ann Radovsky, Les Freshwater, et al. "An Oral Developmental Neurotoxicity Study of Decabromodiphenyl Ether (DecaBDE) in Rats." *Birth Defects Research* 92 (2011): 17–35.

Bijker, Wiebe E., Thomas P. Hughes, and Trevor J. Pinch, eds. *The Social Construction of Technological Systems: New Directions in the Sociology and History of Technology.* Cambridge, Mass.: MIT Press, 1987.

Biomonitoring California. "Biomonitoring California." Accessed April 30, 2015. http://www.biomonitoring.ca.gov.

Birnbaum, Linda S., and Daniele F. Staskal. "Brominated Flame Retardants: Cause for Concern?" *Environmental Health Perspectives* 112, no. 1 (2004): 9–17.

Birnbaum, Linda S., John R. Bucher, Gwen W. Collman, Darryl C. Zeldin, Anne F. Johnson, Thaddeus T. Schug, and Jerrold J. Heindel. "Consortium-Based Science: The NIEHS's Multipronged, Collaborative Approach to Assessing the Health Effects of Bisphenol A." *Environmental Health Perspectives* 120, no. 12 (2012): 1640–1644.

Blais, Matthew. *The Utility of CA TB 117: Does the Regulation Add Value?* San Antonio: Southwest Research Institute, 2012.

Blais, Matthew, and Karen Carpenter. "Flexible Polyurethane Foams: A Comparative Measurement of Toxic Vapors and Other Toxic Emissions in Controlled Combustion Environments of Foams with and without Fire Retardants." *Fire Technology* 51 (2013): 3–18.

Blum, Arlene. *Breaking Trail: A Climbing Life.* New York: Scribner, 2005.

——. "Halogenated Flame Retardants: A Global Concern." Paper presented at the Flame Retardant Dilemma, Berkeley, Calif., February 10, 2012.

Blum, Arlene, and Bruce N. Ames. "Flame-Retardant Additives as Possible Cancer Hazards." *Science* 195, no. 4273 (1977): 17–23.

Blum, Arlene, Marian Deborah Gold, Bruce N. Ames, Christine Kenyon, Frank R. Jones, Eva A. Hett, Ralph C. Dougherty, et al. "Children Absorb Tris-BP Flame Retardant from Sleepwear: Urine Contains the Mutagenic Metabolite, 2,3-Dibromopropanol." *Science* 201, no. 4360 (1978): 1020–1023.

Boeing Company. "Boeing/OIRA Discussion of Proposed EPA Regulations Regarding DecaBDE." Presentation to U.S. Office of Information and Regulatory Affairs, February 16, 2011. http://www.whitehouse.gov/sites/default/files/omb/assets/oira_2070 /2070_02162011-1.pdf.

Boltanski, Luc. *On Critique: A Sociology of Emancipation*. Cambridge: Polity, 2011.

Boudia, Soraya, and Nathalie Jas. "Introduction: The Greatness and Misery of Science in a Toxic World." In *Powerless Science? Science and Politics in a Toxic World*, ed. Soraya Boudia and Nathalie Jas, 1–26. New York: Berghahn, 2014.

Bourdieu, Pierre. *Science of Science and Reflexivity*. Chicago: University of Chicago Press, 2004.

——. "The Peculiar History of Scientific Reason." *Sociological Forum* 6, no. 1 (1991): 3–26.

Bradbury, Judith. "The Policy Implications of Differing Concepts of Risk." *Science, Technology, & Human Values* 14, no. 4 (1989): 380–399.

Brindley, Chaille. "Pallet Wars Round II: State and Federal Pallet Bans Point to Increased Government Scrutiny." *Pallet Enterprise*, June 1, 2010. http://www .palletenterprise.com/articledatabase/view.asp?articleID=3119.

Bromine Science and Environment Forum. "Who We Are." Accessed April 30, 2015. http://www.bsef.com/who-we-are.

——. "Report of Lobbyist Employer, Form 635, 7/01/2007–9/30/2007." Accessed February 5, 2013. http://cal-access.ss.ca.gov/PDFGen/pdfgen.prg?filingid=1294651&amendid=1.

Brown, Elizabeth C., Carolyne R. Hathaway, Julia A. Hatcher, William K. Rawson, and Robert M. Sussman. *TSCA Deskbook*. Washington, D.C.: Environmental Law Institute, 1999.

Brown, Phil. *Toxic Exposures: Contested Illnesses and the Environmental Health Movement*. New York: Columbia University Press, 2007.

Brown, Phil, and Alissa Cordner. "Lessons Learned from Flame Retardant Use and Regulation Could Enhance Future Control of Potentially Hazardous Chemicals." *Health Affairs* 30, no. 5 (2010): 906–914.

Brown, Phil, Sabrina McCormick, Brian Mayer, Stephen Zavestoski, Rachel Morello-Frosch, Rebecca Gasior Altman, and Laura Senier. "'A Lab of Our Own': Environmental Causation of Breast Cancer and Challenges to the Dominant Epidemiological Paradigm." *Science, Technology, & Human Values* 31, no. 5 (2006): 499–536.

Brown, Phil, Rachel Morello-Frosch, and Stephen Zavestoski. *Contested Illnesses: Citizens, Science, and Health Social Movements*. Berkeley: University of California Press, 2012.

Brown, Phil, Stephen Zavestoski, Sabrina McCormick, Joshua Mandelbaum, Theo Luebke, and Meadow Linder. "A Gulf of Difference: Disputes over Gulf War-Related Illnesses." *Journal of Health and Social Behavior* 42 (2001): 235–257.

Bryant, Chris. "Letter from Chris Bryant, American Chemistry Council, to James Gulliford, U.S. EPA, July 17, 2008," 2008. http://www.epa.gov/oppt/vccep/pubs/090508 .pdf.

Butt, Craig, Johanna Congleton, Kate Hoffman, Mingliang Fang, and Heather M. Stapleton. "Metabolites of Organophosphate Flame Retardants and 2-Ethylhexyl Tetrabromobenzoate (EH-TBB) in Urine from Paired Mothers and Toddlers." *Environmental Science & Technology* 48, no. 17 (2014): 10432–10438.

Calabrese, Edward J., and Linda A. Baldwin. "Toxicology Rethinks Its Central Belief." *Nature* 421, no. 6924 (February 13, 2003): 691–692.

——. "Defining Hormesis." *Human and Experimental Toxicology* 21, no. 2 (February 1, 2002): 91–97.

Cal-Access. "Lobbying Activity: Bromine Science and Environmental Forum." Sacramento: California Secretary of State. Accessed February 2, 2013. http://cal-access.ss .ca.gov/Lobbying/Employers/Detail.aspx?id=1254267&view=activity.

California, State of. *AB 302: Polybrominated Diphenyl Ether*, 2003. http://www.leginfo .ca.gov/pub/03–04/bill/asm/ab_0301–0350/ab_302_bill_20030811_chaptered .html.

——. "Chemicals Known to the State to Cause Cancer or Reproductive Toxicity." Sacramento: California EPA, 2012.

——. "SB-1019: Upholstered Furniture: Flame Retardant Chemicals," 2014. http:// leginfo.legislature.ca.gov/faces/billNavClient.xhtml?bill_id=201320140SB1019.

California Department of Consumer Affairs. *Technical Bulletin 117-2012: Requirements, Test Procedure, and Apparatus for Testing the Smolder Resistance of Upholstered Furniture*. Sacramento: California Bureau of Electronic and Appliance Repair, Home Furnishings and Thermal Insulation, 2012. http://www.bearhfti.ca .gov/about_us/tb117_finaldraft.pdf.

——. *Technical Bulletin 117-2013*. Sacramento: California Bureau of Electronic and Appliance Repair, Home Furnishings and Thermal Insulation, 2013. http://www.bhfti .ca.gov/about/laws/propregs.shtml.

California Office of Environmental Health Hazard Assessment. "Evidence on the Carcinogenicity of tris(1,3-Dichloro-2-Propyl) Phosphate," 2011. http://oehha.ca.gov /prop65/hazard_ident/pdf_zip/TDCPP070811.pdf.

Callahan, Patricia, and Sam Roe. "Big Tobacco Wins Fire Marshals as Allies in Flame Retardant Push." *Chicago Tribune*, May 8, 2012, http://www.chicagotribune.com/ct -met-flames-tobacco-20120508-story.html.

Callon, Michel. "Some Elements of a Sociology of Translation: Domestication of the Scallops and the Fishermen of St. Brieuc Bay." In *Power, Action, and Belief: A New Sociology of Knowledge?* ed. J. Law, 196–223. London: Routledge, 1986.

Carolan, Michael. "The Bright- and Blind-Spots of Science: Why Objective Knowledge Is Not Enough to Resolve Environmental Controversies." *Critical Sociology* 34, no. 5 (2008): 725–740.

——. "The Precautionary Principle and Traditional Risk Assessment: Rethinking How We Assess and Mitigate Environmental Threats." *Organization & Environment* 20, no. 1 (2007): 5–24.

Carpenter, David, and Pamela Miller. "Environmental Contamination of the Yupik People of St. Lawrence Island, Alaska." *Journal of Indigenous Research* 1, no. 1 (2011): 1–3.

Carter, Luther J. "Michigan's PBB Incident: Chemical Mix-Up Leads to Disaster." *Science* 192, no. 4236 (1976): 240–243.

Center for Responsive Politics. "American Chemistry Council." Accessed March 26, 2015. https://www.opensecrets.org/orgs/summary.php?id=D000000365.

——. "Sparber & Assoc." Washington, D.C.: Center for Responsive Politics. Accessed February 3, 2013. http://www.opensecrets.org/lobby/firmlbs.php?id=F27638.

Center for Science in the Public Interest. "Harvard University." *Integrity in Science*, 2014. https://www.cspinet.org/integrity/nonprofits/harvard_university .html.

Centers for Disease Control and Prevention. *Fourth National Report on Human Exposure to Environmental Chemicals*. Atlanta: CDC, 2009.

Chemtura. "Chemtura Corporation: Homepage." Accessed February 20, 2013. www .chemtura.com.

Chen, Aimin, Kimberly Yolton, Stephen A. Rauch, Glenys M. Webster, Richard Hornung, Andreas Sjödin, Kim N. Dietrich, and Bruce P. Lanphear. "Prenatal Polybrominated Diphenyl Ether Exposures and Neurodevelopment in U.S. Children through 5 Years of Age: The HOME Study." *Environmental Health Perspectives* 122, no. 8 (2014): 856–862.

Chevrier, Jonathan, Kim Harley, Asa Bradman, Myriam Gharbi, Andreas Sjödin, and Brenda Eskenazi. "Polybrominated Diphenylether (PBDE) Flame Retardants and Thyroid Hormone during Pregnancy." *Environmental Health Perspectives* 118, no. 10 (2010): 1444–1449.

Citizens for Fire Safety. "Citizens for Fire Safety Institute Webpage." Accessed August 31, 2012. www.cffsi.org.

——. "Form 990. Return of Organization Exempt from Income Tax," 2008. http:// www.scribd.com/doc/48001474/fire-safety.

——. "Form 990. Return of Organization Exempt from Income Tax," 2009. http:// www.scribd.com/doc/48001474/fire-safety.

——. "Form 990. Return of Organization Exempt from Income Tax," 2010. http://www .scribd.com/doc/48001474/fire-safety.

Clarke, Lee. "Explaining Choices among Technological Risks." *Social Problems* 35, no. 1 (1988): 22–35.

Clean Production Action. GreenScreen, verson 1.2. 2013. Accessed April 30, 2015. http://
    www.cleanproduction.org/Greenscreen.php.
Collins, Harry M., and Robert Evans. *Rethinking Expertise*. Chicago: University of
    Chicago Press, 2007.
——. "The Third Wave of Science Studies: Studies of Expertise and Experience." *Social
    Studies of Science* 32, no. 2 (2002): 235–296.
Commonweal. "Wingspread Statement on the Precautionary Principle." Accessed
    March 9, 2013. http://www.commonweal.org/programs/wingspread-statement.html.
Consumer Products Safety Commission. *Upholstered Furniture Full Scale Chair
    Tests—Open Flame Ignition Results and Analysis*. Bethesda, Md.: CPSC Divi-
    sion of Combustion and Fire Sciences, 2012. http://www.cpsc.gov/PageFiles/93436
    /openflame.pdf.
——. Flammable Fabrics Act of 1953, 15 U.S.C. §§ 1191–1204 (1953); Pub. L. 83–88; 67
    Stat. 11 (June 30, 1953). http://www.cpsc.gov/PageFiles/98982/FLAMMABLE%20
    FABRICS%20ACT.txt
Cordner, Alissa. "Strategic Science Translation and Environmental Controversies."
    Science, Technology, & Human Values. 40, no. 6 (2015): 915–938.
——. "Defining and Defending Risk: Conceptual Risk Formulas in Environmental
    Controversies." *Journal of Environmental Studies and Sciences*. 5, no. 3 (2015):
    241–250.
Cordner, Alissa, and Phil Brown. "A Multisector Alliance Approach to Environmen-
    tal Social Movements: Flame Retardants and Chemical Reform in the United
    States." *Environmental Sociology* 1, no. 1 (2015): 69–79.
——. "Moments of Uncertainty: Ethical Considerations and Emerging Contami-
    nants." *Sociological Forum* 28, no. 3 (2013): 469–494.
Cordner, Alissa, Phil Brown, and Margaret Mulcahy. "Playing with Fire: The World of
    Flame Retardant Activism and Policy." In *Players and Arenas: The Interactive Dy-
    namics of Protest*, ed. Jan Duyvendak and James M. Jasper, 211–228. Amsterdam:
    Amsterdam University Press, 2014.
Cordner, Alissa, David Ciplet, Phil Brown, and Rachel Morello-Frosch. "Reflexive Re-
    search Ethics for Environmental Health and Justice: Academics and Movement-
    Building." *Social Movement Studies* 11, no. 2 (2012): 161–176.
Cordner, Alissa, Kathryn M. Rodgers, Phil Brown, and Rachel Morello-Frosch. "Fire-
    fighters and Flame Retardant Activism." *New Solutions* 24, no. 4 (2015): 511–534.
Corporate Europe Observatory. "House of Mirrors: Burson-Marsteller Brussels Lob-
    bying for the Bromine Industry," January 2005. http://archive.corporateeurope.org
    /lobbycracy/houseofmirrors.html.
CorpWatch. "Methyl Bromide Working Group." Modified March 31, 1997. http://
    corpwatch.org/article.php?id=906.
Costa, Lucio G., and Gennaro Giordano. "Developmental Neurotoxicity of Polybro-
    minated Diphenyl Ether (PBDE) Flame Retardants." *Neurotoxicology* 28, no. 6
    (2007): 1047–1067.

Coutlakis, Anna. "Introduction to the TSCA New Chemicals Program: Scope of the Program and Authorities." Presentation at Chemical Assessment and Management Workshop, Beijing, People's Republic of China, September 2010.

Crews, Eddy, and Margaret Vaughn. "Memorandum from Eddy Crews, Associated Fire Fighters of Illinois, and Margaret Vaughn, Illinois Fire Fighters Association, to the Members of the Illinois Legislature." March 20, 2007. http://www.mnceh.org /fire-fighter-organizations-across-us-support-banning-or-restricting-toxic-flame -retardant-deca-bde.

Croplife America. "Update on the 2012 CLA Science Forum." Presentation at Science Forum conference, 2012. http://www.croplifeamerica.org/sites/default/files/2012 CLA Science Forum Update on Weight of Evidence.pdf.

Cullen, Lawrence. "Letter from Lawrence Cullen, Counsel for iGPS, to Maria Doa, Environmental Protection Agency." July 31, 2012.

De Carlo, Vincent J. "Studies on Brominated Chemicals in the Environment." *Annals of the New York Academy of Sciences* 320 (1979): 678–681.

deCourcy Hinds, Michael. "Reagan Signs Law on Pajama Makers." *New York Times*, January 1, 1983.

Demko, Donald. "So You Want a Flame-Retarded Formulation . . . : Guidance for the Novice, and Burning Questions You Should Be Prepared to Answer." *Plastics Engineering* 59, no. 1 (2003).

Denison, Richard. "EDF Comments on National Academy of Sciences Workshop on 'Weight of Evidence' in Chemcial Assessments." Washington, D.C.: Environmental Defense Fund, 2013.

——. "States Act While Congress Fiddles." *EDF Health,* January 28, 2013, http://blogs .edf.org/health/2013/01/28/states-act-while-congress-fiddles/.

——. "Stymied at Every Turn: EPA Withdraws Two Draft TSCA Proposals in the Face of Endless Delay at OMB." *EDF Health,* September 6, 2013. http://blogs.edf.org /health/2013/09/06/stymied-at-every-turn-epa-withdraws-two-draft-tsca -proposals-in-the-face-of-endless-delay-at-omb/.

Denison, Richard, and Marianne Engleman. "Letter from Richard Denison and Marianne Engleman to Cass Sunstein and OIRA." Washington, D.C.: Environmental Defense Fund, 2012. http://blogs.edf.org/nanotechnology/files/2012/02/TSCA -Letter-to-OMB-re-CBI-02–29–12-Final.pdf.

Design for the Environment. *Flame Retardant Alternatives for Hexabromocyclododecane (HBCD)*. Washington, D.C.: EPA, 2014. http://www.epa.gov/dfe/pubs/projects /hbcd/hbcd-full-report-508.pdf.

——. *Flame Retardants Used in Flexible Polyurethane Foam: An Alternatives Assessment Update.* Washington, D.C.: EPA, 2015. http://www.epa.gov/saferchoice-flame-retardant-used-polyurethane-foam.

——. *Furniture Flame Retardancy Partnership: Environmental Profiles of Chemical Flame-Retardant Alternatives for Low-Density Polyurethane Foam.* Washington, D.C.: EPA, 2005. http://www.epa.gov/dfe/pubs/flameret/altrep-v1/altrepv1-f1c.pdf.

Desmond, Matthew. "Relational Ethnography." *Theory and Society* 43, no. 5 (2014): 547–579.

Devereux, George. *From Anxiety to Method in the Behavioral Sciences.* The Hague: Mouton, 1966.

DiGangi, Joe, Arlene Blum, Åke Bergman, Cynthia A de Wit, Donald Lucas, David Mortimer, Arnold Schecter, Martin Scheringer, Susan Shaw, and Thomas Webster. "San Antonio Statement on Brominated and Chlorinated Flame Retardants." *Environmental Health Perspectives* 118, no. 12 (2010): A516–A518.

Dingell, John D., and Bart Stupak. "Letter from John D. Dingell and Burt Stupak, Subcommittee on Oversight and Investigations, to Stephen Johnson, EPA," 2008. http://www.ewg.org/sites/default/files/110-ltr_031308_EPA_BPA.pdf.

Dishaw, Laura, Christina Powers, Ian Ryde, Simon Roberts, Frederic Seidler, Theodore Slotkin, and Heather M. Stapleton. "Is the PentaBDE Replacement, Tris(1,3-Dichloro-2-Propyl)phosphate (TDCPP), a Developmental Neurotoxicant? Studies in PC12 Cells." *Toxicology and Applied Pharmacology* 256, no. 3 (2011): 281–289.

Do, Rylee Phuong, Richard W. Stahlhut, Davide Ponzi, Frederick S. Vom Saal, and Julia A. Taylor. "Non-Monotonic Dose Effects of in Utero Exposure to di(2-Ethylhexyl) Phthalate (DEHP) on Testicular and Serum Testosterone and Anogenital Distance in Male Mouse Fetuses." *Reproductive Toxicology* 34, no. 4 (December 2012): 614–621.

Dodson, Robin, Laura J. Perovich, Adrian Covaci, Nele Van den Eede, Alin Ionas, Alin Dirtu, Julia Brody, and Ruthann Rudel. "After the PBDE Phase-Out: A Broad Suite of Flame Retardants in Repeat House Dust Samples from California." *Environmental Science & Technology* 46, no. 24 (2012): 13056–13066.

Dow Chemical Company. "ICL-IP Harnesses Dow's Polymeric Flame Retardant Technology in New Product Development." *Additives for Polymers* 2012, no. 3 (2012): 2.

Edelstein, Michael R. *Contaminated Communities: The Social and Psychological Impacts of Residential Toxic Exposure.* Boulder, Colo.: Westview, 1988.

Egeghy, Peter, Richard Judson, Sumit Gangwal, Shad Mosher, Doris Smith, James Vail, and Elaine Cohen-Hubal. "The Exposure Data Landscape for Manufactured Chemicals." *Science of The Total Environment* 414, no. 1 (2012): 159–166.

Egger, Gerda, Gangning Liang, Ana Aparicio, and Peter A. Jones. "Epigenetics in Human Disease and Prospects for Epigenetic Therapy." *Nature* 429, no. 6990 (2004): 457–463.

Egginton, Joyce. *The Poisoning of Michigan.* New York: Norton, 2009.

Emory University Rollins School of Public Health. "PBB Registry Research Findings." Atlanta: Emory University Rollins School of Public Health, 2013.

Environmental Working Group. "Body Burden: The Pollution in Newborns." Washington, D.C., July 14, 2005. http://www.ewg.org/research/body-burden-pollution-newborns.

——. "EPA Axes Panel Chair at Request of Chemical Industry Lobbyists: 17 Conflicted Reviewers," 2008. http://www.ewg.org/research/epa-axes-panel-chair-request-chemical-industry-lobbyists/17-conflicted-reviewers.

——. "EPA Axes Panel Chair at Request of Chemical Industry Lobbyists: Review Panel Timeline," 2008. http://www.ewg.org/research/epa-axes-panel-chair-request-chemical-industry-lobbyists/review-panel-timeline.

——. "Staff." Accessed April 30, 2015. http://www.ewg.org/about-us/staff.

——. "Form 990. Return of Organization Exempt from Income Tax." 2015. http://www.scribd.com/doc/48001474/fire-safety.

Epseland, Wendy, and Mitchell Stevens. "Commensuration as a Social Process." *American Journal of Sociology* 24 (1998): 313–343.

Epstein, Steven. *Impure Science: AIDS, Activism, and the Politics of Knowledge.* Berkeley: University of California Press, 1996.

Eriksson, Per, Eva Jakobsson, and Anders Fredriksson. "Brominated Flame Retardants: A Novel Class of Developmental Neurotoxicants in Our Environment?" *Environmental Health Perspectives* 109, no. 9 (2001): 903–908.

Eskenazi, Brenda, Jonathan Chevrier, Stephen Rauch, Katherine Kogut, Kim Harley, Caroline Johnson, Celina Trujillo, Andreas Sjodin, and Asa Bradman. "In Utero and Childhood Polybrominated Diphenyl Ether (PBDE) Exposures and Neurodevelopment in the CHAMACOS Study." *Environmental Health Perspectives* 121, no. 2 (2013): 257–262.

Etzion, Dror, and Gerald F. Davis. "Revolving Doors?: A Network Analysis of Corporate Officers and U.S. Government Officials." *Journal of Management Inquiry* 17, no. 3 (2008): 157–161.

European Chemicals Agency. *Guidance on Information Requirements and Chemical Safety Assessment.* Helsinki: European Chemicals Agency, 2008. http://echa.europa.eu/documents/10162/13632/information_requirements_r11_en.pdf.

European Union. "Restriction of the Use of Certain Hazardous Substances (RoHS). Directive 2011/65/EU," 2011.

Evans, James. "Industry Collaboration, Scientific Sharing and the Dissemination of Knowledge." *Social Studies of Science* 40, no. 5 (2010): 757–791.

Evarts, Ben. *Trends and Patterns of U.S. Fire Losses in 2010.* Quincy, Mass.: National Fire Protection Association, 2011.

Eyal, Gil. "For a Sociology of Expertise: The Social Origins of the Autism Epidemic." *American Journal of Sociology* 118, no. 1 (2013): 1–45.

Faber, Daniel. *Capitalizing on Environmental Injustice: The Polluter-Industrial Complex in the Age of Globalization.* Lanham, Md.: Rowman & Littlefield, 2008.

Feiock, Richard, and Christopher Stream. "Environmental Protection Versus Economic Development: A False Trade-Off?" *Public Administration Review* 61, no. 3 (2001): 313–321.

Fischer, Douglas, Kim Hooper, Maria Athanasiadou, Ioannis Athanassiadis, and Ake Bergman. "Children Show Highest Levels of Polybrominated Diphenyl Ethers in a California Family of Four: A Case Study." *Environmental Health Perspectives* 114, no. 10 (2006): 1581–1584.

Freudenburg, William. "Perceived Risk, Real Risk: Social Science and the Art of Probabilistic Risk Assessment." *Science* 242, no. 4875 (1988): 44–49.

——. "Privileged Access, Privileged Accounts: Toward a Socially Structured Theory of Resources and Discourses." *Social Forces* 84, no. 1 (2005): 89–114.

Freudenburg, William, and Susan Pastor. "Public Responses to Technological Risks: Toward a Sociological Perspective." *Sociological Quarterly* 33, no. 3 (1992): 389–412.

Frickel, Scott. "Just Science? Organizing Scientist Activism in the U.S. Environmental Justice Movement." *Science as Culture* 13, no. 4 (2004): 449–469.

Frickel, Scott, Richard Campanella, and M. Bess Vincent. "Mapping Knowledge Investments in the Aftermath of Hurricane Katrina: A New Approach for Assessing Regulatory Agency Responses to Environmental Disaster." *Environmental Science and Policy* 12, no. 2 (2009): 119–133.

Frickel, Scott, and Michelle Edwards. "Untangling Ignorance in Environmental Risk Assessment." In *Powerless Science? Science and Politics in a Toxic World*, ed. Soraya Boudia and Nathalie Jas, 215–233. New York: Berghahn, 2014.

Frickel, Scott, Sahra Gibbon, Jeff Howard, Joanna Kempner, Gwen Ottinger, and David Hess. "Undone Science: Charting Social Movement and Civil Society Challenges to Research Agenda Setting." *Science, Technology, & Human Values* 35, no. 4 (2010): 444–476.

Frickel, Scott, and Neil Gross. "A General Theory of Scientific/Intellectual Movements." *American Sociological Review* 70, no. 2 (2005): 204–232.

Frickel, Scott, and Kelly Moore. *The New Political Sociology of Science: Institutions, Networks, and Power*. Madison: University of Wisconsin Press, 2006.

Frickel, Scott, Rebekah Torcasso, and Annika Y. Anderson. "The Organization of Expert Activism: Shadow Mobilization in Two Social Movements." *Mobilization* 20, no. 2 (2015): 305–323.

Gagnon, Chantal. "Political Translation." In *Handbook of Translation Studies*. Vol. 12, ed. Yves Gambier and Luc van Doorslaer, 252–256. Amsterdam: John Benjamins, 2010.

Gauchat, Gordon. "Politicization of Science in the Public Sphere: A Study of Public Trust in the United States, 1974–2010." *American Sociological Review* 72, no. 2 (2012): 167–187.

George, Alexander, and Andrew Bennett. *Case Studies and Theory Development in the Social Sciences*. Cambridge, Mass.: MIT Press, 2005.

German Toxicology Society. "Guide for Authors." *Toxicology*. Accessed April 30, 2015. http://www.elsevier.com/journals/toxicology/0300–483X/guide-for-authors #54000.

Gibbs, Lois. "Citizen Activism for Environmental Health: The Growth of a Powerful New Grassroots Health Movement." *Annals of the American Academy of Political and Social Science* 584 (2002): 97–109.

Giddens, Anthony. "Risk and Responsibility." *Modern Law Review* 62, no. 1 (1999): 1–10.

Gieryn, Thomas. "Boundary-Work and the Demarcation of Science from Non-Science: Strains and Interests in Professional Ideologies of Scientists." *American Sociological Review* 48, no. 6 (1983): 781–795.

——. *Cultural Boundaries of Science.* Chicago: University of Chicago Press, 1999.

Give Toxics the Boot. Homepage. Accessed April 30, 2015. http://givetoxicstheboot.org/.

Global Information Inc. "US Flame Retardant Demand to Reach 938 Million Pounds in 2016, Growing 4.6% Annually." Press release. October 1, 2012. http://www.ireachcontent.com/news-releases/us-flame-retardant-demand-to-reach-938-million-pounds-in-2016-growing-46-annually-172172821.html.

Gold, Marian Deborah, Arlene Blum, and Bruce N. Ames. "Another Flame Retardant, Tris-(1,3-Dichloro-2-Propyl)-Phosphate, and Its Expected Metabolites Are Mutagens." *Science* 200, no. 19 (1978): 785–787.

Golden, A. L., S. B. Markowitz, and P. J. Landrigan. "The Risk of Cancer in Firefighters." *Occupational Medicine* 10, no. 4 (1995): 803–820.

Goodman, Bryan. "North American Flame Retardant Alliance to Provide All Advocacy Services to Member Companies." Press release. August 31, 2012, American Chemistry Council. http://www.americanchemistry.com/Media/PressReleasesTranscripts/ACC-news-releases/NAFRA-to-Provide-All-Advocacy-Services-to-Members.html.

Goodman, Julie, John Briesemeier, Giffe Johnson, Casey Harbison, Raymond Harbison, Yiliang Zhu, Richard Lee, Hanna Silberberg, Marcia Hardy, and Todd Stedeford. "Fecundability and Serum PBDE Concentrations in Women." *Environmental Health Perspectives* 118, no. 8 (2010): A330.

Goodman, Julie E. "Neurodevelopmental Effects of Decabromodiphenyl Ether (BDE-209) and Implications for the Reference Dose." *Regulatory Toxicology and Pharmacology* 54, no. 1 (June 2009): 91–104.

Gottmann, E., S. Kramer, B. Pfahringer, and C. Helma. "Data Quality in Predictive Toxicology: Reproducibility of Rodent Carcinogenicity Experiments." *Environmental Health Perspectives* 109, no. 5 (May 2001): 509–514.

Gould, Kenneth, David Pellow, and Allan Schnaiberg. "Interogating the Treadmill of Production: Everything You Wanted to Know about the Treadmill but Were Afraid to Ask." *Organization & Environment* 17, no. 3 (2004): 296–316.

Gould, Kenneth, Allan Schnaiberg, and Adam Weinberg. *Local Environmental Struggles: Citizen Activism in the Treadmill of Production.* New York: Cambridge University Press, 1996.

Gradient. "Science and Strategies for Health and the Environment." Accessed April 30, 2015. http://www.gradientcorp.com/who-we-are-and-what-we-do.html.

Gray, George. "Best Practices for Sustainable Alternative Chemical Substitution." In *HESI Annual Meeting.* Alexandria, Va., 2013. http://www.hesiglobal.org/files/Gray-Sustainable Alternative Chemical Solutions.pdf.

——. "George M. Gray Curriculum Vitae." Modified February 2012. https://dtsc.ca.gov/LawsRegsPolicies/upload/Gray_CV.pdf.

———. "Letter from George Gray, U.S. EPA, to Nancy Sandrof, American Chemistry Council," 2007. http://www.ewg.org/research/epa-axes-panel-chair-request-chemical -industry-lobbyists/review-panel-timeline.

Gray, George, Steven Baskin, Gail Charnley, Joshua Cohen, Lois Gold, Nancy Kerkvliet, Harold Koenig, et al. "The Annapolis Accords on the Use of Toxicology in Risk Assessment and Decision-Making: An Annapolis Center Workshop Report." *Toxicological Mechanisms and Methods* 11, no. 3 (2001): 225–231.

Gray, George M., and Joshua T. Cohen. "Rethink Chemical Risk Assessments." *Nature* 489 (2012): 27–28.

Green Science Policy Institute. "About." Accessed April 30, 2015. http://greenscience policy.org/about/.

———. "Another 'Candle Standard' Bites the Dust." Accessed April 6, 2015. http:// greensciencepolicy.org/another-candle-standard-bites-the-dust/.

———. "Bibliography." Accessed April 30, 2015. http://greensciencepolicy.org/bibliography /#health.

Greenberg, Daniel S. *Science, Money, and Politics: Political Triumph and Ethical Erosion.* Chicago: University of Chicago Press, 2001.

Greene, Ronnie. "Ouster of Scientist from EPA Panel Shows Industry Clout." Center for Public Integrity, February 13, 2013. http://www.publicintegrity.org/2013/02/13 /12199/ouster-scientist-epa-panel-shows-industry-clout.

Groenewegen, Peter, and Lois Peters. "The Emergence and Change of Materials Science and Engineering in the United States." *Science, Technology, & Human Values* 27, no. 1 (2002): 112–133.

Gross, Liza. "Special Report: Flame Retardant Industry Spent $23 Million on Lobbying, Campaign Donations." *Environmental Health News*, November 16, 2011. http:// www.environmentalhealthnews.org/ehs/news/2011/money-to-burn.

Gross, Matthias. "The Unknown in Process: Dynamic Connections of Ignorance, Non-Knowledge and Related Concepts." *Current Sociology* 55 (2007): 742–759.

Gulliford, James. "Letter from James Gulliford, U.S. EPA, to Nancy Sandrof, American Chemistry Council." Washington, D.C.: EPA, 2008. http://www.epa.gov/oppt/vccep/ pubs/decaltr.pdf.

Habermas, Jurgen. *Toward a Rational Society: Student Protest, Science, and Politics.* Boston: Beacon, 1970.

Hall, Bill. "Defending against Product De-Selection Attacks: Where Do We Stand?" Presentation at Flexible Vinyl Products Division Conference, Society of the Plastics Industry, Burlington, Vt., July 12, 2011.

Hamby, Chris, and Jim Morris. "EPA Proposal Has Been under Review for 638 Days." Center for Public Intergrity, February 9, 2012. http://www.iwatchnews.org/2012/02 /09/8109/chemicals-Concern-List-Stuck-Omb.

Hannerz, Ulf. "Being There . . . and There . . . and There!" *Ethnography* 4, no. 2 (2003): 201–216.

Hardy, Marcia. "Commment on 'Polybrominated Diphenyl Ethers Contamination of United States Food.'" *Environmental Science & Technology* 39, no. 7 (2005): 2414.

Hardy, Marcia, Marek Banasik, and Todd Stedeford. "Toxicology and Human Health Assessment of Decabromodiphenyl Ether." *Critical Reviews in Toxicology* 38, Suppl. 3 (2009): 1–44.

Hardy, Marcia, John Biesemeier, Marek Banasik, and Todd Stedeford. "Comment on 'Alternate and New Brominated Flame Retardants Detected in U.S. House Dust.'" *Environmental Science & Technology* 42, no. 24 (2008): 9453–9454.

Hardy, Marcia, Michael Mercieca, Dean Rodwell, and Todd Stedeford. "Prenatal Developmental Toxicity of Decabromodiphenyl Ethane in the Rat and Rabbit." *Birth Defects Research Part B: Developmental and Reproductive Toxicology* 89, no. 2 (2010): 139–146.

Harley, Kim G., Amy R. Marks, Jonathan Chevrier, Asa Bradman, Andreas Sjödin, and Brenda Eskenazi. "PBDE Concentrations in Women's Serum and Fecundability." *Environmental Health Perspectives* 118, no. 5 (2010): 699–704.

Harley, Kim, Nicholas Jewell, Raul Aguilar, Amy Marks, Jonathan Chevrier, Asa Bradman, and Brenda Eskenazi. "PBDE Concentrations in Women: Harley et al. Respond." *Environmental Health Perspectives* 118, no. 8 (2010): A330–A331.

Hawthorne, Michael. "Chemical Firm's Champion Now EPA Expert." *Chicago Tribune*, September 10, 2012, http://articles.chicagotribune.com/2012-09-10/news/ct -met-flame-retardants-epa-20120910_1_flame-retardants-chemicals-epa-admini strator-lisa-jackson.

——. "Toxic Flame Retardant May Get a Reprieve." *Chicago Tribune*, December 20, 2012, http://articles.chicagotribune.com/2012-12-20/news/ct-met-flames-chemical -rules-20121220_1_flame-retardants-pbdes-ban-toxic-chemicals.

Healey, Michael J. "The Stuff Upholstery's Made Of." Loudon, Tenn.: Polyurethane Foam Association, 2015. http://www.pfa.org/affiliates/rr.html.

Health Care Without Harm. "Health Care Sector Moves Away from Flame Retardants in Upholstered Furniture." Press release. September 11, 2014. https://noharm -uscanada.org/articles/press-release/us-canada/us-canada/health-care-sector-moves-away -flame-retardants-upholstered.

Healthy Child Healthy World. "Graco Commits to Banning Toxic Flame Retardants from Children's Products." July 16, 2012. http://www.healthychild.org/blog /comments/press_release_graco_commits_to_banning_toxic_flame_retardants _from_children.

Hecht, Gabrielle. *Being Nuclear: Africans and the Global Uranium Trade.* Cambridge, Mass.: MIT Press, 2012.

Hegstad, Maria. "EPA Presses Industry to Provide Flame Retardants' Data for TSCA Review." *Inside EPA*, Feburary 2, 2013.

Heinzerling, Lisa. "Who Will Run the EPA?" *Yale Journal of Regulation* 30 (2013): 39–45.

Herbstman, Julie B., Andreas Sjödin, Matthew Kurzon, Sally A. Lederman, Richard S. Jones, Larry L. Needham, Deliang Tang, Megan Niedzwiecki, Richard Y. Wang, and Frederica Perera. "Prenatal Exposure to PBDEs and Neurodevelopment." *Environmental Health Perspectives* 118, no. 5 (2010): 712–719.

Hess, David. *Alternative Pathways in Science and Industry: Activism, Innovation, and the Environment in an Era of Globalization.* Cambridge, Mass.: MIT Press, 2007.

——. "The Potentials and Limitations of Civil Society Research: Getting Undone Science Done." *Sociological Inquiry* 79, no. 3 (2009): 306–327.

Hites, Ronald A. "Polybrominated Diphenyl Ethers in the Environment and in People: A Meta-Analysis of Concentrations." *Environmental Science & Technology* 38, no. 4 (January 2004): 945–956.

Hoban, Rose. "Q&A with NIEHS Head Linda Birnbaum." *North Carolina Health News*, April 2, 2012. http://www.northcarolinahealthnews.org/2012/04/02/q-a-with-niehs -head-linda-birnbaum-on-bpa/.

Hoffman, Sebastian, and Thomas Hartung. "Toward an Evidence-Based Toxicology." *Human and Experimental Toxicology* 25 (2006): 497–513.

Hoffmann-Riem, Holger, and Brian Wynne. "In Risk Assessment, One Has to Admit Ignorance." *Nature* 416, no. 6877 (March 14, 2002): 123.

Hogue, Cheryl. "Assessing Alternatives to Toxic Chemicals." *Chemical and Engineering News* 91, no. 50 (2013): 19–20. http://cen.acs.org/articles/91/i50/Assessing-Alternatives -Toxic-Chemicals.html.

Hook, Gary E.R. "Responsible Care and Credibility." *Environmental Health Perspectives* 104, no. 11 (1996): 1138–1139.

Horton, Megan K., Sabine Bousleiman, Richard Jones, Andreas Sjodin, Xinhua Liu, Robin Whyatt, Ronald Wapner, and Pam Factor-Litvak. "Predictors of Serum Concentrations of Polybrominated Flame Retardants among Healthy Pregnant Women in an Urban Environment: A Cross-Sectional Study." *Environmental Health* 12 (2013): 1–13.

Hosenball, Mark. "Karl Marx and the Pajama Game." *Mother Jones*, November 1979. http://www.motherjones.com/politics/1979/11/karl-marx-and-pajama-game.

Huber, Peter. *Galileo's Revenge: Junk Science in the Courtroom.* New York: Basic Books, 1991.

Iallonardo, Tony. " 'Safe Chemicals Act of 2011' Introduced Today Legislation Would Protect American Families from Toxic Chemicals." Safer Chemicals Healthy Families, April 14, 2011. http://www.saferchemicals.org/2011/04/safe-chemicals-act-of -2011-introduced-today-legislation-would-protect-american-families-from-toxic -chemicals.html.

International Agency for Research on Cancer. "Painting, Firefighting, and Shiftwork." *IARC Monographs on the Evaluation of Carcinogenic Risks to Humans,* Vol. 98. Lyon, France: IARC, 2010.

International Association of Fire Fighters. "About the IAFF." Accessed April 30, 2015. http://www.iaff.org/about/default.asp.

——. "State Presumptive Disability Laws." Accessed April 30, 2015. http://www.iaff.org /hs/phi/docs/PresumptiveDisabilityChart.pdf.

——. "IAFF Calls on Canadian Government to Ban PBDEs." Press release. November 2, 2007. http://www.iaff.org/canada/Updates/IAFF_PBDE_02112007.htm.

International Electrotechnical Commission. "TC 108 Voting Result." August 8, 2012. http://www.iec.ch/dyn/www/f?p=103:52:0::::FSP_ORG_ID,FSP_DOC_ID,FSP _DOC_PIECE_ID:1311,139509,269491.

Israeli Chemicals Limited. "ICL to Increase Production of Polymeric Flame Retardant at West Virginia Facility." *PR Newswire*, November 13, 2012. http://www.prnewswire .com/news-releases/icl-to-increase-production-of-polymeric-flame-retardant-at -west-virginia-facility-179095971.html.

Jack, Maia. "Principles of Alternatives Assessment." Grocery Manufacturers Association, September 7, 2012. http://www.gmaonline.org/file-manager/20120907_Green %20Chemistry%20AA%20Coalition%20Document_Principles%20of%20Alterna tives%20Assessment.pdf.

Jasanoff, Sheila S. "Contested Boundaries in Policy-Relevant Science." *Social Studies of Science* 17, no. 2 (May 1987): 195–230.

Jayvee Brand, Inc., et al., v. United States of America. No 82-1167. United States Court of Appeals, District of Columbia Circuit Court (1983).

Jones, James J. "TSCA Existing Chemicals Program." Presentation at American Chemistry Council GlobalChem Conference, February 26, 2013, Washington, D.C.

Jones, Rebecca. "Evaluation of Health Hazard Endpoints." Presentation given at Chemical Assessment and Management Workshop, September 2010, Beijing, People's Republic of China, 2010.

Joyce, Kelly. "Is Tuna Safe? A Sociological Analysis of Federal Fish Advisories." In *Mercury Pollution: A Transdisciplinary Treatment*, ed. Sharon L. Zuber and Michael C. Newman, 71–100. Boca Raton, Fla.: CRC, 2011.

Judson, Richard, Ann Richard, David Dix, Keith Houck, Matthew Martin, Robert Kavlock, Vicki Dellarco, et al. "The Toxicity Data Landscape for Environmental Chemicals." *Environmental Health Perspectives* 117, no. 5 (2009): 685–695.

Kaiser, Karen. "Protecting Respondent Confidentiality in Qualitative Research." *Qualitative Health Research* 19, no. 11 (2009): 1632–1641.

Kaiser Permanente. "Kaiser Permanente Commits to Purchasing Furniture Free from Toxic Flame Retardant Chemicals." Press release. June 3. Accessed April 30, 2014. http://share.kaiserpermanente.org/article/kaiser-permanente-commits-to -purchasing-furniture-free-from-toxic-flame-retardant-chemicals/.

Kemmlein, Sabine, Dorte Herzke, and Robin J. Law. "Brominated Flame Retardants in the European Chemicals Policy of REACH—Regulation and Determination in Materials." *Journal of Chromatography A* 1216, no. 3 (2009): 320–333.

Kempler, Joanna, Jon F. Merz, and Charles L. Bosk. "Forbidden Knowledge: Public Controversy and the Production of Nonknowledge." *Sociological Forum* 26, no. 3 (2011): 475–500.

Kerr, Peter. "Demand Increases for Fire-Safe Clothing." *New York Times*, June 11, 1983, http://www.nytimes.com/1983/06/11/style/demand-increases-for-fire-safe-clothing.html.

Kinchy, Abby J. *Seeds, Science, and Struggle: The Global Politics of Transgenic Crops.* Cambridge, Mass.: MIT Press, 2012.

Kinchy, Abby J. "Anti-Genetic Engineering Activism and Scientized Politics in the Case of 'Contaminated' Mexican Maize." *Agriculture and Human Values* 27, no. 4 (2010): 505–517.

Kinchy, Abby J., and Daniel Lee Kleinman. "Organizing Credibility: Discursive and Organizational Orthodoxy on the Borders of Ecology and Politics." *Social Studies of Science* 33, no. 6 (2003): 869–896.

Kirschner, Michael, and Arlene Blum. "When Product Safety and the Environment Appear to Collide." *Conformity* (2009): 12–16.

Klein, Hans K., and Daniel L. Kleinman. "The Social Construction of Technology: Structural Considerations." *Science, Technology, & Human Values* 27, no. 1 (2002): 28–52.

Kleinman, Daniel, and Sainath Suryanarayanan. "Dying Bees and the Social Production of Ignorance." *Science, Technology, & Human Values* 38, no. 4 (May 3, 2012): 492–517.

Kleinman, Daniel, and Steven Vallas. "Science, Capitalism, and the Rise of the 'Knowledge Worker': The Changing Structure of Knowledge Production in the United States." *Theory and Society* 30, no. 4 (2001): 451–492.

Kneiss, Sharon. "Letter from Sharon Kneiss, American Chemistry Council, to George Gray, U.S. EPA Assistant Administrator for Research and Development. May 3, 2007," 2007. http://www.ewg.org/research/epa-axes-panel-chair-request-chemical-industry-lobbyists/review-panel-timeline.

Kriebel, David, and Joel Tickner. "Reenergizing Public Health Through Precaution." *American Journal of Public Health* 91, no. 9 (2001): 1351–1355.

Kriebel, David, Joel Tickner, Paul Epstein, John Lemons, Richard Leveins, Edward Loechler, Margaret Quinn, Ruthann Rudel, Ted Schettler, and Michael Stoto. "The Precautionary Principle in Environmental Science." *Environmental Health Perspectives* 109, no. 9 (2001): 871–876.

Krimsky, Sheldon. *Hormonal Chaos: The Scientific and Social Origins of the Environmental Endocrine Hypothesis.* Baltimore: Johns Hopkins University Press, 2000.

——. "The Funding Effect in Science and Its Implications for the Judiciary." *Journal of Law and Policy* 8, no. 1 (2005): 43–68.

——. "The Weight of Scientific Evidence in Policy and Law." *American Journal of Public Health* 95, Suppl. 1 (2005): S129–S136.

Krimsky, Sheldon, and Dominic Golding. *Social Theories of Risk.* Westport, Conn.: Praeger, 1992.

Lacey, Anthony. "Sequester Order Cuts EPA Less Than Anticipated but Key Programs Hit." *Inside EPA*, March 3, 2013.

Lakoff, Andrew, and Erik Klinenberg. "Of Risk and Pork: Urban Security and the Politics of Objectivity." *Theory and Society* 39, no. 5 (2010): 503–525.

Latour, Bruno. *Science in Action: How to Follow Scientists and Engineers through Society.* Cambridge, Mass.: Harvard University Press, 1987.

Lavoie, Emma, Lauren Heine, Helen Holder, Mark Rossi, Robert Lee, Emily Connor, Melanie Vrable, David DiFiore, and Clive Davies. "Chemical Alternatives Assessment: Enabling Substitution to Safer Chemicals." *Environmental Science & Technology* 44, no. 24 (2010): 9244–9249.

Le Bras, Michel. "Preface." In *Fire Retardancy of Polymers: New Applications of Mineral Fillers*, ed. Michel le Bras, Charles Wilkie, and Serge Bourbigot, v–vii. Cambridge: Royal Society of Chemistry, 2005.

Lee, Stephanie. "Patchwork of Bills Cover Flame Retardants." *San Francisco Chronicle*, March 14, 2013, http://www.sfgate.com/health/article/Patchwork-of-bills-cover -flame-retardants-4353348.php.

Leffall, LaSalle, Margaret Kripke, and Suzanne Reuben. *President's Cancer Panel: Reducing Environmental Cancer Risk*. Bethesda, Md.: U.S. Department of Health and Human Services, 2010.

LeMasters, Grace, Ash Genaidy, Paul Succop, James Deddens, Tarek Sobeih, Heriberto Barriera-Viruet, Kari Dunning, and James Lockey. "Cancer Risk among Firefighters: A Review and Meta-Analysis of 32 Studies." *Journal of Occupational and Environmental Medicine* 48, no. 11 (2006): 1189–1202.

Levchik, Sergei. "Modes of Flame Retardant Action: Can Halogen-Free FRs Replace Brominated FRs?" Presentation at the 12th Workshop on Brominated and Other Flame Retardants. June 6–7, 2011, Boston.

Lubitow, Amy. "The Battle over Bisphenol-A: United States Chemical Policy and the New Networked Environmental Politics." Ph.D., diss., Northeastern University, 2011.

Lyon, Thomas P., and John W. Maxwell. "Astroturf: Interest Group Lobbying and Corporate Strategy." *Journal of Economics and Management Strategy* 13, no. 4 (2004): 561–597.

MacDonald, Sherry. "Hospital Initiative Urges Development of Safer Interior Furnishings." Healthier Hospitals Initiative. Accessed July 11, 2012. http://healthierhospitals .org/media-center/press-releases/hospital-initiative-urges-development-safer -interior-furnishings.

MacGillivray, Brian Hector, Ruth Alcock, and Jerry Busby. "Is Risk-Based Regulation Feasible? The Case of Polybrominated Diphenyl Ethers (PBDEs)." *Risk Analysis* 31, no. 2 (2011): 266–281.

Magnus, David. "Risk Management versus the Precautionary Principle: Agnotology as a Strategy in the Debate over Genetically Engineered Organisms." In *Agnotology: The Making and Unmaking of Ignorance*, ed. Robert Proctor and Londa Schiebinger, 250–265. Stanford, Calif.: Stanford University Press, 2008.

Maine Department of Environmental Protection, and Maine Center for Disease Control and Prevention. *Brominated Flame Retardants: Third Annual Report to the Maine Legislature.* Augusta, Maine, 2007.

Maine, State of. *An Act to Reduce Contamination of Breast Milk and the Environment from Release of Brominated Chemicals in Consumer Products,* 2004. http://www .mainelegislature.org/legis/bills/bills_121st/billtexts/LD179001-1.asp.

——. *Toxic Chemicals in Children's Products: Maine Revised Statutes Title 38, Chapter 16-D.* http://www.mainelegislature.org/legis/statutes/38/title38ch16-Dseco.html.

Marcus, George. "Ethnography in/of the World System: The Emergence of Multi-Sited Ethnography." *Annual Review of Anthropology* 24 (1995): 95–117.

Markowitz, Gerald E., and David Rosner. *Deceit and Denial: The Deadly Politics of Industrial Pollution.* Berkeley: University of California Press, 2002.

Marteinson, Sarah C., Robert J. Letcher, Laura Graham, Sarah Kimmins, Gregg Tomy, Vince P. Palace, Ian J. Ritchie, Lewis T. Gauthier, David M. Bird, and Kim J. Fernie. "The Flame Retardant B-1,2-Dibromo-4-(1,2-Dibromoethyl)cyclohexane: Fate, Fertility, and Reproductive Success in the American Kestrels (*Falco sparverius*)." *Environmental Science & Technology* 46, no. 15 (2012): 8440–8447.

Martell, John. "Letter from John Martell, Matt Vinci, Dennis Sweeney, and Kelly Fox to Cal Dooley. Re: Telling the Truth about Chemical Flame Retardants." July 13, 2012. http://www.saferchemicals.org/PDF/fire_fighters_letter_to_chemical_industry .pdf.

Massachusetts Department of Fire Services. "Comprehensive Model Fire Code," 2014. http://www.mass.gov/eopss/agencies/dfs/dfs2/osfm/fire-prev/comprehensive -model-fire-code.html.

Maxwell, Nancy I. *Understanding Environmental Health.* Boston: Jones and Bartlett, 2009.

Mayer, Brian. *Blue-Green Coalitions: Fighting for Safe Workplaces and Health Communities.* Ithaca, N.Y.: Cornell University Press, 2008.

McCormick, Sabrina. "Democratizing Science Movements: A New Framework for Mobilization and Contestation." *Social Studies of Science* 37, no. 4 (2007): 609–623.

McCright, Aaron M., and Riley E. Dunlap. "Challenging Global Warming as a Social Problem: An Analysis of the Conservative Movement's Counter-Claims." *Social Problems* 47, no. 4 (2000): 499–522.

McCright, Aaron M., and Riley E. Dunlap. "The Politicization of Climate Change and the Polarization in the American Public's Views of Global Warming, 2001–2010." *Sociological Quarterly* 52, no. 2 (2011): 155–194.

McDonald, Thomas A. "Polybrominated Diphenylether Levels among United States Residents: Daily Intake and Risk of Harm to the Developing Brain and Reproductive Organs." *Integrated Environ Assess Management* 1, no. 4 (2005): 343–354.

McGoey, Linsey. "Strategic Unknowns: Towards a Sociology of Ignorance." *Economy and Society* 41, no. 1 (2012): 1–16.

Meeker, John D., and Heather M. Stapleton. "House Dust Concentrations of Organo-phosphate Flame Retardants in Relation to Hormone Levels and Semen Quality Parameters." *Environmental Health Perspectives* 118, no. 3 (2010): 318–323.

Meironyté, Daiva, Koidu Norén, and Åke Bergman. "Analysis of Polybrominated Diphenyl Ethers in Swedish Human Milk. A Time-Related Trend Study, 1972–1997." *Journal of Toxicology and Environmental Health* 58, no. 6 (1999): 329–341.

Merton, Robert King, and Norman William Storer. *The Sociology of Science: Theoretical and Empirical Investigations.* Chicago: University of Chicago Press, 1973.

Meyer, David S. "Protest and Political Opportunities." *Annual Review of Sociology* 30 (2004): 125–145.

Michaels, David. *Doubt Is Their Product: How Industry's Assault on Science Threatens Your Health.* New York: Oxford University Press, 2008.

——. "Manufactured Uncertainty: Contested Science and the Protection of the Public's Health and Environment." In *Agnotology: The Making and Unmaking of Ignorance,* ed. Robert Proctor and Londa Schiebinger, 90–107. Stanford, Calif.: Stanford University Press, 2008.

Michaels, David, and Celeste Monforton. "Manufacturing Uncertainty: Contested Science and the Protection of the Public's Health and Environment." *American Journal of Public Health* 95, Suppl. 1 (2005): S39–48.

Michigan Department of Community Health. "PBBs in Michigan: Frequently Asked Questions 2011 Update." Accessed April 30, 2015. http://www.michigan.gov /documents/mdch_PBB_FAQ_92051_7.pdf.

Miller, Kevin. "DEP Urges Legislative Ban on Fire Retardant." *Bangor Daily News,* February 16, 2007, http://archive.bangordailynews.com/2007/02/16/dep-urges-legislative -ban-on-fire-retardant/.

Mohai, Paul, David Pellow, and J. Timmons Roberts. "Environmental Justice." *Annual Review of Environment and Resources* 34 (2009): 405–430.

Mol, Arthur. *The Refinement of Production: Ecological Modernization Theory and the Chemical Industry.* Utrecht: Van Arkel, 1995.

Mol, Arthur, David A. Sonnenfeld, and Gert Spaargaren. *The Ecological Modernisation Reader: Environmental Reform in Theory and Practice.* London: Routledge, 2009.

Moore, Kelly. *Disrupting Science: Social Movements, American Scientists, and the Politics of the Military, 1945–1975.* Princeton, N.J.: Princeton University Press, 2008.

——. "Organizing Integrity: American Science and the Creation of Public Interest Organizations, 1955–1975." *American Journal of Sociology* 101, no. 6 (1996): 1592–1627.

Moore, Kelly, Daniel Kleinman, David Hess, and Scott Frickel. "Science and Neoliberal Globalization: A Political Sociological Approach." *Theory and Society* 40, no. 5 (2011): 505–532.

Moore, Marshall. "Testimony Regarding Oversight of EPA Authorities and Actions to Control Exposures to Toxic Chemicals," U.S. Senate Committee Hearing on Environment and Public Works. July 24, 2014.

Morello-Frosch, Rachel, Julia Brody, Phil Brown, Rebecca Gasior Altman, Ruthann Rudel, and Carla Perez. "Toxic Ignorance and Right-to-Know in Biomonitoring Results Communication: A Survey of Scientists and Study Participants." *Environmental Health* 8, no. 1 (2009): 6.

Morello-Frosch, Rachel, Stephen Zavestoski, Phil Brown, Rebecca Gasior Altman, Sabrina McCormick, and Brian Mayer. "Embodied Health Movements: Responses to a 'Scientized' World." In *The New Political Sociology of Science: Institutions, Networks, and Power*, ed. Scott Frickel and Kelly Moore, 244–271. Madison: University of Wisconsin Press, 2006.

Myers, Daniel, and Daniel Cress. *Authority in Contention*. Boston: Elsevier, 2004.

Nader, Laura. "Up the Anthropologist—Perspectives Gained from Studying Up." In *Reinventing Anthropology*, ed. Dell Hymes, 284–311. New York: Pantheon, 1974.

National Association of State Fire Marshals. "National Furniture Flammability Standard." 2013. http://www.firemarshals.org/pdf/NASFM_upholstered_furniture_flammability_resolution_2013-4.pdf.

——. "News." *NASFM News*, February 2008. http://www.firemarshals.org/pdf/Feb-08-News.pdf.

——. "Policy Regarding Corporate Support of the National Association of State Fire Marshals." Accessed February 5, 2013. http://www.firemarshals.org/organization/finance.html.

National Cancer Institute. "Dictionary of Cancer Terms." Accessed February 15, 2013. http://www.cancer.gov/dictionary.

National Caucus of Environmental Legislators. "Status: PBDE Legislation," July 13, 2007. www.ncel.net/articles/PBDE.Legislation.Laws.Website.doc.

National Commission on Fire Prevention and Control. *America Burning*. Washington, D.C., 1973.

National Fire Protection Association. "History." Accessed April 30, 2015. http://www.nfpa.org/safety-information/for-consumers/causes/smoking/coalition-for-fire-safe-cigarettes/history.

——. "Home Fires." Modified September 2014. http://www.nfpa.org/research/reports-and-statistics/fires-by-property-type/residential/home-fires.

——. "States That Have Passed Fire-Safe Cigarette Laws." August 26, 2011. http://www.nfpa.org/safety-information/for-consumers/causes/smoking/coalition-for-fire-safe-cigarettes/states-that-have-passed-fire-safe-cigarette-laws.

National Institute on Money in State Politics. "Lobbyist Link: Bromine Science and Environmental Forum." Helena, Mont.: National Institute on Money in State Politics, 2013.

National Research Council. *A Framework to Guide Selection of Chemical Alternatives*. Washington, D.C.: NRC, 2015. http://www.nap.edu/catalog/18872/a-framework-to-guide-selection-of-chemical-alternatives.

——. *Risk Assessment in the Federal Government: Managing the Process.* Washington, D.C.: NRC, 1983.

——. *Science and Decisions: Advancing Risk Assessment.* Washington, D.C.: Committee on Improving Risk Analysis Approaches Used by the U.S. EPA. National Academies Press, 2009.

——. *Toxicity Testing in the 21st Century: A Vision and a Strategy.* Washington, D.C.: Committee on the Toxicity Testing and Assessment of Environmental Agents. National Academies Press, 2007.

National Safety Council. *Injury Facts.* Itasca, Ill.: NSC, 2011.

National Toxicology Program. *Report on Carcinogens*, 12th ed. Research Triangle Park, N.C.: U.S. Department of Health and Human Services, 2011.

Nelson, Gordon. "Testimony Regarding the Consumer Choice Fire Safety Act before the California State Senate Committee on Business, Professions, & Consumer Protection." April 25, 2011. http://24.104.59.141/channel/viewvideo/2353.

New Jersey Department of Environmental Protection. *Strategies for Addressing Cumulative Impacts in Environmental Justice Communities.* Trenton: New Jersey Department of Environmental Protection, 2009. http://www.nj.gov/dep/ej/docs/ejac_impacts_report200903.pdf.

New York, State of. "A6195–2011: Prohibits the Sale of Child Products Containing Tris." 2011.

Nguyen, Nhan. "Overview of Occupational Exposure and Environmental Release Assessment." Presentation given at the Chemical Assessment and Management Workshop, Beijing, People's Republic of China, September 2010.

Noren, Koidu, and Daiva Meironyte. "Certain Organochlorine and Organobromine Contaminates in Swedish Human Milk in Perspective of Past 20–30 Years." *Chemosphere* 40 (2000): 1111–1123.

Nylund, Kerstin, Lillemor Asplund, Bo Jansson, Per Jonsson, Kerstin Litzen, and Ulla Sellstrom. "Analysis of Some Polyhalogenated Organic Pollutants in Sediment and Sewage Sludge." *Chemosphere* 24, no. 12 (1992): 1721–1730.

O'Brien, David. *What Process Is Due? Courts and Science-Policy Disputes.* New York: Russel Sage Foundation, 1987.

O'Fallon, Liam, and Allen Dearry. "Community-Based Participatory Research as a Tool to Advance Environmental Health Sciences." *Environmental Health Perspectives* 110, Suppl. 2 (2002): 155–159.

Oregon Department of Environmental Quality. "Fact Sheet: Sources of Polychlorinated Biphenyls." Accessed April 30, 2015. http://www.deq.state.or.us/lq/cu/nwr/PortlandHarbor/docs/SourcePCBs.pdf.

Oreskes, Naomi, and Erik M. Conway. *Merchants of Doubt: How a Handful of Scientists Obscured the Truth on Issues from Tobacco Smoke to Global Warming.* New York: Bloomsbury, 2010.

Panofsky, Aaron L. "Field Analysis and Interdisciplinary Science: Scientific Capital Exchange in Behavior Genetics." *Minerva* 49, no. 3 (July 30, 2011): 295–316.

Parthasarathy, Shobita. "Breaking the Expertise Barrier: Understanding Activist Strategies in Science and Technology Policy Domains." *Science and Public Policy* 37, no. 5 (June 1, 2010): 355–367.

Pataki, Gyorgy. "Ecological Modernization as a Paradigm of Corporate Sustainability." *Sustainable Development* 17, no. 2 (2009): 82–91.

Patisaul, Heather, Simon Roberts, Natalie Mabrey, Katherine McCaffrey, Robin Gear, Joe Braun, Scott Belcher, and Heather M. Stapleton. "Accumulation and Endocrine Disrupting Effects of the Flame Retardant Mixture Firemaster 550 in Rats: An Exploratory Assessment." *Journal of Biochemical Molecular Toxicology* 27, no. 2 (2012): 124–136.

Peeples, Lynne. "Firefighters Sound Alarm on Toxic Chemicals." *Huffington Post*, March 27, 2014. http://www.huffingtonpost.com/2014/03/27/firefighters-toxic -chemicals-regulation-flame-retardants_n_5034976.html.

Perrow, Charles. *Normal Accidents: Living with High-Risk Technologies*. New York: Basic Books, 1984.

Petreas, Myrto, Jianwen She, F. Reber Brown, Jennifer Winkler, Gayle Windham, Evan Rogers, Guomao Zhao, Rajiv Bhatia, and M. Judith Charles. "High Body Burdens of 2,2′,4,4′-Tetrabromodiphenyl Ether (BDE-47) in California Women." *Environmental Health Perspectives* 111, no. 9 (March 10, 2003): 1175–1179.

Petryna, Adriana. *Life Exposed: Biological Citizens after Chernobyl*. Princeton, N.J.: Princeton University Press, 2002.

Phosphorous, Inorganic, and Nitrogen Flame Retardants Association. "About Us." Accessed August 10, 2015. http://pinfa.org/index.php/about-us/who-is-pinfa.

Plankey-Videla, Nancy. "Informed Consent as Process: Problematizing Informed Consent in Organizational Ethnographies." *Qualitative Sociology* 35, no. 1 (2012): 1–21.

Plautz, Jason. "EPA to Withdraw 2 High-Profile Regulations." *Greenwire*. September 6, 2013. http://www.eenews.net/stories/1059986842.

Polanyi, Michael. "The Republic of Science: Its Political and Economic Theory." *Minerva* 38 (2000): 1–32.

Pope, Arden, and Douglas Dockery. "Health Effects of Fine Particulate Air Pollution: Lines That Connect." *Journal of the Air and Waste Management Association* 56 (2006): 742.

Proctor, Robert, and Londa Schiebinger. *Agnotology: The Making and Unmaking of Ignorance*. Stanford, Calif.: Stanford University Press, 2008.

QSR International. NVivo, version 10. 2010.

Raffensperger, Carolyn, and Joel Tickner. *Protecting Public Health and the Environment: Implementing the Precautionary Principle*. Washington, D.C.: Island Press, 1999.

Reich, Michael R. "Environmental Politics and Science: The Case of PBB Contamination in Michigan." *American Journal of Public Health* 73, no. 3 (1983): 302–313.

Renn, Ortwin. *Risk Governance: Coping with Uncertainty in a Complex World*. London: Earthscan, 2008.

Reporterlink. "World Flame Retardants Market." Modified March 7, 2013. http://www
.prnewswire.com/news-releases/world-flame-retardants-market-196039651.html.

Rice, Deborah C., Elizabeth A. Reeve, Aleece Herlihy, R. Thomas Zoeller, W. Douglas
Thompson, and Vincent P. Markowiski. "Developmental Delays and Locomotor
Activity in the C57BL6/J Mouse Following Neonatal Exposure to the Fully-
Brominated PBDE, Decabromodiphenyl Ether." Neurotoxicology and Teratology
29, no. 4 (2007): 511–520.

Rice, Deborah C., W. Douglas Thompson, Elizabeth A. Reeve, Kristen D. Onos, Mina
Assadollahzadeh, and Vincent P. Markowski. "Behavioral Changes in Aging but
Not Young Mice after Neonatal Exposure to the Polybrominated Flame Retardant
decaBDE." Environmental Health Perspectives 117, no. 12 (December 2009):
1903–1911.

Richards, Ira. Principles and Practice of Toxicology in Public Health. Boston: Jones and
Bartlett, 2008.

Rizzuto, Pat. "Law Center Says Strict Chemical Controls Foster Innovation, Markets,
Protect People." Bloomberg BNA, February 18, 2013.

R.J. Reynolds. "Strategic Plan 1996." Accessed April 30, 2015. http://legacy.library.ucsf
.edu/tid/bus60doo.

RnR Market Research. "Flame Retardant Industry to Touch $10.340 Million by 2019."
Press release. Accessed August 8, 2014. http://beforeitsnews.com/science-and
-technology/2014/08/flame-retardant-industry-to-touch-10340-million-by
-2019-2712888.html.

Roberts, J. Timmons. "Emerging Global Environmental Standards: Prospects and Perils." In Globalization and the Evolving World Society, ed. Proshanta Nandi and
Shahid Shahidullah, 144–163. Boston: Leiden, 1998.

Roe, Sam. "Doubts Cast on New Research Touted by Fire-Retardant Lobby." Chicago
Tribune, December 30, 2012. http://articles.chicagotribune.com/2012-12-30/news/ct
-met-flames-southwest-study-20121230_1_flame-retardants-vytenis-babrauskas
-paper.

Ruckelshaus, William. "Risk in a Free Society." Risk Analysis 4, no. 3 (1984): 157–162.

Rudel, Ruthann A., David E. Camann, John D. Spengler, Leo R. Korn, and Julia G.
Brody. "Phthalates, Alkylphenols, Pesticides, Polybrominated Diphenyl Ethers, and
Other Endocrine-Disrupting Compounds in Indoor Air and Dust." Environmen-
tal Science & Technology 37, no. 20 (2003): 4543–4553.

Rudel, Ruthann A., and Elizabeth Newton. "Letter to the Editor: Exposure Assessment
for Decabromodiphenyl Ether (decaBDE) Is Likely to Underestimate General U.S.
Population Exposure." Journal of Children's Health 2, no. 2 (2004): 171–173.

Rudel, Ruthann A., and Laura J. Perovich. "Endocrine Disrupting Chemicals in Indoor
and Outdoor Air." Atmospheric Environment 43, no. 1 (2009): 170–181. doi:10.1016/j.
atmosenv.2008.09.025.

Sachs, Noah M. "Rescuing the Strong Precautionary Principle from Its Critics." University of Illinois Law Review, no. 4 (2011): 1285–1338.

Safer Chemicals Healthy Families. "National Stroller Brigade Today at U.S. Capital."
  February 5, 2013. http://www.saferchemicals.org/2012/05/national-stroller-brigade
  -today-at-us-capitol.html.
——. "Presentation of Findings from a Survey of 825 Voters in 75 Swing Congressional
  Districts." Washington, D.C.: Mellman Group (2010). http://www.saferchemicals
  .org/PDF/resources/schf-poll-final.pdf.
——. "Safer Chemicals, Healthy Families: Who We Are." Accessed April 30, 2015. http://
  saferchemicals.org/who-we-are.
Sandman, Peter. "Hazard versus Outrage in the Public Perception of Risk." In
  *Effective Risk Communication: The Role and Responsibility of Government and
  Nongovernment Organizations*, ed. V. T. Covello, David McCallum, and Maria
  Pavlova, 45–49. New York: Plenum, 1989.
Sarewitz, Daniel. "How Science Makes Environmental Controversies Worse." *Environ-
  mental Science and Policy* 7 (2004): 385–403.
Sass, Jennifer. "Credibility of Scientists: Conflict of Interest and Bias." *Environmental
  Health Perspectives* 114, no. 3 (2006): A147–148.
Sasso, Alissa. "ECHA Raises Its Sights: Several Recent Additions to the REACH Can-
  didate List Set Precedents," Environmental Defense Fund, January 29, 2013. http://
  blogs.edf.org/nanotechnology/2013/01/29/echa-raises-its-sights-several-recent
  -additions-to-the-reach-candidate-list-set-precedents.
Schecter, Arnold, Olaf Papke, Kuang-Chi Tunk, Daniele F. Staskal, and Linda Birn-
  baum. "Polybrominated Diphenyl Ethers Contamination of United States Food."
  Environmental Science & Technology 38, no. 20 (2004): 5306–5311.
Schmidt, Charles. "Uncertain Inheritance: Transgenerational Effects of Environmen-
  tal Exposures." *Environmental Health Perspectives* 121, no. 10 (2013): 298–303.
Schnaiberg, Allan, and Kenneth Alan Gould. *Environment and Society: The Enduring
  Conflict*. New York: St. Martin's, 1994.
Schreder, Erika. "Hidden Hazards in the Nursery." Seattle: Washington Toxics Coali-
  tion/Safer States, 2012.
Scruggs, Caroline E., Leonard Ortolano, Megan R. Schwartzman, and Michael P.
  Wilson. "The Role of Chemical Policy in Improving Supply Chain Knowledge
  and Product Safety." *Journal of Environmental Studies and Sciences* 4 (2015):
  132–141.
Serbaroli, Joseph. *Plastics 101: A Primer on Flame Retardants for Thermoplastics*.
  Ampacet Technical Services Report, 2005. www.ampacet.com/usersimage/File
  /tutorials/FlameRetardants.pdf.
Shane, Scott. "Tobacco Industry Tied to Firefighters: Donations Seen as Way to Weaken
  Support for Fire-Safe Cigarettes." *Baltimore Sun*, February 16,1999. http://articles
  .baltimoresun.com/1999-02-16/news/9902160065_1_tobacco-fire-safe-cigarettes
  -cigarette-manufacturers.
Shapiro, Nicholas. "Un-Knowing Exposure: Toxic Emergency Housing, Strategic Inclusivity and Governance in the US Gulf South." In *Knowledge, Technology, and*

Law, ed. Emilie Cloatre and Martyn Pickersgill, 189–205. Hoboken, N.J.: Taylor and Francis, 2014.

Shaw, Susan, Arlene Blum, Roland Weber, Kuruthachalam Kannan, David Rich, Donald Lucas, Catherine P. Koshland, Dina Dobraca, Sarah Hanson, and Linda S. Birnbaum. "Halogenated Flame Retardants: Do the Fire Safety Benefits Justify the Risks?" *Reviews on Environmental Health* 25, no. 4 (2010): 261–305.

Shaw, Susan D., Michelle L. Berger, Jennifer H. Harris, Se Hun Yun, Qian Wu, Chunyang Liao, Arlene Blum, Anthony Stefani, and Kurunthachalam Kannan. "Persistent Organic Pollutants Including Polychlorinated and Polybrominated Dibenzo-p-Dioxins and Dibenzofurans in Firefighters from Northern California." *Chemosphere* 91, no. 10 (2013): 1386–1394.

She, Jianwen, Myrto Petreas, Jennifer Winkler, Patria Visita, Michael McKinney, and Dianne Kopec. "PBDEs in the San Francisco Bay Area: Measurements in Harbor Seal Blubber and Human Breast Adipose Tissue." *Chemosphere* 46, no. 5 (2002): 697–707.

Shin, Annys. "Fighting for Safety." *Washington Post*, January 26, 2008. http://www .washingtonpost.com/wp-dyn/content/article/2008/01/25/AR2008012503170.html.

Shostak, Sara. *Exposed Science: Genes, the Environment, and the Politics of Population Health*. Berkeley: University of California Press, 2013.

Shwed, Uri, and Peter Bearman. "The Temporal Structure of Scientific Consensus Formation." *American Sociological Review* 75, no. 6 (2010): 817–840.

Singla, Veena. "HBCD Alternatives Assessment: Narrow Focus Misses Large Problems." Green Science Policy Institute, October 10, 2013. http://greensciencepolicy .org/hbcd-alternatives-assessment-narrow-focus-misses-large-problems/.

Sismondo, Sergio. *An Introduction to Science and Technology Studies*. 2nd ed. Malden, Mass.: Blackwell, 2010.

Sjodin, Andreas, and Hakan Carlsson. "Flame Retardants in Indoor Air at an Electronics Recycling Plant and at Other Work Environments." *Environmental Science & Technology* 35, no. 3 (January 2001): 448–454.

Sjödin, Andreas, Lee-Yang Wong, Richard S. Jones, Annie Park, Yalin Zhang, Carolyn Hodge, Emily Dipietro, Cheryl McClure, Wayman Turner, Larry L. Needham, and Donald G. Patterson, Jr. "Serum Concentrations of Polybrominated Diphenyl Ethers (PBDEs) and Polybrominated Biphenyl (PBB) in the United States Population: 2003–2004." *Environmental Science & Technology* 42, no. 4 (February 15, 2008): 1377–1384.

Slovic, Paul. "Perception of Risk." *Science* 236, no. 4799 (1987): 280–285.

Small, Chanley M., Deanna Murray, Metrecia L. Terrell, and Michele Marcus. "Reproductive Outcomes among Women Exposed to a Brominated Flame Retardant in Utero." *Archives of Environmental and Occupational Health* 66, no. 4 (2011): 201–208.

Smith, Richard. "Medical Journals Are an Extension of the Marketing Arm of Pharmaceutical Companies." *PLoS Medicine* 2, no. 5 (2005): 364–366.

Smith, Ted, David A. Sonnenfeld, and David Pellow. *Challenging the Chip: Labor Rights and Environmental Justice in the Global Electronics Industry*. Philadelphia: Temple University Press, 2006.

Snow, David. "Social Movements as Challenges to Authority: Resistance to an Emerging Conceptual Hegemony." In *Authority in Contention*, ed. Daniel Myers and Daniel Cress, 3–26. Boston: Elsevier, 2004.

Snow, David, Burke Rochford Jr., Steven K. Worden, and Robert D. Benford. "Frame Alignment Processes, Micromobilization, and Movement Participation." *American Sociological Review* 51, no. 4 (1986): 464–481.

Spring Mills, Inc. v. CPSC. "Springs Mills, Inc. v. Consumer Products Safety Commission." 434 F. Supp. 416, 1977.

Stapleton, Healther, Susan Klosterhaus, Sarah Eagle, Jennifer Fuh, John D. Meeker, Arlene Blum, and Thomas Webster. "Detection of Organophosphate Flame Retardants in Furniture Foam and U.S. House Dust." *Environmental Science & Technology* 43, no. 19 (2009): 7490–7495.

Stapleton, Heather M., Joseph Allen, Shannon Kelly, Alex Konstantinov, Susan Klosterhaus, Deborah Watkin, Michael McClean, and Thomas Webster. "Alternate and New Brominated Flame Retardants Detected in U.S. House Dust." *Environmental Science & Technology* 42, no. 18 (2008): 6910–6916.

Stapleton, Heather M., Susan Klosterhaus, Alex Keller, P. Lee Ferguson, Saskia van Bergen, Ellen Cooper, Thomas Webster, and Arlene Blum. "Identification of Flame Retardants in Polyurethane Foam Collected from Baby Products." *Environmental Science & Technology* 45, no. 12 (2011): 5323–5331.

Stapleton, Heather M., Smriti Sharma, Gordon Getzinger, P. Lee Ferguson, Michelle Gabriel, Thomas Webster, and Arlene Blum. "Novel and High Volume Use Flame Retardants in US Couches Reflective of the 2005 PentaBDE Phase Out." *Environmental Science & Technology* 46, no. 24 (2012): 13432–13439.

Stec, Anna. "Influence of Fire Retardants on Fire Toxicity." Presentation at the American Chemical Society Meeting, March 27, 2013, San Diego, Calif.

Stedeford, Todd. "Todd Stedeford." LinkedIn. Accessed April 30, 2015. http://www.linkedin.com/pub/todd-stedeford/5/16/321.

Steingraber, Sandra. *Living Downstream: An Ecologist Looks at Cancer and the Environment*. Reading, Mass.: Addison-Wesley, 1997.

Steinzor, Rena, and Michael Patoka. "The Bottleneck." *Environmental Forum*, January/February 2012, 36–40.

Stirling, Andy, and David Gee. "Science, Precaution, and Practice." *Public Health Reports* 117, no. 6 (2002): 521–533.

Stockholm Convention. "Reports and Decisions." 2014. http://chm.pops.int/Convention/ConferenceoftheParties(COP)/Decisions/tabid/208/Default.aspx.

——. "Stockholm Convention." Chatelaine, Switzerland: Secretariat of the Stockholm Convention, 2012.

Superfund Research Program. "Community Engagement and Research Translation," National Institute of Environmental Health Sciences, 2014. http://www.niehs.nih .gov/research/supported/dert/programs/srp/outreach/index.cfm.

Susser, Mervyn. "Editorial: Goliath and Some Davids in the Tobacco Wars." *American Journal of Public Health* 87, no. 10 (1997): 1593–1594.

Szasz, Andrew. *Shopping Our Way to Safety: How We Changed from Protecting the Environment to Protecting Ourselves*. Minneapolis: University of Minnesota Press, 2007.

Taylor-Gooby, Peter, and Jens Zinn. "Current Directions in Risk Research: New Developments in Psychology and Sociology." *Risk Analysis* 26, no. 2 (2006): 397–411.

Thomas, Pat. "Flame Retardants Are the New Lead." *The Ecologist*, June 13, 2014.

Tickner, Joel A. "From Reaction to Prevention." Presentation at the American Bar Association 39th National Spring Conference on the Environment—Chemicals Regulation. June 11, 2010, Baltimore.

Tickner, Joel A., Jessica N. Schifano, Ann Blake, Catherine Rudisill, and Martin J. Mulvihill. "Advancing Safer Alternatives Through Functional Substitution." *Environmental Science & Technology* 49, no. 2 (2015): 742–749.

Tierney, Kathleen. "Toward a Critical Sociology of Risk." *Sociological Forum* 14, no. 2 (1999): 215–242.

Tolich, Martin. "Internal Confidentiality: When Confidentiality Assurances Fail Relational Informants." *Qualitative Sociology* 27, no. 1 (2004): 101–106.

Tuncak, Baskut. "How Stronger Laws Help Bring Safer Chemicals to Market." Washington, D.C.: Center for International Environmental Law, 2013.

Union of Concerned Scientists. *Voices of Scientists at the EPA: Human Health and the Environment Depend on Independent Science*. Cambridge, Mass., 2008. http://www .ucsusa.org/assets/documents/scientific_integrity/epa-survey-brochure.pdf.

U.S. Census. "Historical Census of Housing Table: House Heating Fuel." *Census of Housing*, 2011.

U.S. Congress. "Toxic Substances Control Act." *15 U.S.C 2601–2692*, 1976.

U.S. Consumer Products Safety Commission. *Children's Sleepwear Regulations: 16 CFR 1615 & 1616*. U.S., 2001. http://www.cpsc.gov//PageFiles/103092/regsumsleepwear.pdf.

——. "News from CPSC: CSPS Bans TRIS-Treated Children's Garments." Release 77-030. Washington, D.C.: Office of Information and Public Affairs, 1977. http:// www.cpsc.gov/cpscpub/prerel/prhtml77/77030.html.

U.S. Department of Energy. "Average Material Consumption for a Light Vehicle," Table 4.16. In *Transportation Energy Data Book*. 31st ed. Washington, D.C.: DOE, 2012.

U.S. Department of Justice. Freedom of Information Act. Home Page. Washington, D.C.: DOJ, 2013.

U.S. Department of Transportation. *Federal Motor Vehicle Safety Standard 302: Flammability of Interior Materials*. Washington, D.C.: U.S. Department of Transportation, 1972. http://www.nhtsa.gov/cars/rules/import/fmvss/index.html.

U.S. Environmental Protection Agency. "A Dictionary of Technical and Legal Terms Related to Drinking Water." Washington, D.C.: EPA. Accessed April 30, 2015. http://nepis.epa.gov/Exe/ZyPURL.cgi?Dockey=20001QWV.txt.

——. "An Alternatives Assessment for the Flame Retardant Decabromodiphenyl Ether (DecaBDE)." Washington, D.C.: EPA, Design for the Environment, 2014. http://www.epa.gov/oppt/dfe/pubs/projects/decaBDE/deca-report-complete.pdf.

——. "Analog Identification Method." Washington, D.C.: EPA, 2012.

——. "Certain Polybrominated Diphenylethers: Significant New Use Rule and Test Rule. 40 CFR Parts 721, 795, and 799." *Federal Register* 77, no. 63 (2012): 19862–19898.

——. "Comments on the Design for the Environment (DfE) Program Alternatives Assessment for the Flame Retardant Decabromodiphenyl Ether." Washington, D.C.: EPA, 2012.

——. *Community Air Screening How-To Manual, EPA-744-B-04-001.* Research Triangle Park, N.C.: EPA, Office of Air Quality Planning and Standards, 2004. http://www.epa.gov/oppt/cahp/pubs/howto.htm.

——. "DecaBDE Phase-Out Initiative." Washington, D.C.: EPA, 2009.

——. "Design for the Environment Program Alternatives Assessment Criteria for Hazard Evaluation." Washington, D.C.: EPA, Office of Pollution Prevention and Toxics, 2011. http://www.epa.gov/dfe/alternatives_assessment_criteria_for_hazard_eval.pdf.

——. "Draft Toxicological Reviews of Polybrominated Diphenyl Ethers (PBDEs): In Support of the Summary Information in the Integrated Risk Information System (IRIS)." *Federal Register* 71, no. 246 (2006). http://www.epa.gov/fedrgstr/EPA-RESEARCH/2006/December/Day-22/r21969.htm.

——. "Endocrine Disruptor Screening Program for the 21st Century: EDSP21 Work Plan." Washington, D.C.: EPA, Office of Chemical Safety and Pollution Prevention, 2011. http://www.epa.gov/endo/pubs/edsp21_work_plan_summary _overview_final.pdf.

——. "EPA Announces Chemicals for Risk Assessment in 2013, Focus on Widely Used Flame Retardants." Press release. Washington, D.C.: EPA, 2013. http://yosemite.epa.gov/opa/admpress.nsf/bd4379a92ceceeac8525735900400c27/c6be79994c3fd08785257b3b0054e2fa!OpenDocument.

——. "EPA Releases Formerly Confidential Chemical Information." Washington, D.C.: EPA, 2011.

——. "EPA Settles Case Involving 3M Voluntary Disclosures of Toxic Substances Violations." Washington, D.C.: EPA, 2006.

——. "EPA Settles PFOA Case against DuPont for Largest Environmental Administrative Penalty in Agency History." Washington, D.C.: EPA, 2005.

——. "EPA's Budget and Spending," 2014. http://www2.epa.gov/planandbudget/budget.

——. *Essential Principles for Reform of Chemicals Management Legislation.* Washington, D.C.: EPA, 2009.

——. "Existing Chemicals Action Plans." Washington, D.C.: EPA, 2012.

——. *External Peer Review: Toxicological Review for Polybrominated Diphenyl Ethers (PBDEs) Human Health Assessment* (August 2007), 2007. http://www.ewg.org /research/epa-axes-panel-chair-request-chemical-industry-lobbyists/review-panel -timeline.

——. "External Peer Review: Toxicological Review for Polybrominated Diphenyl Ethers (PBDEs) Human Health Assessment (March 2007)," 2007. http://www.ewg .org/research/epa-axes-panel-chair-request-chemical-industry-lobbyists/review -panel-timeline.

——. *Flame Retardants in Printed Circuit Boards*. Washington, D.C.: EPA, 2014. http:// www2.epa.gov/sites/production/files/2015–01/documents/pcb_updated_draft _report.pdf.

——. "Good Laboratory Practices Standards." Washington, D.C.: EPA, 2013.

——. "Green Chemistry." Washington, D.C.: EPA, 2013.

——. *Guidelines for Exposure Assessment*. Washington, D.C.: EPA, Risk Assessment Forum, 1992.

——. "Hexabromocyclododecane (HBCD) Action Plan," 2010. http://www.epa.gov /oppt/existingchemicals/pubs/actionplans/RIN2070-AZ10_HBCD action plan _Final_2010–08–09.pdf.

——. "IRIS Glossary," Washington, D.C.: EPA, 2011. http://ofmpub.epa.gov/sor_inter- net/registry/termreg/searchandretrieve/glossariesandkeywordlists/search.do? details=&glossaryName=IRIS%20Glossary.

——. "Letter from George Gray, U.S. EPA, to Nancy Sandrof, American Chemistry Council," 2008. http://www.ewg.org/research/epa-axes-panel-chair-request -chemical-industry-lobbyists/review-panel-timeline.

——. "New Chemicals: Assessing Risk." Washington, D.C.: EPA, 2010. http://www.epa .gov/oppt/newchems/pubs/assess.htm.

——. "New Chemicals: Chemicals Categories Report." Washington, D.C.: EPA, 2012.

——. "Office of Children's Health Protection." Washington, D.C.: EPA, 2013. http:// www2.epa.gov/children.

——. "Persistent Bioacccumulative and Toxic (PBT) Chemical Program." Washington, D.C.: EPA, 2011.

——. "Pesticides: Glossary." Washington, D.C.: EPA, Office of Pesticide Programs, 2012.

——. *Polybrominated Diphenyl Ethers (PBDEs) Action Plan*. Washington, D.C.: EPA, 2009. http://www.epa.gov/oppt/existingchemicals/pubs/actionplans/pbdes_ap_2009 _1230_final.pdf.

——. *Polybrominated Diphenyl Ethers (PBDEs) Project Plan*. Washington, D.C.: EPA, 2006. http://www.epa.gov/oppt/pbde/pubs/pbdestatus1208.pdf.

——. *Polyhalogenated Dibenzo-P-Dioxins/Dibenzofurans; Testing and Reporting Re- quirements*, Washington, D.C.: EPA, 1987.

——. *Polymer Exemption Guidance Manual*. Washington, D.C.: EPA, 1997. http:// www.epa.gov/oppt/newchems/pubs/polyguid.pdf.

——. "Press Release October 21, 1976: Train Sees New Toxic Substances Law as 'Preventive Medicine.'" Washington, D.C.: U.S. EPA, 1976. http://www2.epa.gov/aboutepa /train-sees-new-toxic-substances-law-preventive-medicine.

——. *Response to Comments on the Draft Alternatives Assessment for Decabromodiphenyl Ether (decaBDE)—January, 2014.* Washington, D.C., EPA, 2014. http://www .epa.gov/oppt/dfe/pubs/projects/decaBDE/140127-response-to-comments-on -decabde-report.pdf.

——. "Risk Assessment Portal: Basic Information." Washington, D.C.: U.S. EPA, 2013. http://www.epa.gov/risk/.

——. "Safer Choice." Washington, D.C.: U.S. EPA, 2015. http://www2.epa.gov /saferchoice.

——. "Summary of Accomplishments: New Chemicals Program Activities through September 30, 2012." Washington, D.C.: U.S. EPA, 2012.

——. "Summary of the Toxic Substances Control Act: 15 U.S.C. §2601 et Seq.," Washington, D.C., EPA, 2014. http://www2.epa.gov/laws-regulations/summary -toxic-substances-control-act.

——. "ToxCast: Screening Chemicals to Predict Toxicity Faster and Better." Washington, D.C.: EPA, 2013.

——. *Toxicological Review of Decabromodiphenyl Ether (BDE-299) in Support of Summary Information on the Integrated Risk Information System.* EPA 635-R-07-008F. Washington, D.C.: EPA, Integrated Risk Information System, 2008. www.epa .gov/ncea/iris/toxreviews/0035tr.pdf.

——. "Toxics Release Inventory Release Reports." Washington, D.C.: EPA, 2015.

——. "Tris(2,3-Dibromopropyl) Phosphate Significant New Use Rule." *Federal Register* 52, no. 2703 (1987): 326.

——. "TSCA Chemical Substance Inventory." Washington, D.C.: EPA, 2011.

——. "TSCA Section 5(b)(4) Concern List." Washington, D.C.: EPA, 2012.

——. "TSCA Work Plan Chemicals." Washington, D.C.: EPA, 2014. http://www.epa .gov/oppt/existingchemicals/pubs/workplans.html.

——. "TSCA Work Plan for Chemical Assessments: 2014 Update." Washington, D.C.: EPA, 2015. http://www.epa.gov/oppt/existingchemicals/pubs/TSCA_Work_Plan _Chemicals_2014_Update-final.pdf.

——. "TSCA Workplan Chemicals: Methods Document." Washington, D.C.: EPA, 2012. http://www.epa.gov/oppt/existingchemicals/pubs/wpmethods.pdf.

——. "VCCEP Status: Octabromodiphenyl Ether." Washington, D.C.: EPA, 2010. http://www.epa.gov/oppt/vccep/pubs/chem23.html.

——. "VCCEP Status: Pentabromodiphenyl Ether." Washington, D.C.: EPA, 2010. http://www.epa.gov/oppt/vccep/pubs/chem22.html.

——. "Voluntary Children's Chemical Evaluation Program (VCCEP)." Washington, D.C.: EPA, 2010. http://www.epa.gov/oppt/vccep/.

——. "What Is the TSCA Chemical Substance Inventory?" Washington, D.C.: EPA, 2010.

U.S. Government Accountability Office. *Chemical Regulation: Options Exist to Improve EPA's Ability to Assess Health Risks and Manage Its Chemical Review Program.* Washington, D.C.: GAO, 2005. http://www.gao.gov/products/GAO-05-458.

U.S. House of Representatives. "Science Under Siege: Scientific Integrity at the Environmental Protection Agency." Washington, D.C.: U.S. Government Printing Office, 2008.

U.S. Office of Information and Regulatory Affairs. "Reginfo.gov: EPA/OCSPP TSCA Chemicals of Concern List." Washington, D.C.: Office of Information and Regulatory Affairs, 2013.

——. "View Rule: RIN 2070-AJ08." Reginfo.gov, 2014. http://reginfo.gov/public/do/eAgendaViewRule?pubId=201210&RIN=2070-AJ08.

U.S. Office of the Inspector General. *EPA's Voluntary Chemical Evaluation Program Did Not Achieve Children's Health Protection Goals.* Report No. 11-P-0379. Washington, D.C.: EPA, 2011.

U.S. Office on Smoking and Health. *Smoking and Health: A Report of the Surgeon General.* Washington, D.C., 1979. http://profiles.nlm.nih.gov/NN/B/C/M/D/.

U.S. Senate. "Oversight of EPA Authorities and Actions to Control Exposures to Toxic Chemicals. Committee on Environment and Public Works." July 24, 2012.

Van Wyke, Ben. "Ethics and Translation." In *Handbook of Translation Studies.* Volume 1, ed. Yves Gambier and Luc van Doorslaer, 111–115. Amsterdam: John Benjamins, 2010.

Vandenberg, Laura N., Theo Colborn, Tyrone B. Hayes, Jerrold J. Heindel, David R. Jacobs, Duk-Hee Lee, Toshi Shioda, Ana M. Soto, Frederick S. vom Saal, Wade V. Welsons, R. Thomas Zoeller, and John Peterson Myers. "Hormones and Endocrine-Disrupting Chemicals: Low-Dose Effects and Nonmonotonic Dose Responses." *Endocrine Reviews* 33, no. 3 (2012): 378–455.

VECAP. "The Voluntary Emissions Control Action Programme." Accessed April 30, 2015. http://www.vecap.info.

Viberg, Henrik, Anders Fredriksson, and Per Eriksson. "Neonatal Exposure to Polybrominated Diphenyl Ether (PBDE 153) Disrupts Spontaneous Behavior, Impairs Learning and Memory, and Decreases Hippocampal Cholinergic Receptors in Adult Mice." *Toxicology and Applied Pharmacology* 192, no. 2 (2003): 95–106.

Viberg, Henrik, Anders Fredriksson, Eva Jakobsson, Ulrika Orn, and Per Eriksson. "Neurobehavioral Derangements in Adult Mice Receiving Decabrominated Diphenyl Ether (PBDE 209) during a Defined Period of Neonatal Brain Development." *Toxicological Sciences* 76, no. 1 (2003): 112–120.

Vogel, Sarah A. *Is It Safe? BPA and the Struggle to Define the Safety of Chemicals.* Berkeley: University of California Press, 2013.

Vogel, Sarah A., and Jody A. Roberts. "Why the Toxic Substances Control Act Needs an Overhaul, and How to Strengthen Oversight of Chemicals in the Interim." *Health Affairs* 30, no. 5 (2011): 898–905.

Voith, Melody. "Stronger Demand in Fourth Quarter." *Chemical and Engineering News* 88, no. 5 (2010): 9.

Vom Saal, Frederick, and Claude Hughes. "An Extensive New Literature Concerning Low-Dose Effects of Bisphenol A Shows the Need for a New Risk Assessment." *Environmental Health Perspectives* 113, no. 8 (2005): 926–933.

Wagner, Wendy, and David Michaels. "Equal Treatment for Regulatory Science: Extending the Controls Governing the Quality of Public Research to Private Research." *American Journal of Law and Medicine* 30 (2004): 119–154.

Walker, Edward T. "Industry-Driven Activism." *Contexts* 9, no. 2 (2010): 45–49.

——. "Privatizing Participation: Civic Change and the Organizational Dynamics of Grassroots Lobbying Firms." *American Sociological Review* 74, no. 1 (2009): 83–105.

Walker, Edward T., Andrew W. Martin, and John D. McCarthy. "Confronting the State, the Corporation, and the Academy: The Influence of Institutional Targets on Social Movement Repertoires." *American Journal of Sociology* 114, no. 1 (2008): 35–76.

Walker, Edward T., and Christopher M. Rea. "The Political Mobilization of Firms and Industries." *Annual Review of Sociology* 40 (2014): 281–304.

Washington State Department of Ecology. "Rule Development—Children's Safe Products Reporting Rule," 2012. http://www.ecy.wa.gov/programs/swfa/rules/ruleChildSafe.html.

Weart, Spencer. "Global Warming: How Skepticism Became Denial." *Bulletin of the Atomic Scientists* 67, no. 1 (2011): 41–50.

Welke, B. Y. "The Cowboy Suit Tragedy: Spreading Risk, Owning Hazard in the Modern American Consumer Economy." *Journal of American History* 101, no. 1 (2014): 97–121.

Williams, Christopher. "Polymeric Flame Retardants: Possible Less Hazardous Alternatives to Decabromodiphenyl Ether?" Presented at the 13th Workshop on Brominated and Other Flame Retardants, Winnipeg, Canada, June 4–5, 2012.

Williams, Rebecca. "The Environment Report: Is Fire Safety Putting Us at Risk?" University of Michigan, 2010. Accessed April 30, 2015. http://environmentreport.org/fire_safety.php.

Wilson Center. "Wilson Center Experts: Peter O'Rourke." Accessed April 30, 2015. http://www.wilsoncenter.org/staff/peter-orourke.

Wing, Steve. "The Limits of Epidemiology." *Journal of Radiological Protection* 1, no. 2 (1994): 74–86.

Winner, Langdon. *The Whale and the Reactor.* Chicago: University of Chicago Press, 1986.

Wisconsin Department of Natural Resources. "Asbestos: History and Uses." Modified August 31, 2007. http://web.archive.org/web/20071228042633/http://www.dnr.state.wi.us/org/aw/air/reg/asbestos/asbes3.htm.

Woodhouse, Edward. "Nanoscience, Green Chemistry, and the Privileged Position of Science." In *The New Political Sociology of Science: Institutions, Networks, and Power,* ed. Scott Frickel and Kelly Moore, 148–181. Madison: University of Wisconsin Press, 2006.

Wright, Eric O. "Interrogating the Treadmill of Production: Some Questions I Still Want to Know about and Am Not Afraid to Ask." *Organization and Environment* 17, no. 3 (2004): 317–322.

Zinn, Jens. "Introduction: The Contribution of Sociology to the Discourse on Risk and Uncertainty." In *Social Theories of Risk and Uncertainty: An Introduction*, ed. Jens Zinn, 1–17. Malden, Mass.: Blackwell, 2008.

Zota, Ami, Gary Adamkiewicz, and Rachel Morello-Frosch. "Are PBDEs an Environmental Equity Concern? Exposure Disparities by Socioeconomic Status." *Environmental Science & Technology* 44, no. 15 (2010): 5691–5692.

Zota, Ami, Ruthann Rudel, Rachel Morello-Frosch, and Julia Brody. "Elevated House Dust and Serum Concentrations of PBDEs in California: Unintended Consequences of Furniture Flammability Standards?" *Environmental Science & Technology* 42, no. 21 (2008): 8158–8164.

INDEX

Note: italicized page numbers indicate figures (*f*) and tables (*t*).

furniture flammability standards, 185
furniture manufacturers, 2, 25, 28, 46

Good Laboratory Practices (GLP), 94,
105–106, 148, 159–160
Goodman, Julie E., 152–153
Graco, 182, 209
Gradient, 151–152
Gray, George, 121–123, 149–150
Great Lakes Solutions (Chemtura), 23–24,
26, 30, 37–39
green building movement, and high-
density foam insulations, 21
green chemistry principles, 71, 144, 212–213
greening of capitalism, 9–10
Greenpeace, 182
Green Science Policy Institute, 2, 167, 182,
275n3
guideline studies. See Good Laboratory
Practices (GLP)

halogenated flame retardant chemicals
(organo-halogens), 22–24, 48, 117, 168
Harvard Center for Risk Analysis (HCRA),
149–150
hazard, as component of risk, 52–53, 69
hazard assessments, 68–70, 80–81, 111, 183
hazard calls, in DfE's alternatives
assessment process, 139–140, 148, 161
hazard-centric risk formula, 69–71, 79–80,
85–86, 142–144, 212
Health and Environmental Sciences
Institute (HESI), 150
health outcomes, complexity of, 6–7, 207
Healthy Hospitals Initiative, 171
Hess, David, 11–12, 181
Hewlett-Packard (HP), as Greenpeace
target, 182
hexabromocyclododecane (HBCD),
46, 50
high-throughput screening, 95–96
hormesis, defined, 62
hormone-disrupting chemicals, nonlinear
dose–response relationships, 62–63. See
also exposure-proxy risk formula

house fires, 28–31. See also fire incidence
rates in U.S.
household exposure studies of PBDE levels
in U.S., 35–36

ignorance, 13, 90–91, 262n10
industrial chemicals, ubiquity of, 7, 211,
222
Industrial Chemistry branch, EPA, 130–131
industry science, reputation of, 219
Integrated Risk Information System (IRIS),
EPA, 59, 121, 124
interdisciplinary research, 7, 103, 154–155
International Association of Fire Fighters
(IAFF), 171–172
inverted quarantine, 192
in vitro assays, high-throughput screening
of, 95–96
IRIS (Integrated Risk Information System),
EPA, 59, 121, 124
Israeli Chemicals Limited Industrial
Products (ICL-IP), 23–24

junk science, 151–152, 272–273n84

Kaiser Permanente, 171
kestrels, effects of β-Tech on, 87–88
knowledge production, scientific
uncertainty and, 90

labeling requirements, 216
Leno, Mark, 174–175
level of proof, 97–98
linear dose–response model, challenges to,
61. See also conceptual risk formulas

Maine, 120, 214, 271n64
market, industrial, for PBDEs and other
flame retardants, 36
market campaigns, 181–184, 209, 214
market forces, in product development and
safety, 155–157, 163–164, 279–280n60
medical professions, collaboration with
flame retardant coalition, 171
methodological uncertainty, 93–94